T0215631

INTERNAL GRAVITY WAVES

The study of internal gravity waves provides many challenges: they move along interfaces as well as in fully three-dimensional space, and they do so at relatively fast temporal and small spatial scales – making them difficult to observe and resolve in weather and climate models. The equations describing their evolution are well established but their solution poses various mathematical challenges associated with singular boundary value problems and large amplitude dynamics.

This book provides the first comprehensive treatment of the theory for small- and large-amplitude internal gravity waves, whether existing as interfacial waves in a layered fluid or as internal waves in continuously stratified fluid. Over 120 schematics, numerical simulations and images from laboratory experiments illustrate the theory and mathematical techniques, while the 130 exercises allow the reader to test their understanding of the theory and its applications. This is an invaluable single resource for academic researchers and graduate students studying the motion of waves within the atmosphere and oceans, and more generally for mathematicians, physicists and engineers interested in the peculiar properties of propagating, growing and breaking waves.

BRUCE R. SUTHERLAND holds a B.Math. from the University of Waterloo and a Ph.D. in Physics from the University of Toronto where, under the supervision of W. R. Peltier, he ran numerical simulations of internal waves generated by shear instability. As a Research Associate working with P. F. Linden at the University of Cambridge, he helped develop the laboratory method known as the synthetic schlieren technique, which was applied to the examination of internal waves generated by turbulence. Now a Professor in the departments of Physics and of Earth and Atmospheric Sciences, and an Adjunct Professor in Mathematical and Statistical Sciences, at the University of Alberta, he continues to develop theories and to run laboratory experiments and numerical simulations that examine the generation, propagation and breaking of internal gravity waves.

INTERNAL GRAVITY WAVES

B. R. SUTHERLAND

Departments of Physics and of Earth & Atmospheric Sciences
University of Alberta

CAMBRIDGE
UNIVERSITY PRESS

University Printing House, Cambridge CB2 8BS, United Kingdom

One Liberty Plaza, 20th Floor, New York, NY 10006, USA

477 Williamstown Road, Port Melbourne, VIC 3207, Australia

314-321, 3rd Floor, Plot 3, Splendor Forum, Jasola District Centre, New Delhi-110025, India

79 Anson Road, #06-04/06, Singapore 079906

Cambridge University Press is part of the University of Cambridge.

It furthers the University's mission by disseminating knowledge in the pursuit of education, learning and research at the highest international levels of excellence.

www.cambridge.org
Information on this title: www.cambridge.org/9781108457088

© Bruce R. Sutherland 2010

First published 2010
First paperback edition 2018

A catalogue record for this publication is available from the British Library

ISBN 978-0-521-83915-0 Hardback
ISBN 978-1-108-45708-8 Paperback

To Brendan, Cameron and Samantha

Contents

Preface

Why write a book on internal gravity waves when so many other books cover the subject already? The textbooks listed in the appendix include at least some discussion of internal gravity waves. Some focus upon interfacial waves, which are internal gravity waves at interfaces; some focus upon internal waves, which exist in continuously stratified fluid. Different books emphasize different dynamics such as mechanisms for generation, propagation in non-uniform media, nonlinear evolution and stability. Textbooks on geophysical fluid dynamics (e.g. Gill (1982), Vallis (2006)) understandably devote only a chapter to the subject because, although internal waves are non-negligible in their influence upon global weather and ocean circulation patterns, they are by no means dominant. Internal waves are noise, if sometimes irritatingly loud. Textbooks on the theory of waves and instability (e.g. Whitham (1974), Lighthill (1978), Drazin and Reid (1981), Craik (1985)) examine how non-uniform media and nonlinearity affect the evolution of inter-facial and internal waves. But these books can be daunting to graduate students lacking strong mathematical backgrounds. Textbooks on stratified fluid dynam-ics (e.g. Turner (1973), Baines (1995)) help to provide physical insight into the dynamics of internal gravity waves through a combination of theory and laboratory experiments, though sometimes without providing the mathematical details. Some textbooks are devoted to the subject of internal gravity waves (e.g. Miropol'sky (2001), Nappo (2002), Vlasenko *et al.* (2005)), but these focus either on atmospheric or oceanic waves.

The approach taken here is to provide the physics and mathematics describing internal gravity waves in a way that is accessible to students who have been exposed to multivariable calculus and ordinary differential equations. An understanding of partial differential equations, though useful, is not necessary. A background in atmosphere–ocean science and fluid dynamics is not assumed. Chapter 1 covers this material at an introductory level, presenting only those details that are necessary for modelling internal gravity waves and the environment in which they exist. This chapter also introduces the mathematical description of waves and their properties.

Chapter 2 describes the structure and evolution of periodic, small-amplitude interfacial waves, beginning with a detailed description of surface waves. Although surface waves are not internal gravity waves, they are part of everyone's common experience thus making it easier to draw the link between mathematical theory and reality. We will find that surface waves are a special case of internal gravity waves at the interface between two fluids. They occur in the limit where the upper layer density (that of air) is much smaller than the lower layer density (that of water). The discussion goes on to describe waves at the interface between fresh and salty water or between hot and cold fluid, whether a gas or liquid. In the presence of shear an otherwise flat interface may become unstable to undular disturbances. The influence of interfacial waves upon the growth and structure of the instability is also discussed in this chapter.

Whereas interfacial waves occur where the density decreases rapidly with height over a negligibly small distance, internal waves move vertically through a fluid whose (effective) density decreases continuously with height. The rate of this decrease determines a fundamental quantity used in the description of internal waves known as the buoyancy frequency. This is derived for liquids and gases at the start of Chapter 3. Thereafter, the equations for periodic, small-amplitude internal waves in uniformly stratified fluid are derived and solved. This chapter includes a discussion of the peculiar behaviour of internal waves near sloping boundaries and describes how their structure is affected by rotation and relatively rapid density changes with height.

Chapter 4 introduces the mathematics necessary to model waves of non-negligibly small amplitude. The changes in frequency and structure of finite-amplitude interfacial and internal waves are examined. Special attention is drawn to the case of finite-amplitude interfacial waves in shallow water which can take the form of hump-shaped, solitary waves. The chapter also describes the various forms of instability associated with waves including modulational instability, parametric subharmonic instability, overturning and shear instability.

Internal gravity waves are generated by flow over topography, convective storms, imbalance of large-scale circulations, thunderstorm outflows, river plumes and so on. Of these, the first generation mechanism is best understood theoretically and is the focus of Chapter 5. This begins with the classic problem of internal waves generated by an oscillating cylinder. The mathematics of this section is more advanced than elsewhere but is included in part to illustrate how this conceptually simple problem is challenging to model mathematically in a way that gives meaningful physical results. The rest of the chapter discusses the generation of interfacial and internal waves by steady and oscillatory (tidal) flow over hills. The generation of internal waves by non-rigid sources such as plumes, gravity currents and turbulence is becoming better understood as a result of high-resolution numerical simulations and laboratory experiments. But it is beyond the scope of this book to discuss such recent and on-going research.

In Chapter 6 the propagation of waves in non-uniform media is described. This includes the description of interfacial waves approaching a slope and of internal waves in non-uniformly stratified shear flows. In parts of the ocean and atmosphere, internal waves exhibit a somewhat universal relationship between their amplitude, frequency and spatial scale. The chapter closes with an empirical description of these waves.

Although references are not included in the text, the appendix lists other text-books and articles that the reader can use to follow-up on various topics. The journal articles are organized by subject matter, more or less following the order of presentation in the book. It is hoped that this style will help the reader follow the history of research into each subject up until the time of writing. In some cases, this organization also serves to emphasize links between the theory of internal gravity waves in both layered and continuously stratified fluids.

Many colleagues have helped guide the structure and content of this book. In particular, I would like to thank Joan Alexander, Eric D'Asaro, Oliver Buhler, Colm-cille Caulfield, Kathleen Dohan, Morris Flynn, David Fritts, Jody Klymak, Eric Kunze, Jennifer MacKinnon and Rob Pinkel for their illuminating insights and stimulating discussions. I am particularly grateful to Joseph Ansong, Geoffrey Brown, Heather Clark, Hayley Dosser, Kate Gregory, Amber Holdsworth, Justine McMillan, James Munroe and Joshua Nault for their constructive criticism and support. Finally, I wish to acknowledge the hard work and ingenuity of undergraduate students Kyle Holland and Cara Kozack who helped prepare many of the figures.

1

Stratified fluids and waves

1.1 Introduction

Outside of theoretical interest in their peculiar properties, one of the main motivations for understanding the dynamics of internal gravity waves is that they occur naturally in the atmosphere and oceans. In the atmosphere, through transporting horizontal momentum from the ground upwards, internal waves influence wind speeds and consequently the thermal structure of the atmosphere. By contrast, internal waves in the ocean are primarily important as they affect mixing through the transport of energy. Although internal waves do not play a dominant role in the evolution of weather and climate, their influence is non-negligible: numerical simulations that do not include the effects of internal waves predict wind speeds and temperature in the atmosphere incorrectly and they do not account for the observed levels of turbulent diffusivity in the oceans. At the mesoscale in the atmosphere internal wave breaking is a source of clear-air turbulence and in the ocean, internal solitary waves influence biological activity over continental shelves through mass transport and mixing.

This chapter begins with a brief introduction to stratified fluids and internal gravity waves and then gives an overview of the structure of the atmosphere and oceans with mention of internal gravity wave phenomena in these fluids. In the following sections, we derive the equations describing the motion and thermodynamics of fluids and then make approximations relevant to internal gravity wave dynamics. The derivations are sometimes heuristic, aiming to provide physical intuition rather than emphasizing rigour. Finally, we review the mathematics used to describe the structure and evolution of periodic waves, wavepackets and modes.

1.2 Stratified fluids

A stratified fluid tends to move in horizontal layers. ('*Strata*' from modern Latin means 'layers'.) Vertical motions are inhibited because the density and pressure

1

change with height in such a way that it costs energy to move against or with the
direction of gravity. Put another way, vertically displaced fluid in a stratified fluid
feels a buoyancy restoring force acting in a direction opposite to the displacement.
In this sense, the force acts like a spring, which feels a contracting force when
stretched and an expanding force when compressed.

Sometimes a stratified fluid is said to be 'stably stratified' to distinguish it from
an 'unstably stratified fluid', such as boiling water, in which buoyancy forces cause
relatively light fluid to rise and relatively dense fluid to sink. At the conceptual
boundary between these two classes of stratification is 'uniform', 'homogeneous'
or 'neutrally stratified' fluid: vertically displaced fluid experiences no buoyancy
forces. We are primarily interested in stably stratified fluids in this book because
only these support internal gravity waves.

Both liquids and gases can be stratified. We will consider these separately below.

1.2.1 Stratified liquids and the ocean

In preparing a salad dressing, oil and vinegar form a stratified fluid with oil floating
in a layer above the more dense vinegar. Many cocktails are also stratified fluids
with dense sugary or creamy liqueurs resting at the bottom of the glass and lighter
aperitifs layered on top, as shown in Figure 1.1.

As well as describing the sequential stacking of slabs of fluid, stratified fluids
also include liquids whose density decreases continuously with height. For example,

Fig. 1.1. Tasty stratified beverages.

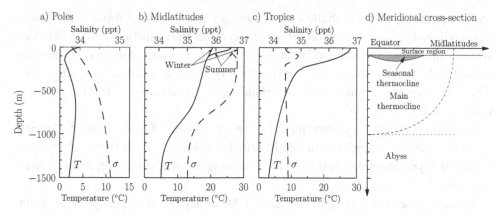

Fig. 1.2. Schematic showing vertical temperature (solid line) and salinity (dashed line) profiles of the ocean a) near the poles, b) at midlatitudes and c) near the equator. d) The vertical and meridional density structure of the ocean between the equator and poles showing the surface region, seasonal thermocline, main thermocline and the abyss. The seasonal thermocline is most pronounced in summer. In all plots the vertical axis is depth, as indicated in a).

saline solutions are 'continuously stratified' if the fluid becomes gradually less salty from bottom to top. Likewise, water is stratified by temperature if it becomes cooler with depth. The fluid is said to be 'strongly stratified' if the density decreases relatively rapidly with height, whereas it is 'weakly stratified' if the density decreases slowly with height.

The vertical structure of the ocean is effectively subdivided into layers according to the strength of its stratification. These are the surface mixed regions, the strongly stratified thermocline and the underlying weakly stratified abyss. These are illustrated schematically in Figure 1.2. Temperature, salinity and pressure change with depth but, except near estuaries, it is the temperature variation that is most important in determining the density change. In freshwater lakes, temperature alone determines the stratification.

Wind stress, surface wave breaking and convection lead to active turbulence near the ocean surface. This surface mixed region is unstratified or weakly stratified and generally extends downwards as far as 100 m depending upon latitude and season. The top 10 m of the ocean, which is most turbulent as a result of wind stress and surface wave breaking, is the surface skin region.

The thermocline region includes the seasonal thermocline and below it the main thermocline. Both extend from midlatitudes in the southern hemisphere to midlatitudes in the northern hemisphere. The seasonal thermocline is situated between approximately 50 and 150 m depth, depending upon season. Its stratification is strongest and its depth is shallowest in summer when vertical mixing is strongly inhibited by the large density contrast between the warm surface and cooler deep waters. In winter, surface cooling drives convection which deepens the thermocline, weakening its stratification and potentially, in lakes, eliminating it altogether. The main thermocline in the ocean, extending to approximately 1000 m depth, is unaffected by the seasons.

Strictly speaking a 'thermocline' is a highly stratified layer in which the temperature decreases rapidly with depth; this definition ignores salinity changes. A stratified layer resulting strictly from increasing saltiness with depth is called a 'halocline'.

The abyss lies between approximately 1000 m depth and the ocean floor, about 5000 m below the surface. This part of the ocean is weakly stratified and nearly quiescent, except where tidal flows move over the oceanic ridges and continental margins.

1.2.2 Stratified gases and the atmosphere

Gases as well as liquids form stratified fluids and share many of the properties of stratified liquids: just as warm water floats on cold, so does warm air hover near the ceiling and cold air near the floor of a room. When cold air formed by dry-ice creates clouds that pour off a stage it exposes a stratified flow of cold moisture-laden air in a relatively warm ambient.

On larger vertical scales the temperature of air can decrease with height and still be stably stratified. Thus cold air at a mountain top does not necessarily rush downslope into a warm valley. If it is pushed downward, it is compressed by higher pressures acting to increase its density and temperature. The compressed air continues downslope only if its increased temperature is nonetheless colder than its surroundings. If the temperature of the air is constant with height, vertically displaced air always feels a force that tends to restore it to its original position – it is stably stratified.

Generally, it is insufficient to say that a gas is stably stratified if its density decreases with height. Its stability depends primarily upon ambient pressure and temperature variations, as discussed in Section 1.7.1.

Like the ocean, the atmosphere is subdivided into layers depending upon the relative strength of the stratification at different altitudes. Immediately above the Earth's surface is the troposphere. Above that the stratosphere and mesosphere comprise what is called the middle atmosphere. Above these is the upper atmosphere composed of the thermosphere, which contains the ionosphere near its base. These are illustrated in Figure 1.3. Although the air masses evolve smoothly from one layer to the next, for conceptual convenience we separate the layers by boundaries.

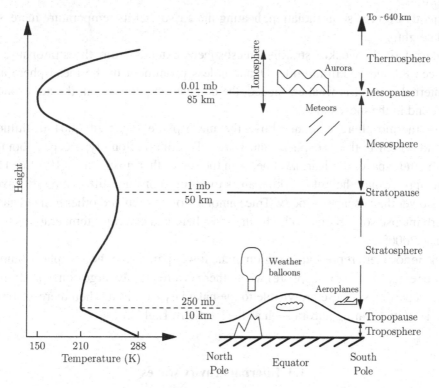

Fig. 1.3. Schematic representation of the main four layers of the atmosphere, which are determined by thermal variations with height.

Thus the top of the troposphere is the tropopause, the top of the stratosphere is the stratopause and the top of the mesosphere is the mesopause.

The troposphere, where our weather occurs, is neutrally stable or weakly stratified for the most part. Its relative homogeneity results from convection: during the day the sun warms the ground and the air above it mixes to about 7 km altitude near the poles and approximately 16 km near the equator.

In a neutrally stratified atmosphere, the temperature decreases with height at a rate of about 10°C per kilometre. This is known as the 'dry adiabatic lapse rate'. It means that if air containing no water vapour is lifted upwards by 1 km then its temperature will drop by 10°C as a result of cooling when it expands in the lower pressure surroundings.

The stratosphere extends from the tropopause to about 50 km altitude. It gets its name because it is relatively strongly stratified. One reason why commercial aeroplanes fly in the lower stratosphere is because vertical updrafts and turbulence are inhibited by the stable stratification. The stratification is strong due to chemistry: ozone, which is produced in large quantities in the equatorial stratosphere, efficiently absorbs solar radiation, heating the air so that its temperature increases with height.

The relatively weakly stratified mesosphere extends from the stratopause to between 80 and 85 km altitude. Ozone is less prominent in the mesosphere and so internal heating is weaker. Indeed, the coldest temperatures in the atmosphere are found in the mesosphere.

The thermosphere extends above the mesosphere to about 600 km altitude, beyond which is the exosphere and space. The internationally accepted boundary to outer space, the Kármán line, is in the lower thermosphere at 100 km. This is the approximate height at which air is too thin to provide lift to aircraft travelling slower than escape velocity. Trace amounts of oxygen and other ionized gases absorb intense solar radiation in the thermosphere and can raise temperatures to as high as 2000°C.

The ionosphere represents a region in the lower part of the thermosphere (and to a lesser extent in the mesosphere) where there is a relatively large concentration of charged particles. These are visible to the naked eye at night when there is strong solar activity forming aurora such as the Northern Lights.

1.3 Internal gravity waves

Surface waves on the ocean move up and down due to buoyancy. Water lifted upwards is heavier than its surroundings and falls downwards. It then overshoots its equilibrium position and experiences a restoring force that lifts the downward-displaced surface upwards again. This spring-like motion is felt collectively by the fluid which oscillates in space as well as time. Because buoyancy forces, and ultimately gravity, are necessary for the existence of surface waves, they are sometimes called gravity waves.

Likewise, the interface between warm and cold fluid or between fresh and salt water can oscillate forming what is called an 'internal gravity wave': a gravity wave that moves *within* a fluid. In a continuously stratified fluid, internal gravity waves are not confined to an interface and can in fact propagate vertically as well as horizontally through the fluid. Internal gravity waves at an interface and in continuously stratified fluid are shown in Figure 1.4.

Whether at an interface or moving within a continuously stratified fluid, internal gravity waves are sometimes more laconically called 'gravity waves' or 'internal waves'. The former terminology is avoided since it is often specifically applied to surface waves. The latter terminology is used most often in this book to describe waves moving due to buoyancy in continuously stratified fluid. We use the term 'interfacial waves' to describe waves at the interface between hot and cold or fresh

a) Interfacial waves

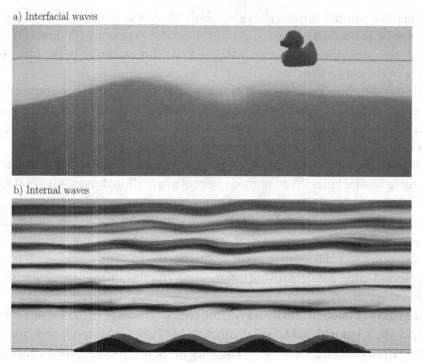

b) Internal waves

Fig. 1.4. a) An interfacial wave between fresh and underlying (dyed) salt water moving beneath a nearly flat surface, and b) internal waves in uniformly stratified fluid visualized by the displacement of dye lines spaced apart by 5 cm. The waves in b) result from the rightward translation of a model set of hills at the bottom of the image.

and salty water. Waves at a water–air interface are specifically referred to as 'surface waves' or 'water waves'.

In addition to buoyancy forces, internal gravity waves with periods comparable to a day feel the effects of the Earth's rotation through Coriolis forces. This class of waves we call 'inertial waves' if they occur at an interface and 'inertia gravity waves' if they occur in continuously stratified fluid. The terminology arises not because the waves possess inertia but because Coriolis forces arise from apparent accelerations occurring within a non-inertial (rotating) frame of reference.

1.4 Co-ordinate systems

In most circumstances fluid motions will be described in Cartesian co-ordinates with x and y locating horizontal co-ordinates and z the vertical. We denote a position in three-dimensional space by $\vec{x} \equiv (x, y, z)$. Velocities are correspondingly denoted by $\vec{u} \equiv (u, v, w)$.

Throughout we use arrows above symbols to denote vectors. The magnitude of the vector is denoted by $|\vec{u}| \equiv (u^2 + v^2 + w^2)^{1/2}$. Vectors with unit length are denoted by a superimposed hat so that $\hat{x} \equiv \vec{x}/|\vec{x}|$, for example, is the unit vector pointing in the x-direction.

To describe the motion of internal gravity waves on the Earth, one could employ spherical co-ordinates, but the motions are confined sufficiently close to the Earth's surface and with a sufficiently small horizontal extent compared to the Earth's radius that a Cartesian co-ordinate system suffices.

In this system, the z-axis is oriented upwards, pointing parallel to the net gravitational force that includes the combined effect of gravitational and centripetal forces on the rotating Earth. This is discussed in more detail in Section 1.10.

It is typical in geophysical fluid dynamics to orient the x-axis in the zonal direction, pointing from west to east, and to orient the y-axis in the meridional direction, pointing from south to north. Likewise, the velocity components u, v and w correspond to the zonal (eastward), meridional (northward) and vertical (upward) velocity, respectively.

In the study of internal gravity waves, the effective change of the Earth's rotation with latitude is ignored and so the distinction between north–south and east–west becomes irrelevant (see Exercises). The only thing that matters when considering the evolution of large-scale, slowly evolving internal gravity waves is the sign and magnitude of the local effects of rotation. This is discussed in more detail in Section 1.10.3. Here it suffices to say that we can arbitrarily orient the x- and y-axes. This is convenient in the consideration of plane waves for which the x-axis can be oriented in the horizontal direction of wave propagation thus rendering their description two- or even one-dimensional in space.

1.5 Lagrangian and Eulerian frames of reference

For conceptual convenience, it is sometimes useful to describe how a fluid evolves by imagining that it is composed of 'fluid parcels', which are infinitesimally small fluid elements. A parcel is not a real object like a molecule. Effectively, it represents the collective properties of fluid of a sufficiently large volume that atomic-level details can be ignored, but not so large that the properties of the parcel vary within its volume. Thus a fluid parcel can be assigned a density, velocity and so on.

The evolution of a fluid parcel can be described mathematically in a frame of reference that moves with the parcel. This is the Lagrangian description of a fluid. We denote the Lagrangian representation of a fluid parcel's properties with an 'L' subscript. For example, $C_L(t; \vec{x}(t))$ is a property (such as the concentration of injected dye, density, horizontal velocity, etc.) of a fluid parcel situated at $\vec{x}(t_0)$ at time t_0 and which has moved to position $\vec{x}(t_1)$ at time t_1, as illustrated in Figure 1.5.

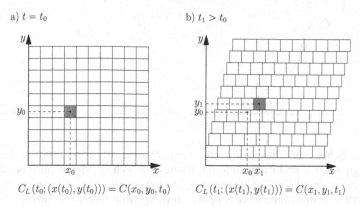

a) $t = t_0$ b) $t_1 > t_0$

$C_L(t_0; (x(t_0), y(t_0))) = C(x_0, y_0, t_0)$ $C_L(t_1; (x(t_1), y(t_1))) = C(x_1, y_1, t_1)$

Fig. 1.5. Schematics illustrating the relationship between Eulerian and Lagrangian co-ordinates. A fluid parcel located initially at $(x_0, y_0) \equiv (x(t_0), y(t_0))$, as shown in a), moves to position $(x_1, y_1) \equiv (x(t_1), y(t_1))$ at time t_1, as shown in b). In Lagrangian co-ordinates the properties of this parcel are tracked while it moves; in Eulerian co-ordinates the properties of the fluid passing a fixed position are tracked.

The semicolon in the argument to C_L is meant to indicate that knowing the position of the parcel as time evolves is not necessary in the Lagrangian frame. Only its initial position, $\vec{x}(t_0)$, is necessary for labelling the fluid parcel with which we are dealing at later times. Thus $C_L(t; \vec{x}(t))$ is a function of time alone and so we may denote changes to C_L in the Lagrangian frame by the ordinary time derivative, d/dt.

This is one reason why the Lagrangian description of a fluid is convenient. The statement that a conserved quantity s does not change following the motion of a fluid parcel is given simply by the differential equation

$$\frac{ds_L}{dt} = 0. \tag{1.1}$$

Though appealing for its simple description of conservation laws, the Lagrangian frame is often impractical for analytical and numerical calculations because it does not explicitly specify the differences between a parcel and its surroundings. To work out how density varies spatially at some time, one would have to trace the paths of all surrounding fluid parcels back to time $t = t_0$, when their properties were first prescribed.

The Eulerian frame is often more practical. Rather than focusing upon the evolution of a fluid parcel as it moves with a flow, one examines instead how the properties of the fluid change in time at a fixed position. The changes occur not only because the properties of the parcel vary in time but also because a train of parcels with different properties moves past that point.

Stratified fluids and waves

a) $C_L(t)$ b) $C(x,t)$

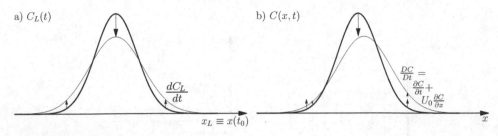

Fig. 1.6. Schematic illustrating the time change of concentration in a) a Lagrangian frame and b) an Eulerian frame in which the diffusing blob advects rightwards at speed U_0. The initial concentration is indicated by the thick solid line and its value a short time later is indicated by the thin line. Solid arrows indicate time variations in concentration due to diffusion. Dotted arrows indicate changes due to advection.

In the Eulerian frame we denote by $C(\vec{x},t)$ the property C of fluid at position $\vec{x}$ and at time t. Here $\vec{x}$ is independent of t and correspondingly we have dropped the L subscript employed in our Lagrangian definition. Translating from the Lagrangian to Eulerian frame is done through straightforward application of the chain rule of (partial) derivatives, which has the effect of switching from a moving to a stationary frame of reference.

To demonstrate this, consider the example in one spatial dimension as illustrated in Figure 1.6. Here $C_L(t)$ represents the concentration, for example of injected dye, associated with fluid parcels located initially at $x(t_0)$. The concentration is assumed to change in time due to diffusion, so that the peak concentration of the central fluid parcel decreases while the concentration of dye associated with parcels at the flanks increases. The rate of change of concentration is given by dC_L/dt. Even if the background flow is moving, in the Lagrangian frame the value of C_L is determined by their initial and not instantaneous position. The value of concentration in both space and time is revealed in the Eulerian frame, shown in Figure 1.6b. In this example, we assume the background flow moves at constant horizontal speed U_0. So, at a fixed position, x, the concentration $C(x,t)$ changes, both because of the instantaneous change in time of concentration $\partial_t C$, and because of the advection of the concentration gradient $U_0 \partial_x C$. Explicitly, after a short time the concentration in the Eulerian frame is related to that in the Lagrangian frame by

$$\frac{d}{dt} C_L(t; x(t_0) + U_0 t) = \frac{\partial C}{\partial t} + U_0 \frac{\partial C}{\partial x}. \tag{1.2}$$

The concentration changes at a point, both because the concentration of a parcel passing the point changes in time (the first term on the right-hand side of (1.2)), and because the surrounding parcels that move into and away from the point have different concentrations (the second term on the right-hand side of (1.2)).

Suppose the fluid parcels move at non-constant speed. In the Lagrangian frame, the change in concentration of a parcel situated at position x at time t is $dC_L(t; x(t))/dt$. In the Eulerian frame, by the chain rule, the time derivative becomes

$$\frac{d}{dt} C_L(t; x(t)) = \frac{\partial C}{\partial t} + \frac{dx}{dt} \frac{\partial C}{\partial x}.$$

The ordinary derivative dx/dt equals the instantaneous horizontal velocity, u, of the fluid, which itself is a function of position and time.

More generally, the Lagrangian time derivative of a property s of a fluid parcel moving in three dimensions is transformed into an Eulerian frame by writing

$$\frac{d\, s_L}{dt} = \frac{\partial s}{\partial t} + u \frac{\partial s}{\partial x} + v \frac{\partial s}{\partial y} + w \frac{\partial s}{\partial z} \equiv \frac{Ds}{Dt}. \tag{1.3}$$

In the last expression we have introduced a mathematically convenient notation for the material derivative:

$$\frac{D}{Dt} \equiv \frac{\partial}{\partial t} + u \frac{\partial}{\partial x} + v \frac{\partial}{\partial y} + w \frac{\partial}{\partial z} = \frac{\partial}{\partial t} + \vec{u} \cdot \nabla, \tag{1.4}$$

in which $\nabla \equiv (\partial_x, \partial_y, \partial_z)$ is the gradient operator.

The expression $\vec{u} \cdot \nabla s$ represents the 'advection terms' (sometimes also called the 'convective terms') of the material derivative. These are the terms that capture changes at a point due to fluid parcels moving through it at speed $\vec{u}$. The advection terms make the equations of motion nonlinear and so make analytic solutions difficult, if not impossible, to find.

Analogous to (1.1), which states that s is conserved in the Lagrangian frame, in the Eulerian frame the conservation law is given by the partial differential equation

$$\frac{Ds}{Dt} \equiv \frac{\partial s}{\partial t} + \vec{u} \cdot \nabla s = 0. \tag{1.5}$$

1.6 Equations of state

Liquids and gases are both treated as fluids because their motion is governed by nearly identical equations. They differ significantly in the equation of state which, for example, relates how the fluid's density changes in response to variations of temperature, pressure and, in the case of ocean water, salinity.

1.6.1 Equations of state for air

Throughout this book we will treat the atmosphere as an ideal gas. Thus the equation of state requires that its density, ϱ, is proportional to pressure, P, and is inversely proportional to temperature T (measured in Kelvin). Explicitly, we assume

$$P = \varrho R_a T, \tag{1.6}$$

in which $R_a = 287\,\text{J/(kg K)}$ is the ideal gas constant for dry air. Additional formulae that describe how properties change with varying concentrations of water vapour and other gases are neglected here.

1.6.2 *Equations of state for sea water*

Pressure does not significantly affect the density of water except in the abyss. Temperature and salinity pose the dominant influence upon density although the formulae describing this relationship are nontrivial. For example, Figure 1.7a shows that the density of fresh water is a nonlinear function of temperature, having a maximum around 4°C.

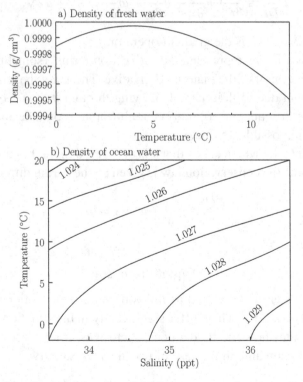

Fig. 1.7. a) Sketch of the density of fresh water as a function of temperature. b) Contours of density in g/cm^3 ($10^3\,\text{kg/m}^3$) as a function of temperature and salinity given for ranges typical of the world's oceans. Salinity is given in parts by weight of salt per thousand parts of water.

Likewise, at fixed temperature, the density is a nontrivial function of salt concentration. Salinity, S, is defined as the number of grams of salt dissolved in one kilogram of solvent. Though dimensionless, it is sometimes represented in units of $^o/_{oo}$ or 'ppt', meaning parts per thousand. Figure 1.7b plots constant density contours as a function of temperature and salinity for values representative of the ocean.

The equation of state used to produce this plot was established by the Joint Panel on Oceanographic Tables and Standards in 1981. It consists of three polynomials involving 41 empirical constants.

Provided the temperature and salinity do not vary too much, it is sufficient to make the linear approximation

$$\varrho = \varrho_0 \left[1 - \alpha_T (T - T_0) + \alpha_S (S - S_0) \right]. \tag{1.7}$$

Here ϱ_0 is the density at temperature T_0 and salinity S_0, α_T is the thermal expansion coefficient and α_S is the haline expansion coefficient. At 20°C for fresh water $\varrho_0 \simeq 0.99823 \, \text{g/cm}^3$, $\alpha_T \simeq 2.1 \times 10^{-4} \, \text{K}^{-1}$ and $\alpha_S \simeq 7.4 \times 10^{-4} \, \text{ppt}^{-1}$.

Note that if 34 g of salt is added to 1 kg of water, its density is not $1.034 \, \text{g/cm}^3$, but is approximately $1.025 \, \text{g/cm}^3$. The density is smaller because the volume of water expands somewhat upon incorporating the salt. In the abyss, pressure must also be accounted for in determining the density. This is discussed in Section 1.7.3.

1.7 Conservation of internal energy

The 'internal energy' of a fluid parcel usually refers to its heat energy, which is established using thermodynamics. The internal energy of sea water depends upon both temperature and salinity. In either case, the internal energy is distinct from mechanical energy, which is comprised of kinetic and potential energy (see Section 1.14).

Heat energy is conserved if it is neither absorbed nor radiated away during its evolution. Salinity is conserved if its diffusion is neglected. Conditions for the conservation of internal energy will be formulated separately for gases and liquids.

1.7.1 Thermodynamics of dry air

We have seen that the density of air depends both upon temperature and pressure through the ideal gas law (1.6). According to this equation, if air is compressed by increasing the pressure while keeping the temperature fixed, then its density increases. However, if heat energy is not given off or absorbed during the process of compression, then the air temperature will also increase and so the density will be somewhat less.

Another relationship between density and pressure is necessary to determine this temperature increase and hence the density change. Specifically, we assume that compression and expansion is an 'adiabatic' process, meaning that internal energy is conserved.

The internal energy per unit mass of a gas at temperature T (measured in Kelvin) is

$$E_I \equiv C_v T, \tag{1.8}$$

in which C_v is the specific heat of gas at constant volume and E_I is measured in units of energy per unit mass. So, for example, with $C_v = 718 \, \text{J}/(\text{kg K})$ for dry air, one would have to add 718 J of energy to 1 kg of air in order to raise its temperature by one degree.

Because its volume is held fixed the pressure of the heated gas increases. If the volume is allowed to expand so that its pressure returns to that before it was heated, the gas will cool slightly. Thus additional energy is required to heat 1 kg of air by one degree while keeping the pressure, not volume, constant. This energy, which includes both the internal energy and the work of the expanding gas, is given by $C_p T$, in which C_p is the specific heat at constant pressure. It turns out that C_p is related to C_v through

$$C_p = R_a + C_v. \tag{1.9}$$

Energy conservation requires that E_I changes in response to the work done on the gas through pressure. In one dimension, work energy is force times distance. Extending this to three dimensions we derive the differential expression

$$dE_I = -P d\left(\frac{1}{\varrho}\right). \tag{1.10}$$

The differential on the right-hand side is the volume per unit mass. If pressure acts to change the volume with the mass held fixed, the resulting change in energy per unit mass is dE_I.

Using (1.8), (1.10) and the ideal gas law (1.6) to eliminate T, we formulate the following differential equation relating P and ϱ:

$$\frac{dP}{d\varrho} = \gamma \frac{P}{\varrho}. \tag{1.11}$$

Here, $\gamma \equiv 1 + R_a/C_v \simeq 7/5$, as is typical for such diatomic gases as oxygen and nitrogen that comprise the majority of air.

If the density is ϱ_0 when the pressure is P_0, the solution of (1.11) is

$$\varrho = \varrho_0 \left(\frac{P}{P_0}\right)^{1/\gamma}. \tag{1.12}$$

Likewise, the temperature changes with pressure according to

$$T = T_0 \left(\frac{P}{P_0} \right)^{\kappa}, \tag{1.13}$$

in which

$$\kappa = 1 - 1/\gamma \simeq 2/7. \tag{1.14}$$

Sometimes (1.13) is written in terms of the 'Exner pressure', defined by

$$\Pi \equiv \left(\frac{P}{P_0} \right)^{\kappa}. \tag{1.15}$$

Thus the equation adopts the simple algebraic form $T = T_0 \Pi$.

The exponents being positive in (1.12) and (1.13) confirm that both density and temperature increase as the pressure increases. But the rate of density increase is not linear with pressure, as it would be if the air temperature was kept constant.

1.7.2 Potential temperature of a gas

The relationship between temperature and pressure given by (1.13) is used to define the 'potential temperature':

$$\vartheta \equiv T \left(\frac{P}{P_0} \right)^{-\kappa}. \tag{1.16}$$

In terms of the Exner pressure, $\vartheta = T/\Pi$.

If a parcel of air has temperature T and pressure P, then its potential temperature is the temperature it would have if brought adiabatically to the reference pressure P_0, typically taken to be that at sea level. This is illustrated schematically in Figure 1.8.

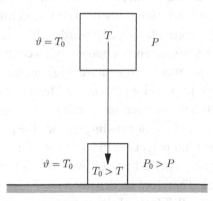

Fig. 1.8. Schematic illustrating the compressional heating of a fluid parcel as it moves adiabatically downwards to higher pressure surroundings. The potential temperature does not change following this motion.

As the parcel descends it is compressed due to the higher pressure at the surface. This heats the parcel increasing its temperature from T to T_0. So the potential temperature of the parcel both before and after it is displaced is T_0.

Potential temperature is defined to be a conserved quantity: the potential temperature for a parcel of air does not change following its adiabatic motion although its temperature, pressure and density may change. That is,

$$\frac{D\vartheta}{Dt} = 0. \tag{1.17}$$

This expresses the law of conservation of internal energy for a gas.

If we know the temperature of a fluid parcel at some height z_1 and the pressure $P(z)$ is prescribed, we can determine the temperature the parcel would have if displaced to any vertical level. From (1.16), its potential temperature is $\vartheta = T(z_1)(P(z_1)/P_0)^{-\kappa}$ and so at its displaced position z_2 the temperature would be

$$T_2 = \vartheta/(P(z_2)/P_0)^{-\kappa} = T(z_1)(P(z_1)/P(z_2))^{-\kappa}.$$

Knowing the new temperature, the ideal gas law (1.6) can be used to determine its new density:

$$\varrho_2 = P(z_2)/(R_a T_2).$$

Now, let us ignore the motion of fluid parcels and instead look at a snapshot of the atmosphere. At any vertical position z we can record the temperature $T(z)$ and pressure $P(z)$. Therefore, from (1.16) we can compute the vertical profile of the potential temperature $\vartheta(z)$. The resulting profiles are shown in Figure 1.9, for the case of an isothermal gas and for an atmosphere whose temperature decreases linearly with height in such a way that the potential temperature is constant with height.

In the former case the atmosphere is said to be stably stratified. This is because air brought downwards would be compressed and so have a higher temperature than the surrounding air at that level. Hence it would be less dense than its surroundings and would feel a buoyancy force pushing it upwards. Likewise, if displaced upwards it would cool and so would experience a buoyancy force directed downwards.

Generally, where the potential temperature $\vartheta(z)$ increases with height the air is stably stratified. If a fluid parcel with potential temperature $\vartheta_1 \equiv \vartheta(z_1)$ moves downwards to z_2, it still has potential temperature ϑ_1 but this is larger than the potential temperature $\vartheta(z_2)$ of its surroundings. Since the pressure of the surrounding air and the displaced fluid parcel are the same at z_2, from (1.16) we know that the temperature T_2 of the displaced parcel must be higher than the temperature $T(z_2)$ of its surroundings. Therefore its density $\varrho_2 = P(z_2)/(R_a T_2)$ is smaller than the density $\varrho(z_2) = P(z_2)/(R_a T(z_2))$ of the surrounding air.

These considerations show that the temperature may decrease with height but the atmosphere can still be stably stratified. The special case in which the atmosphere

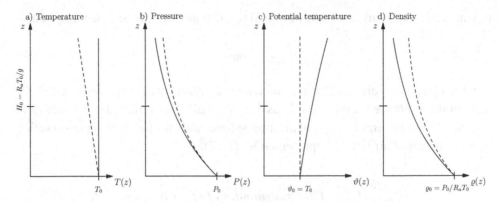

Fig. 1.9. Schematic showing vertical profiles of a) temperature and the corresponding b) hydrostatic pressure, c) potential temperature and d) density corresponding to an isothermal gas (solid line) and a gas whose density decreases at the adiabatic lapse rate (dashed line). All four plots indicate the density scale height, H_0, based upon the surface temperature, T_0. This has a value of the order of 8 km for air at room temperature.

is neutrally stable occurs if the potential temperature is a constant with height. For example, suppose the air between sea level and 100 m above the ground has a uniform potential temperature of 300 K. This means the air at the ground has an actual temperature of 300 K (or about 27°C). The air 100 m above has a lower temperature because the pressure is lower there. However, if this air was pushed to the ground, we know its pressure and temperature would increase to match those of the surrounding air. The ideal gas law therefore predicts that the parcel's density would be the same as its surroundings and so it would not experience any buoyancy forces.

With the additional consideration of the balance between vertical pressure gradient and buoyancy forces, known as hydrostatic balance (see Section 1.11 and Exercises), we find that the atmosphere is neutrally stable if the temperature decreases linearly with height at a rate of $\Gamma \simeq 9.8$ K/km. This is known as the 'adiabatic lapse rate'. It explains why, for example, mountain tops may be covered with snow on a summer day. It may be 20°C at the ground, but two kilometres up it is close to freezing.

From (1.12) we could correspondingly define the potential density as a function of the density and pressure to be

$$\varrho_{\,a,\mathrm{pot}} \equiv \varrho \left(\frac{P}{P_0} \right)^{-1/\gamma}.$$

This would represent the density of a fluid parcel if brought adiabatically to the ground. In combination with the ideal gas law (1.6) and (1.14), it turns out that the potential density is inversely proportional to the potential temperature. Explicitly

$$\varrho_{a,\text{pot}} = \varrho_0 \frac{T_0}{\vartheta},$$
(1.18)

in which ϱ_0 and T_0 are the density and temperature, respectively, measured at the level of the reference pressure P_0. Thus the potential temperature alone is sufficient to characterize the stratification of the atmosphere, and internal energy conservation of a dry parcel of air is well represented by (1.17).

1.7.3 Thermodynamics of sea water

Except in the deep ocean, the density of sea water changes negligibly as the surrounding pressure varies. Thus, neglecting salt and heat diffusivity, the law of conservation of internal energy for sea water can simply be represented by

$$\frac{D\varrho}{Dt} = 0.$$
(1.19)

A more precise expression that accounts for the enormous pressures in the abyss is given by

$$\frac{D\varrho_{\text{pot}}}{Dt} = 0,$$
(1.20)

in which

$$\varrho_{\text{pot}} = \varrho + \varrho_0 \frac{z}{H_\varrho}$$
(1.21)

is the 'potential density'. Here z (which is negative) measures the depth of the fluid parcel below the surface, ϱ_0 is a characteristic value of the density at the surface and H_ϱ is the density scale height (see Section 1.11). For the ocean $H_\varrho \simeq 200\,\text{km}$, which demonstrates that the fluid parcel would increase in density by a factor of about $1/40$ if it descended adiabatically from the surface to the ocean bottom at $5\,\text{km}$ depth.

Unlike for air, the potential temperature of sea water is not dependent upon the potential density. Temperature varies even more weakly with pressure than density, the potential temperature assuming the form

$$\vartheta_w = T + T_0 \frac{z}{H_T},$$
(1.22)

in which the temperature scale height is $H_T \simeq 2000\,\text{km}$. This is significantly larger than the density scale height, H_ϱ.

From vertical profiles of temperature and salinity in the ocean, and using the equation of state for sea water, one can reconstruct a vertical profile of potential

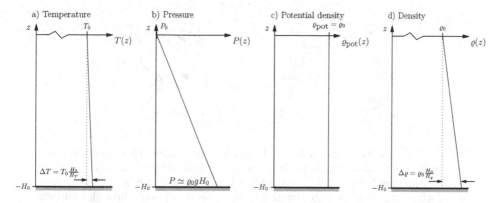

Fig. 1.10. Schematic showing vertical profiles of a) temperature, b) pressure, c) potential density, which is taken to be constant and d) density corresponding to a liquid of constant salinity and great depth.The temperature and density increase with depth at a rate given respectively by the temperature scale height $H_T \simeq$ 2000 km and the density scale height $H_\varrho \simeq 200$ km. Both are much larger than the ocean depth $H_0 \simeq 5$ km.

density. A specific example is illustrated in Figure 1.10, which shows the profiles of temperature and density resulting from a constant potential density profile extending over the ocean depth. As in the atmosphere, the temperature can decrease from the bottom to the top of the ocean but still be stably stratified.

1.8 Conservation of mass

The law of conservation of mass states that a parcel of fluid does not gain or lose mass during its motion. The partial differential equation for this law is commonly called the 'continuity equation'.

Many introductory fluid dynamics texts derive the general formula by starting with a one-dimensional box model, as illustrated in Figure 1.11a. Mass per unit time, $\partial_t(\Delta x A \varrho)$, increases in the box if more mass enters the left side (at rate $(Au\varrho)|_{x-\Delta x/2}$) than leaves the right side (at rate $(Au\varrho)|_{x+\Delta x/2}$). Taking the difference and letting $\Delta x \to 0$ gives the differential formula for the rate of change of density in time:

$$\frac{\partial \varrho}{\partial t} = -\frac{\partial}{\partial x}(u\varrho). \qquad (1.23)$$

This is the continuity equation in one dimension.

Note that the cross-sectional area, A, does not enter into the resulting expression. The negative sign on the right-hand side of (1.23) indicates that the density increases if the mass flux, ϱu, decreases with x.

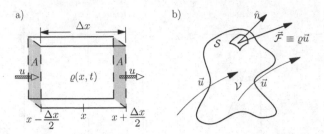

Fig. 1.11. Schematics used to derive the mass conservation law a) in one dimension and b) in three dimensions. In a) the flow passes through a cross-sectional area A with speed $u(x - \Delta x/2)$ on the left-hand side of the box, and exits with speed $u(x + \Delta x/2)$ on the right-hand side. In b) the flow passes through a surface area S containing a volume $\mathcal{V}$. The unit normal vector to the surface, $\hat{n}$, always points outwards.

An alternative derivation that generalizes this result to more dimensions makes use of the 'divergence theorem' (also sometimes called Gauss' theorem) from vector calculus. The derivation is more elegant than box models in the sense that it does not explicitly assume a Cartesian co-ordinate system.

Consider flow through an arbitrary volume, as depicted in Figure 1.11b. This closed blob, which is fixed in space, has an interior denoted by $\mathcal{V}$ and its surface is denoted by S. At any point on S we can define a unit normal vector $\hat{n}$ that points outside the volume. The component of flow that moves into the blob at any point on S is $-\vec{u} \cdot \hat{n}$. The dot-product removes that component of flow moving parallel to the surface and the negative sign is included because we are assigning inward flow to be positive.

The mass flux, which is the mass per unit time advected across a unit area is

$$\vec{\mathcal{F}}_\varrho \equiv \varrho \vec{u}. \tag{1.24}$$

The component of this flux entering the volume per unit area of the surface is $-\vec{\mathcal{F}}_\varrho \cdot \hat{n}$. So the total mass entering the volume per unit time is the surface integral $\iint_S -\vec{\mathcal{F}}_\varrho \cdot \hat{n} \, dS$. Mass conservation requires this integral to equal the increase in mass per unit time within the blob as given by the volume integral of ϱ. That is

$$\frac{\partial}{\partial t} \iiint_\mathcal{V} \varrho \, dV = \iint_S -\vec{\mathcal{F}}_\varrho \cdot \hat{n} \, dS. \tag{1.25}$$

This is where the divergence theorem comes in. The integral over the closed surface on the right-hand side of (1.25) is equal to the volume integral of $-\nabla \cdot \vec{\mathcal{F}}_\varrho = -\nabla \cdot (\varrho \vec{u})$, the negative divergence of the mass flux. Combining everything into one integral we have

$$\iiint_{\mathcal{V}} \frac{\partial \varrho}{\partial t} + \nabla \cdot (\varrho \vec{u}) \, dV = 0.$$

This result holds for any volume within the fluid. It is the arbitrariness of the choice of $\mathcal{V}$ and the assumption that the integrand is continuous which leads us to conclude that the integrand itself must be identically zero everywhere in space and for all time. Thus we have the law of conservation of mass:

$$\frac{\partial \varrho}{\partial t} + \nabla \cdot (\varrho \vec{u}) = 0. \tag{1.26}$$

This is the general form of the continuity (mass conservation) equation.

This conservation equation is readily put into flux form by moving the divergence expression to the right-hand side of the equation. In Cartesian co-ordinates, this becomes

$$\frac{\partial \varrho}{\partial t} = -\nabla \cdot \vec{\mathcal{F}}_\varrho = -\left[\frac{\partial}{\partial x}(u\varrho) + \frac{\partial}{\partial y}(v\varrho) + \frac{\partial}{\partial z}(w\varrho) \right]. \tag{1.27}$$

In one dimension, we recover (1.23).

1.9 Incompressibility

Using the vector identity $\nabla \cdot (\varrho \vec{u}) = \vec{u} \cdot \nabla \varrho + \varrho \nabla \cdot \vec{u}$, (1.26) can alternatively be written

$$\frac{D\varrho}{Dt} = -\varrho \nabla \cdot \vec{u}. \tag{1.28}$$

Here D/Dt is the material derivative defined by (1.4). Thus, although the mass of a fluid parcel may not change, (1.28) states that its density can change following its motion. For example, the density of a fluid parcel would increase if its mass remained fixed while it was compressed due to a converging flow field ($\nabla \cdot \vec{u} < 0$), thus making its volume decrease.

If we further insist that the fluid is incompressible, then we require that its density, not just its mass, does not change following the motion of a fluid parcel. That is

$$\frac{D\varrho}{Dt} = 0. \tag{1.29}$$

Strictly speaking, this is the 'incompressibility condition'. However, it is often more practical to combine (1.28) and (1.29) so that the incompressibility condition is expressed as the diagnostic equation

$$\nabla \cdot \vec{u} = 0. \tag{1.30}$$

For flow restricted to two dimensions, the incompressibility condition in the form (1.30) means that one can define a scalar function ψ, known as the streamfunction, from which the components of $\vec{u}$ can be determined from derivatives of ψ. As such, the streamfunction provides a mathematically useful representation of a vector function by a scalar function similar to the way that a conservative force field can be represented by the gradient of a scalar potential.

For flows restricted to the x–z plane, $\vec{u} = \nabla \times (\psi \hat{y})$ so that

$$u = -\frac{\partial \psi}{\partial z} \text{ and } w = \frac{\partial \psi}{\partial x}. \tag{1.31}$$

It follows from a standard vector identity that (1.30) is automatically satisfied if $\vec{u}$ is defined as the curl of a function.

The definition of the streamfunction extends to non-Cartesian geometries if symmetry dictates that the flow has only two degrees of freedom. For example, one can represent the velocity $\vec{u} = (u_R, 0, w)$ of axisymmetric flow in a cylindrical geometry with no azimuthal flow ($u_\theta = 0$) by $\vec{u} = \nabla \times (\psi \hat{\theta})$ in which $\psi \equiv \psi(R, z)$.

Sometimes the implicit definition of the streamfunction is given in a left- rather than right-handed co-ordinate system. For example, for motion in the x–y plane one could define $\vec{u} = \hat{z} \times \nabla \psi$, in which case $u = -\partial_y \psi$ and $v = \partial_x \psi$. As long as this formulation is consistent in the equations, the results will be the same. But pulling the unit vector outside the gradient operator can be a problem in non-Cartesian co-ordinate systems, when the unit vector is itself a function of position. Then (1.30) may not be satisfied.

Given ψ, it is straightforward to compute $\vec{u}$ as in (1.31). In the converse problem, ψ is found from $\vec{u}$ by integration. For example, suppose $\vec{u}(x, y) = \Omega(-y, x)$. This is the velocity field corresponding to solid-body rotation at frequency Ω in the x–y plane (see Section 1.13.1). Assuming $\vec{u} = (\partial_y \psi, -\partial_x \psi)$ then, matching the x-component of the two expressions, we find $\psi(x, y) = -\Omega y^2/2 + f(x)$, in which f at this point is an arbitrary function of x resulting from integrating in y. Substituting this into the y-components gives $-f'(x) = \Omega x$. The right-hand side is independent of y, as must occur because $\nabla \cdot \vec{u} = 0$. Hence $f(x) = -\Omega x^2/2 + C$, in which C is a constant of integration. Arbitrarily setting $\psi(0, 0) = 0$, the streamfunction for flow describing solid-body rotation is given by $\psi(x, y) = -\Omega(x^2 + y^2)/2$. Lines of constant ψ are circles.

Generally, contours of constant ψ are called streamlines. These represent the paths followed by fluid parcels in steady flow, as is evident from the vector identity $\vec{u} \cdot \nabla \psi = 0$.

Plots showing contours of constant ψ are particularly useful for visualizing flow, as shown for example in Figure 1.12. The lines illustrate the paths followed by fluid

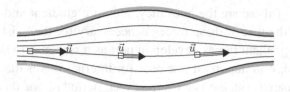

Fig. 1.12. Schematic showing the motion of a fluid parcel following a streamline. Its associated velocity vector lies parallel to streamlines and the speed is faster where streamlines are more closely packed.

parcels. The flow is faster where the lines are closer together, indicating that the across-flow gradient of ψ is larger.

Another consequence of the incompressibility condition written in the form (1.30) is that for any scalar function s, we have

$$\nabla \cdot (s\vec{u}) = \vec{u} \cdot \nabla s. \tag{1.32}$$

In particular, if s is a conserved quantity, then we can rewrite the conservation law (1.5), $Ds/Dt = 0$, into the 'flux-form'

$$\frac{\partial s}{\partial t} = -\nabla \cdot (s\vec{u}). \tag{1.33}$$

This says that s changes at a fixed position in time because the flux of s, $\vec{\mathcal{F}}_s = s\vec{u}$, diverges about that point.

1.10 Conservation of momentum

Momentum conservation gives rise to what is effectively Newton's force law, $F = ma$: force equals mass times acceleration. More precisely, the conservation law states that the momentum density (momentum per unit volume) of a fluid parcel changes in time due to the sum of the forces acting on it.

The change in time of the momentum density is taken following the motion of the fluid parcel and so is represented by the material derivative: $D(\varrho\vec{u})/Dt$. Applying the product rule of derivatives and assuming the fluid is incompressible so that ϱ satisfies (1.29), we find

$$D(\varrho\vec{u})/Dt = \varrho D\vec{u}/Dt. \tag{1.34}$$

This equation is analogous to the statement from Newtonian mechanics that the change in time of a particle's momentum equals the particle's mass times its acceleration.

A wide variety of forces can act upon a fluid parcel to accelerate or decelerate it. Only those relevant to the motion of internal waves will be considered here. The most significant of these are the buoyancy, pressure gradient and Coriolis forces. The presence of the first of these forces is necessary for the existence of internal waves. Pressure gradient forces develop in part as a response to buoyancy forces. Coriolis forces play an important role only for internal waves that evolve on long time-scales compared with the Earth's rotational period of one day.

In laboratory experiments, viscous damping forces may also play a significant role, though generally they are negligible for internal waves on atmospheric and oceanic scales. Numerical simulations of internal waves often include viscous effects as well in order to damp small-scale noise that would otherwise render the code numerically unstable. Although for the most part viscosity will be neglected in the theory of internal waves, here we include the terms in the momentum equation that describe the diffusion of momentum by viscous stresses.

Thus we have the law of conservation of momentum

$$\varrho \frac{D\vec{u}}{Dt} = -\nabla P + \vec{g}\varrho - (2\vec{\Omega}) \times (\varrho\vec{u}) + \mu\nabla^2\vec{u}. \qquad (1.35)$$

This is a particular example of the Navier–Stokes equations written for an incompressible fluid. In the absence of viscosity, the momentum equations are sometimes referred to as the Euler equations, though Euler himself did not explicitly include buoyancy and Coriolis forces.

Equation (1.35) says that the sum of the pressure gradient, buoyancy, Coriolis and viscous damping forces acts to accelerate or decelerate a fluid parcel. The formulation and interpretation of each of the four forces is discussed below.

1.10.1 The buoyancy force

All fluids on Earth feel the force of gravity. This force acting per unit volume on a fluid parcel is

$$\vec{F}_{\text{buoyancy}} = \vec{g}\varrho. \qquad (1.36)$$

Here, $\vec{g}$ is a downward pointing vector whose magnitude is the effective acceleration of gravity. Explicitly, in a Cartesian co-ordinate system $\vec{g} = -g\hat{z}$.

To be more precise, $\vec{g}$ includes not only the effect of gravitational attraction but also of centripetal acceleration, which is negligible near the poles and strongest at the equator. However, the Earth bulges moderately at the equator so that gravity exerts a stronger inward radial force there. The combination conspires so that the sum of the gravitational and centripetal forces gives an almost constant inward force around the globe.

Thus in most geophysical applications we can take the magnitude of $\vec{g}$ to hold the constant value $g \simeq 9.80\,\text{m/s}^2$.

1.10.2 The pressure gradient force

Simply put, the 'pressure gradient force' drives fluid from regions of high to low pressure. The larger the pressure change, the larger the magnitude of the force. The mathematical expression for this force is

$$\vec{F}_{\text{pressure}} = -\nabla P. \tag{1.37}$$

The negative sign ensures the force acts from high to low pressure. Pressure itself has units of force acting per unit area; $\vec{F}_{\text{pressure}}$ has units of force per unit volume.

We have used an upper-case P in (1.37) to represent the total pressure in the fluid. The total pressure at a point is set predominantly by the 'hydrostatic pressure', which is the cumulative weight per unit horizontal area of fluid lying above the point, as illustrated in Figure 1.13. This factor can be subtracted off and the resulting difference between the total and hydrostatic pressure is denoted by p, the dynamic pressure (see Section 1.11).

Dynamic pressure changes are established by the motion of the fluid. However, the fluid also moves in response to the pressure field. Thus one should not think of the influence of pressure upon a fluid's motion as a cause-and-effect relationship. The pressure and flow field change in concert.

1.10.3 The Coriolis force

The Coriolis force is not a physically applied force but an apparent force that acts upon objects moving in an accelerating (non-inertial) frame of reference. In this case the acceleration is due not to a change in speed but to a change of direction. This force significantly affects the large-scale motions of the atmosphere and ocean on the rotating Earth.

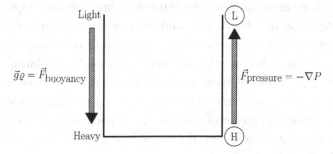

Fig. 1.13. Schematic illustrating how the buoyancy force (a downward force due to the weight of fluid) is equal and opposite to the pressure-gradient force (an upward force pointing from high to low pressure) for a fluid in hydrostatic balance.

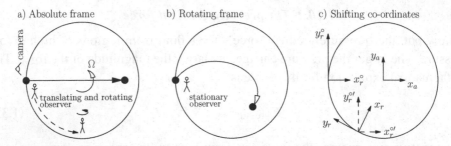

Fig. 1.14. Thought experiment of sliding puck (solid circle) on a rotating table seen a) by a camera in the absolute frame and b) by an observer in the rotating frame of reference (indicated by the stick figure). Note the camera sees the observer on the table rotate and translate along a circular arc. c) Co-ordinate systems used to shift from an absolute to rotating frame.

To illustrate the action of the Coriolis force, imagine the circumstance depicted in Figure 1.14 in which a puck slides at constant speed, c, across the diameter of a frictionless table that rotates counter-clockwise at frequency Ω. A camera records the event from the sidelines in what is called the laboratory, inertial or absolute frame of reference. The camera observes that the puck follows a straight line path across the table, as shown in Figure 1.14a. In contrast, to an observer standing on the table, the puck appears to follow a curved path, as shown in Figure 1.14b. Thus, in the rotating frame of reference it looks as though an external force is responsible for the deflection when, in fact, the camera demonstrates that no deflection occurs in an absolute frame of reference.

This thought experiment can be used to formulate the mathematical description of the Coriolis force. Geometric manipulation allows us to translate from a co-ordinate system (x_a, y_a) in the absolute frame to one in the rotating frame (x_r, y_r). Knowing the puck's position in time measured in the absolute frame, one can then find its position in time, and hence its apparent acceleration, in the rotating frame.

Shifting from one co-ordinate system to the other involves two steps, as illustrated in Figure 1.14c. First, the observer's co-ordinate system translates along a circular arc so that what is originally measured in $(x_r^{\,o}, y_r^{\,o})$ co-ordinates is later measured in $(x_r^{\,o\prime}, y_r^{\,o\prime})$ co-ordinates. Second, the observer rotates while translating along the arc so the $(x_r^{\,o\prime}, y_r^{\,o\prime})$ co-ordinates must be rotated to (x_r, y_r) co-ordinates. The details, which are left as an exercise, reveal two conceptually distinct accelerations. One, which depends upon the apparent rotation of the puck about the observer, is the centripetal acceleration. The other, which is independent of the puck's position on the table, is the Coriolis acceleration.

Large-scale geophysical flows on the Earth are predominately horizontal, so we can adapt the rotating table analogy above to the motion of the atmosphere and ocean. In general, the Coriolis force is

$$\vec{F}_{\text{Coriolis}} = -2\vec{\Omega} \times (\varrho\vec{u}), \tag{1.38}$$

in which $\vec{\Omega}$ has magnitude Ω (measured in radians per unit time) and the vector is directed parallel to the rotation axis.

At the North Pole the horizontal components of $\vec{u}$ are perpendicular to $\vec{\Omega}$. Thus we have

$$\vec{F}_{\text{Coriolis}} = -\varrho(-2\Omega_e v, 2\Omega_e u, 0),$$

in which $\Omega_e = 2\pi/\text{day} \simeq 1.5 \times 10^{-4}\,\text{s}^{-1}$ is the angular frequency of the Earth's rotation.

More generally, at a latitude ϕ (with $\phi = 90°$ at the North Pole and $\phi = -90°$ at the South Pole)

$$\vec{F}_{\text{Coriolis}} = -\varrho(-fv, fu, 0), \tag{1.39}$$

in which f is the 'Coriolis parameter' defined by

$$f \equiv 2\Omega_e \sin\phi. \tag{1.40}$$

In Section 1.13.1 we will show that the vorticity of an object rotating at angular frequency Ω is $\zeta = 2\Omega$. Thus an astronaut looking down upon the North Pole sees the planet itself has vorticity equal to the Coriolis frequency $f = 2\Omega_e$.

In the northern hemisphere $f > 0$ and in the southern hemisphere $f < 0$. This means, in particular, that an astronaut looking down upon the North Pole sees anticlockwise rotation, as in Figure 1.15a, but sees clockwise rotation when looking down upon the South Pole, as in Figure 1.15b. Motion is deflected rightwards by

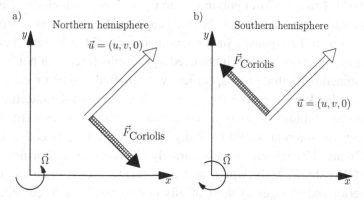

Fig. 1.15. The Coriolis force a) deflects motion to the right in the northern hemisphere, where the sense of background rotation is counter-clockwise and b) deflects motion to the left in the southern hemisphere, where the sense of background rotation is clockwise.

Coriolis forces in the northern hemisphere, consistent with the illustration shown in Figure 1.15b. In the southern hemisphere Coriolis forces act to deflect motion leftwards.

About a fixed latitude, f corresponds to an approximately constant frequency, f_0, which is called the 'inertial frequency' or the 'Coriolis frequency'. The former definition, which we use here, gets its name because it is related to the rotational frequency recorded by a stationary observer in outer space – someone in an inertial (non-accelerating) frame of reference. If the meridional scale of the motion is sufficiently small that $f \simeq f_0$ can be treated as a constant in the equations of motion, then the flow is said to exist on the 'f-plane'. In the atmosphere and oceans at midlatitudes a characteristic value of the inertial frequency is $f_0 \simeq 10^{-4}\,\mathrm{s}^{-1}$.

Generally, this is a good approximation for internal waves with horizontal wavelengths of the order of hundreds of kilometres or less and which do not propagate north and south by more than approximately $10°$ latitude.

In some circumstances, particularly near the equator or for large-scale flows at midlatitudes, the variations of f with latitude result in non-negligible effects that can be captured by a linear approximation:

$$f \simeq f_0 + \beta y. \tag{1.41}$$

Here $f_0 \equiv 2\Omega_e \sin\phi_0$ is the Coriolis parameter computed at a characteristic latitude, ϕ_0, the variable y measures the meridional distance from this latitude, and

$$\beta = 2\Omega_e \cos\phi_0 / R_e \tag{1.42}$$

is the rate of change of the Coriolis parameter with meridional distance about ϕ_0. In the denominator, R_e is the radius of the planet. At the equator on Earth, $\beta \simeq 2.3 \times 10^{-11}\,(\mathrm{ms})^{-1}$. Flow evolving in circumstances in which the effects of the Earth's curvature are non-negligible but well captured by this linear approximation is said to exist on a 'beta-plane'. Such phenomena are said to feel the 'beta-effect'.

Internal gravity waves that are influenced by Coriolis forces but not by the beta-effect are sometimes referred to specifically as inertial waves or inertia gravity waves. In this book we use the former terminology to describe interfacial waves and the latter terminology to describe internal waves in continuously stratified fluid.

The theory for internal waves typically neglects the beta-effect. Waves additionally influenced by this effect are separately characterized as 'planetary waves' (also called 'Rossby waves'), which are waves that feel the curvature of the planet through meridional changes in the Coriolis parameter. Though consideration of planetary waves is beyond the scope of this book, they are discussed in standard textbooks in geophysical fluid dynamics.

1.10.4 The viscous damping force

One can think of viscosity acting in a fluid in the same way that friction acts to retard the motion of a particle sliding on a rough surface. A fast-moving fluid parcel is slowed by an adjacent fluid parcel that moves more slowly. Conversely, since for every action there is an equal and opposite reaction, the speed of the slow parcel increases. Thus the diffusion of momentum by viscosity has a tendency to make the fluid move with a more uniform speed. Only the action of external forces and stresses exerted by boundaries drives the fluid away from this constant velocity state. Thus, for example, a viscous fluid between two horizontal plates adopts a linearly changing velocity profile between the lower and upper plate, if each moves horizontally at a different but steady speed.

This example guides us to develop a formula for the effect of viscosity upon a fluid. We assume that the stress exerted by fluid parcels rubbing together establishes a velocity gradient proportional to the applied stress. The flux of momentum in the x-direction resulting from fast fluid accelerating slower fluid and slower fluid retarding faster fluid is thus given by

$$\vec{\mathcal{F}}_{\mu x} = -\mu \nabla u. \tag{1.43}$$

Similarly the fluxes in the y- and z-directions are written by replacing u with v and w, respectively, in (1.43). The proportionality constant, μ, is called the 'molecular viscosity'. This is an empirically measured property of the fluid, which depends upon its chemical composition and other factors such as temperature. A fluid is said to be 'Newtonian' if viscous stresses are proportional to velocity gradients and if μ can be treated as approximately constant. This is a good description of most atmospheric and oceanic phenomena. Typical values of the molecular viscosity of water and air are $\mu \simeq 10^{-3}$ Pa s and $\mu \simeq 2 \times 10^{-5}$ Pa s, respectively.

The negative sign on the right-hand side of (1.43) indicates that momentum is transported down-gradient from fast to slow. One might compare this formula with Fick's law for the diffusion of heat, in which the flux of heat energy is directed from hot to cold at a rate proportional to the temperature gradient.

A fluid parcel accelerates if the divergence in the momentum flux due to viscous stress is non-zero. That is, viscous forces acting on the three components of velocity are represented by

$$\vec{F}_{\mu} = -\nabla \cdot \vec{\mathcal{F}}_{\mu} \simeq \mu \nabla^2 \vec{u}, \tag{1.44}$$

in which we have assumed that μ is approximately constant.

Whereas the rate of momentum diffusion is determined by the molecular viscosity, the resulting deceleration of fluid by viscous stresses is determined by the kinematic viscosity, defined by

$$\nu = \mu/\varrho_0, \tag{1.45}$$

in which ϱ_0 is the density of the fluid. Typical values of the kinematic viscosity of water and air at standard temperature and pressure are $\nu \simeq 10^{-6} \, \mathrm{m^2/s}$ and $\nu \simeq 2 \times 10^{-5} \, \mathrm{m^2/s}$, respectively.

A measure of the relative importance of viscosity upon the evolution of a fluid is given by the Reynolds number:

$$\mathrm{Re} = \frac{\mathcal{U}\mathcal{L}}{\nu}, \tag{1.46}$$

in which $\mathcal{U}$ and $\mathcal{L}$ are characteristic velocity and length scales of the fluid. If the Reynolds number is large (typically greater than $10\,000$) then viscous effects are negligibly small and the fluid is effectively inviscid.

For example, $\mathcal{L}$ could represent the horizontal wavelength and $\mathcal{U}$ the maximum vertical velocity of a wave. An internal wave created in a tank of salt-stratified water might have $\mathcal{L} = 10\,\mathrm{cm}$ and $\mathcal{U} = 0.1\,\mathrm{cm/s}$, in which case $\mathrm{Re} = 100$. Though viscosity does not play a dominant role in setting the structure of such waves, viscosity will act non-negligibly to attenuate them over time.

In contrast, the Reynolds number is of the order of millions or more for internal waves in the atmosphere and ocean with characteristic length scales greater than tens of metres and velocity scales greater than tens of centimetres per second. Even in turbulent fluids, which have much larger effective viscosities, the Reynolds number is so large that viscous effects are unimportant. Therefore, in the theory of internal waves it is usual to neglect viscous damping forces.

There are some exceptions. As will be shown in Section 5.2, it is sometimes necessary to include viscous effects to avoid mathematical singularities that arise in the theory of waves generated by oscillating bodies. Also, numerical models often include viscosity in order to damp small-scale numerical noise. But, with Reynolds numbers of 1000 or greater, the evolution of the internal waves they simulate is negligibly affected by viscosity over sufficiently short simulation times.

1.11 Hydrostatic balance

In stationary fluid, Coriolis forces are absent and the material derivative is identically zero. Therefore the momentum conservation equation, (1.35), predicts that the pressure is constant at each horizontal level and varies vertically according to

$$\frac{dP}{dz} = -g\varrho. \tag{1.47}$$

This is the statement of hydrostatic balance.

Note that ϱ in (1.47) must be a function of z alone. Otherwise P would vary horizontally, which cannot be the case for a fluid at rest. Letting P_{00} be the pressure at an arbitrary level z_0, the hydrostatic pressure is given explicitly by integrating (1.47) to give

$$P(z) = P_{00} - g \int_{z_0}^{z} \varrho \, dz = P_{00} + g \int_{z}^{z_0} \varrho \, dz. \tag{1.48}$$

The latter formula is useful in applications for which the reference pressure level is at the top of the domain. Equation (1.48) reveals that the hydrostatic pressure is determined by the weight per unit horizontal area of the fluid above z.

Setting $z_0 = 0$ at the ocean surface, the pressure in the ocean interior is $P(z) = P_{00} + g \int_{z}^{0} \varrho \, dz$. If the density $\varrho = \varrho_0$ is constant with depth, we have

$$P(z) = P_{00} - g\varrho_0 z. \tag{1.49}$$

Thus the pressure increases linearly with depth, as illustrated in Figure 1.16a. In particular, (1.49) shows that the pressure increases by approximately 1 atmosphere with every 10 m depth; the weight of the top 10 m of the ocean is equal to the weight of the entire atmosphere above.

Assuming the atmosphere is isothermal having temperature T_0, the ideal gas law (1.6) can be combined with (1.47) to give differential equations for ϱ or P alone.

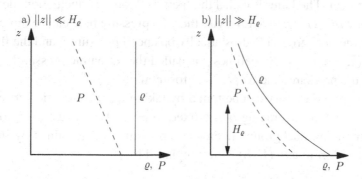

Fig. 1.16. Hydrostatic background pressure associated with background density that a) is approximately constant with height (as is the case of the ocean and in the atmosphere over heights that are small compared with the density scale height, H_ϱ) and b) decreases exponentially with height (as is the case of the atmosphere over large heights compared with H_ϱ).

Thus we find

$$\varrho = \varrho_0 \exp(-z/H_\varrho), \tag{1.50}$$

and

$$P(z) = P_{00} \exp(-z/H_\varrho), \tag{1.51}$$

as illustrated in Figure 1.16b.

Both the pressure and density decrease exponentially with height over an e-folding depth H_ϱ, a quantity known as the 'density scale height', or simply the 'scale height'. Explicitly, in an isothermal atmosphere

$$H_\varrho = \frac{R_a T_0}{g}. \tag{1.52}$$

This formula follows by applying the hydrostatic balance condition (1.47) to (1.51) and using the ideal gas law (1.6) at $z = 0$.

Taking $T_0 \simeq 288\,$K (about 15°C), we find $H_\varrho \simeq 8.4\,$km, a typical value for the density scale height near the ground. Near the tropopause the temperature is in fact cooler and the local value of the scale height is smaller: $H_\varrho \simeq 6.4\,$km. A typical value for the troposphere is taken to be $H_\varrho \simeq 7.4\,$km.

In a moving fluid, the pressure and density change very little if at all from their values when the fluid is stationary. It is therefore convenient to subdivide the density field into a static, z-dependent function and a dynamic spatially and temporally varying function. The former is called the background density and is denoted by $\bar\rho(z)$. The latter is called the 'perturbation' or 'fluctuation' density, and is denoted by $\rho(\vec{x}, t) \equiv \varrho - \bar\rho$. Likewise the total pressure field can be represented as the sum of the background hydrostatically balanced pressure $\bar p$ and the fluctuation pressure $p(\vec{x}, t) \equiv P - \bar p$. Sometimes p is called the 'dynamic pressure', this being the part of the pressure field responsible for motion.

The background density can be found by taking a characteristic vertical profile of ϱ by horizontally averaging ϱ at a fixed time or also by averaging over time. The corresponding background hydrostatic pressure is determined by integrating the analogous formula to (1.48) with ϱ replaced by $\bar\rho$:

$$\frac{d\bar p}{dz} = -g\bar\rho. \tag{1.53}$$

This is the equation for 'background hydrostatic balance'.

Substituting $P = \bar{p} + p$ and $\varrho = \bar{\rho} + \rho$ in the general formula for the momentum conservation equations (1.35), writing the buoyancy and Coriolis forces explicitly using (1.36) and (1.39), neglecting viscosity and invoking background hydrostatic balance using (1.53), we have

$$\varrho \left(\frac{Du}{Dt} - fv \right) = -\frac{\partial p}{\partial x},$$

$$\varrho \left(\frac{Dv}{Dt} + fu \right) = -\frac{\partial p}{\partial y}, \quad (1.54)$$

$$\varrho \frac{Dw}{Dt} = -\frac{\partial p}{\partial z} - g\rho.$$

These describe the law of conservation of momentum with the relevant Coriolis terms given explicitly on the left-hand side of the horizontal momentum equations. By invoking background hydrostatic balance, only the dynamic pressure and density fluctuations responsible for the buoyancy-driven motion remain explicitly on the right-hand side of the equations.

1.12 The Boussinesq and anelastic approximations

The internal energy, mass and momentum conservation laws give a coupled set of partial differential equations that describe a wide variety of fluid motions. In particular, they capture the dynamics not only of internal gravity waves but also of sound waves. For the development of internal gravity wave theory, it is useful to simplify the equations in a way that removes the dynamics of sound while keeping the dynamics of internal gravity waves. The resulting simplified set of equations is easier to manipulate into a form that, subject to further approximations, may be solved analytically. These equations are also useful in numerical simulations of internal gravity waves: without having to resolve the fast dynamics of sound, numerical simulations can take longer time steps without becoming numerically unstable.

Eliminating sound waves is done through what are called the 'Boussinesq' and 'anelastic' approximations. The former assumes that the background density varies vertically by a relatively small amount, as in the ocean; the latter allows the background density to vary by a large fraction of itself, as in the atmosphere over vertical distances comparable to the density scale height. The anelastic approximation is usually derived for a gas. Here we also include its counterpart for a liquid, which we will refer to as a 'non-Boussinesq liquid'.

Before discussing these approximate equations, we briefly review the theory of sound waves.

1.12.1 The speed of sound

In a stationary homogeneous fluid, the density can change in response to pressure variations. Assuming the response is adiabatic, the fields are related by

$$\frac{DP}{Dt} = c_s^2 \frac{D\varrho}{Dt},$$
(1.55)

in which

$$c_s^2 \equiv \frac{dP}{d\varrho}.$$
(1.56)

This denotes how pressure changes adiabatically due to density fluctuations.

Thus the continuity equation in the form of (1.28) can be written

$$\frac{1}{c_s^2} \frac{DP}{Dt} = -\varrho \nabla \cdot \vec{u}.$$
(1.57)

The motion of sound waves generally involves relatively fast and small fluctuations, allowing us to neglect Coriolis forces and the advection terms in the momentum and continuity equations, (1.54) and (1.28), respectively. Combining the resulting equations by eliminating velocity (see Exercises) gives the following formula describing the evolution of pressure changes:

$$\frac{\partial^2 P}{\partial t^2} = c_s^2 \nabla^2 P.$$
(1.58)

This is the classic 'wave equation' in which we now identify c_s as the speed of sound.

For a dry adiabatic gas, the sound speed is given explicitly in terms of pressure and density by combining (1.6), (1.11) and (1.56) to give

$$c_s = \sqrt{\frac{\gamma P}{\varrho}} = \sqrt{\gamma R_a T}.$$
(1.59)

Near the ground at room temperature the speed of sound in air is $c_s \simeq 350\,\text{m/s}$.

The complicated nature of the equation of state for sea water means that an analytic formula for the speed of sound in the ocean is not written as succinctly as (1.59). However, measurements show that a typical value is approximately $1500\,\text{m/s}$, much larger than that in air.

1.12.2 Boussinesq liquids

The Boussinesq approximation can be invoked if the density of the stratified fluid in which internal waves propagate varies little over the depth of the fluid. This is a good approximation for internal waves in the ocean, whose density varies by only a few per cent from top to bottom. The typical statement of the Boussinesq approximation is that density fluctuations are negligible except insofar as they affect buoyancy forces.

To see this, consider the vertical component of the momentum equation in (1.54). Dividing through by the total density gives

$$\frac{Dw}{Dt} = -\frac{1}{\varrho}\frac{\partial p}{\partial z} - \frac{g}{\varrho}\rho. \tag{1.60}$$

Assuming the background density increases over a depth H from the characteristic density ϱ_0 to a moderately larger value $\varrho_0 + \Delta\varrho$, we can write

$$\bar{\rho} = \varrho_0[1 + \bar{\epsilon}\bar{r}(z)], \tag{1.61}$$

in which $\bar{\epsilon} \equiv \Delta\varrho/\varrho_0$ and $r(z)$ is a non-increasing function of z that varies between 0 and 1. An example is shown in Figure 1.17a in which the background density increases by 3% from top to bottom of the domain.

The fluctuation density ρ depends upon the vertical displacements associated with the disturbance and also upon the rate of change of the background density. Fluid lifted upwards by a small amount δz is more dense than its surroundings by $\delta\rho \sim -\bar{\rho}'\delta z$. So the characteristic fluctuation density, for example due to an internal wave with vertical displacement amplitude A, is $A\Delta\varrho/H$. Assuming that the amplitude is much smaller than the domain height, the fluctuation density is much smaller than $\Delta\varrho$.

The total density can therefore be written

$$\varrho = \varrho_0[1 + \bar{\epsilon}\bar{r}(z) + \epsilon r(\vec{x},t)], \tag{1.62}$$

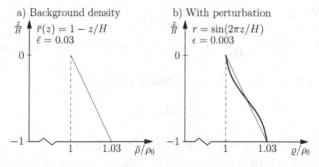

Fig. 1.17. a) Background density represented by (1.62) with $\bar{\epsilon} = 0.03$ and $\bar{r} = 1 - z/H$. Here there is no superimposed perturbation: $\epsilon = 0$. b) As in a) but also with superimposed (thick solid line) sinusoidal perturbation, $r = \sin(2\pi z/H)$, having relative amplitude given by $\epsilon = 0.003$.

in which $\epsilon = (\Delta \varrho / \varrho_0) A / H \ll \bar{\epsilon}$, and the nondimensional fluctuation density r has a largest magnitude on the order of unity. (For example, see Figure 1.17b.)

Over the depth of the thermocline in the ocean we might have $\bar{\epsilon} \simeq 0.01$. Even for internal waves with amplitudes of tens of metres, ϵ would be orders of magnitude smaller.

Substituting (1.62) into (1.60) and using the approximation $(1+x)^{-1} \simeq 1-x$, if x is small, the leading order terms in (1.60) become

$$\frac{Dw}{Dt} \simeq -\frac{1}{\varrho_0} \frac{\partial p}{\partial z} - \frac{g}{\varrho_0} \rho. \tag{1.63}$$

Although ρ is small, this approximate representation of the buoyancy term is still much larger than the terms we have neglected. If we neglect the buoyancy term entirely, we have thrown away the dynamics that govern the evolution of internal waves.

Likewise, in the horizontal momentum equations we can approximate $\varrho \simeq \varrho_0$. Putting these results together in (1.54) gives the Boussinesq form of the momentum equations on the f-plane in which the total density is replaced with the characteristic density on the left-hand side of the equations:

$$\varrho_0 \left(\frac{Du}{Dt} - f_0 v \right) = -\frac{\partial p}{\partial x},$$

$$\varrho_0 \left(\frac{Dv}{Dt} + f_0 u \right) = -\frac{\partial p}{\partial y}, \tag{1.64}$$

$$\varrho_0 \frac{Dw}{Dt} = -\frac{\partial p}{\partial z} - g\rho.$$

Consistent within the Boussinesq approximation the fluid is typically taken to be incompressible. To see this, we substitute (1.62) into the continuity equation, as given by (1.28):

$$\bar{\epsilon} \frac{d\bar{r}}{dz} + \epsilon \frac{Dr}{Dt} + \epsilon r \nabla \cdot \vec{u} = -[1 + \bar{\epsilon}\bar{r} + \epsilon r] \nabla \cdot \vec{u}. \tag{1.65}$$

The leading order term in (1.65) is independent of $\bar{\epsilon}$ and ϵ:

$$\nabla \cdot \vec{u} = 0. \tag{1.66}$$

This is the mass conservation law for a Boussinesq fluid.

Substituting (1.66) into (1.57), we see that the statement of incompressibility is equivalent to assuming that the speed of sound is so fast with respect to the velocity scale of internal waves, that c_s is effectively infinite. By requiring the fluid to be incompressible, we have in this way filtered out the physics of sound waves in the equations of motion.

Table 1.1. *Laws of conservation of momentum, internal energy and mass that are appropriate to describe the motion of internal waves in Boussinesq and non-Boussinesq liquids and gases. Here the equations for a liquid ignore the thermodynamics associated with large pressures in the oceanic abyss. The Boussinesq equations for a gas are given in terms of the fluctuation potential temperature θ. These may be recast in terms of density through the relations $\bar{\theta} = -(\vartheta_0/\varrho_0)\bar{\rho}$, and $\theta = -(\vartheta_0/\varrho_0)\rho$.*

Boussinesq liquid	Boussinesq gas
$\varrho_0\left(\dfrac{Du}{Dt} - f_0 v\right) = -\dfrac{\partial p}{\partial x}$	$\varrho_0\left(\dfrac{Du}{Dt} - f_0 v\right) = -\dfrac{\partial p}{\partial x}$
$\varrho_0\left(\dfrac{Dv}{Dt} + f_0 u\right) = -\dfrac{\partial p}{\partial y}$	$\varrho_0\left(\dfrac{Dv}{Dt} + f_0 u\right) = -\dfrac{\partial p}{\partial y}$
$\varrho_0\dfrac{Dw}{Dt} = -\dfrac{\partial p}{\partial z} - g\rho$	$\varrho_0\dfrac{Dw}{Dt} = -\dfrac{\partial p}{\partial z} + g\dfrac{\varrho_0}{\vartheta_0}\theta$
$\dfrac{D\rho}{Dt} = -w\dfrac{d\bar{\rho}}{dz}$	$\dfrac{D\theta}{Dt} = -w\dfrac{d\bar{\theta}}{dz}$
$\nabla \cdot \vec{u} = 0$	$\nabla \cdot \vec{u} = 0$

Non-Boussinesq liquid	Anelastic gas
$\bar{\rho}\left(\dfrac{Du}{Dt} - f_0 v\right) = -\dfrac{\partial p}{\partial x}$	$\dfrac{Du}{Dt} - f_0 v = -\dfrac{\partial}{\partial x}\left(\dfrac{p}{\bar{\rho}}\right)$
$\bar{\rho}\left(\dfrac{Dv}{Dt} + f_0 u\right) = -\dfrac{\partial p}{\partial y}$	$\dfrac{Dv}{Dt} + f_0 u = -\dfrac{\partial}{\partial y}\left(\dfrac{p}{\bar{\rho}}\right)$
$\bar{\rho}\dfrac{Dw}{Dt} = -\dfrac{\partial p}{\partial z} - g\rho$	$\dfrac{Dw}{Dt} = -\dfrac{\partial}{\partial z}\left(\dfrac{p}{\bar{\rho}}\right) + \dfrac{g}{\bar{\theta}}\theta$
$\dfrac{D\rho}{Dt} = -w\dfrac{d\bar{\rho}}{dz}$	$\dfrac{D\theta}{Dt} = -w\dfrac{d\bar{\theta}}{dz}$
$\nabla \cdot \vec{u} = 0$	$\nabla \cdot (\bar{\rho}\vec{u}) = 0$

The law of conservation of internal energy for a liquid (1.19) is unchanged in the Boussinesq approximation. So the approximation is consistent in the sense that the internal energy equation (1.19) is identical to the continuity (mass conservation) equation for an incompressible fluid in the form (1.29).

Explicitly, in terms of the background and fluctuation density, (1.19) becomes

$$\frac{D\rho}{Dt} = -w\frac{d\bar{\rho}}{dz}. \tag{1.67}$$

In the deep ocean, ϱ would be replaced with the potential density ϱ_{pot}, defined by (1.21).

The laws of conservation of momentum, mass and internal energy for a liquid in the Boussinesq approximation are given by (1.64), (1.66) and (1.67). For motion in three dimensions, these form a coupled set of five equations in the five fields of velocity $\vec{u}$, dynamic pressure p, and fluctuation density ρ. For convenience, the equations are reproduced in Table 1.1.

A consequence of the Boussinesq equations is that they are invariant upon reflection in z if the background density gradient $\overline{\rho}'$ is symmetric in z (see Exercises). In particular, this is the case in a uniformly stratified fluid for which $\overline{\rho}'$ is constant. Thus there is no dynamical distinction between upward and downward-propagating Boussinesq internal waves. This feature is particularly useful in performing laboratory experiments: it is often more convenient to generate downward-propagating waves from the surface of a tank filled with salt-stratified fluid than it is to generate upward-propagating waves from the bottom. Because the fluid is Boussinesq (salt water can be at most 20% more dense than fresh water) the behaviour of upward-propagating waves is reproduced simply by flipping images of the experiment upside-down.

1.12.3 Non-Boussinesq liquids

For internal waves that propagate through a stratified fluid over vertical distances comparable to or larger than a density scale height of the fluid, the Boussinesq approximation is no longer strictly applicable. Nonetheless, one can derive approximate equations that capture the relatively slow-time evolution of internal waves while filtering the fast-time dynamics of sound waves. As will be shown in Section 3.7, non-Boussinesq internal waves grow in amplitude as they propagate upwards into substantially less dense fluid. This is a consequence of momentum conservation: velocities must increase as the background density decreases in order that the flux of momentum remains constant. Such an increase is sometimes referred to as 'anelastic growth'. Thus there is a dynamical difference between upward- and downward-propagating non-Boussinesq internal waves.

A variety of models have been developed to capture the effects of anelastic growth while filtering vertically propagating sound waves. Somewhat unconventionally, we will begin by deriving the equations for a non-Boussinesq liquid and follow by deriving the so-called 'anelastic equations', which describe the motion of a non-Boussinesq gas.

Non-Boussinesq liquids do not occur naturally, but have been used in the laboratory to simulate anelastic effects in gases. In particular, Section 3.7 shows that the dispersion relation for small-amplitude internal waves in a non-Boussinesq liquid is the same as that for anelastic waves in air. One advantage of examining the non-Boussinesq equations for a liquid is that their derivation and form differs only slightly from those for a Boussinesq liquid. This helps to provide better insight into the dynamical difference between Boussinesq and non-Boussinesq waves.

In a non-Boussinesq fluid, the background density varies substantially over the vertical scale of the domain and so we can no longer assume that $\bar{\epsilon}$ in (1.62) is small. Still we can assume that the fluctuations are so small that $\epsilon \ll 1$. Therefore we can approximate $\varrho \simeq \bar{\rho}$ on the right-hand side of the momentum equations:

$$\bar{\rho}\left(\frac{Du}{Dt} - f_0 v\right) = -\frac{\partial p}{\partial x},$$

$$\bar{\rho}\left(\frac{Dv}{Dt} + f_0 u\right) = -\frac{\partial p}{\partial y}, \qquad (1.68)$$

$$\bar{\rho}\frac{Dw}{Dt} = -\frac{\partial p}{\partial z} - g\rho.$$

These differ from the corresponding Boussinesq momentum equations (1.64) only in that the background density $\bar{\rho}(z)$, rather than the characteristic density ϱ_0, multiplies the acceleration terms.

Again the internal energy equation is unchanged and is given by (1.67). For consistency, we approximate the continuity equation as in the Boussinesq approximation for a liquid by taking the fluid to be incompressible so that the velocity field satisfies (1.66).

Equations (1.68), (1.66) and (1.67) constitute five equations in the five unknowns $\vec{u}$, p and ρ. They are listed in Table 1.1 for comparison with the corresponding Boussinesq equations. They differ from the equations for a Boussinesq liquid only by the presence of $\bar{\rho}(z)$, instead of the characteristic density ϱ_0, in front of the material derivatives in the three components of the momentum equation.

1.12.4 Anelastic gases

For internal waves in air, the equations of motion need to include the thermodynamics of an ideal gas such that a fluid parcel expands and contracts as a consequence of moving into lower and higher pressure, respectively. We wish to do this in a way that accounts for the substantial decrease in the background density with height, while still filtering sound waves. The resulting equations are sometimes ambiguously said to correspond to those of a compressible atmosphere. But the density and pressure changes they model do not arise from compression due to sound waves: air may expand and contract when forced vertically up and down by an internal wave, but it does so silently.

There is more than one way to approximate the equations of motion of a non-Boussinesq gas. Here we will derive what are referred to as the 'anelastic equations'.

Because the energy equation for an ideal gas involves potential temperature, not density, we begin by recasting the momentum equations in terms of the background and fluctuation potential temperatures, $\bar{\theta}$ and θ, respectively. Using the definition of potential temperature (1.16) and the ideal gas law (1.6), the background potential temperature may be related directly to the background density $\bar{\rho}$ through

$$\frac{\bar{\theta}}{\vartheta_0} = \frac{\varrho_0}{\bar{\rho}} \left(\frac{\bar{p}}{P_0} \right)^{1/\gamma}, \tag{1.69}$$

in which we have defined $\vartheta_0 \equiv T_0$, the temperature taken at the reference pressure level. Assuming that density and pressure fluctuations due to internal waves are small compared with the background density and pressure, respectively, the fluctuation potential temperature can be related to the fluctuation density and pressure by

$$\theta \simeq \bar{\theta} \left(-\frac{\rho}{\bar{\rho}} + \frac{1}{\gamma} \frac{p}{\bar{p}} \right). \tag{1.70}$$

To derive the equations of motion for an anelastic gas, we begin by dividing both sides of the momentum equations (1.68) by $\bar{\rho}(z)$. The right-hand side of the resulting vertical momentum equation can then be rewritten as follows:

$$-\frac{1}{\bar{\rho}} \frac{\partial p}{\partial z} - g\frac{\rho}{\bar{\rho}} \simeq -\frac{\partial}{\partial z} \left(\frac{p}{\bar{\rho}} \right) + g\frac{\theta}{\bar{\theta}} + \frac{1}{\bar{\theta}} \frac{d\bar{\theta}}{dz} \frac{p}{\bar{\rho}}, \tag{1.71}$$

in which we have related the background density gradient to the background potential temperature gradient through taking a z-derivative of (1.69), and we have used the expression for hydrostatic balance (1.53) together with the approximate relationship between the fluctuation potential temperature, pressure and density (1.70).

The background potential temperature varies relatively slowly with height in the atmosphere, having a scale height of the order of a hundred kilometres. Therefore the last term on the right-hand side of (1.71) can be neglected.

The manipulation of the horizontal momentum equations is more straightforward. Because the background density is a function of z alone, it can be taken inside the horizontal pressure gradient. Thus, we can write the momentum equations on the f-plane in the following form:

$$\frac{Du}{Dt} - f_0 v = -\frac{\partial}{\partial x} \left(\frac{p}{\bar{\rho}} \right),$$

$$\frac{Dv}{Dt} + f_0 u = -\frac{\partial}{\partial y} \left(\frac{p}{\bar{\rho}} \right), \tag{1.72}$$

$$\frac{Dw}{Dt} = -\frac{\partial}{\partial z} \left(\frac{p}{\bar{\rho}} \right) + \frac{g}{\bar{\theta}} \theta.$$

We manipulate the mass conservation equation into a form that removes sound waves but which allows for the adiabatic expansion of a gas as it moves upwards through a gas whose background density decreases significantly with height. The Boussinesq approximation to the continuity equation (1.28) gives rise to the incompressibility condition (1.66). For an anelastic gas, the approximation used in (1.65) is too strong because the scale of the background density variations, as measured by $\overline{\epsilon}$, can be of order unity. However, the fluctuation terms of order ϵ are still negligible. Thus the mass conservation equation (1.26) is approximated by

$$\nabla \cdot (\overline{\rho}\vec{u}) = 0. \tag{1.73}$$

This equation constitutes the anelastic approximation for a gas.

Alternatively, (1.73) can be written

$$\nabla \cdot \vec{u} = \frac{1}{H_\varrho} w, \tag{1.74}$$

in which

$$H_\varrho \equiv \left(-\frac{\overline{\rho}'}{\overline{\rho}} \right)^{-1} \tag{1.75}$$

is the general definition of the density scale height. If the background density decreases exponentially, H_ϱ is the constant e-folding depth, defined implicitly by (1.50).

The internal energy equation for a gas (1.17) is unchanged in the anelastic approximation. Explicitly writing ϑ respectively as the sum of the background and fluctuation potential temperature, $\overline{\theta}(z)$ and $\theta(\vec{x}, t)$, gives

$$\frac{D\theta}{Dt} = -w \frac{d\overline{\theta}}{dz}. \tag{1.76}$$

Because the internal energy equation involves potential temperature, not density, it poses no inconsistency with the approximate mass conservation law (1.73). For a non-Boussinesq liquid, it is inappropriate to use (1.73) instead of (1.66) because ϱ, not ϑ, plays the key variable in the internal energy equation. It would be inconsistent with the continuity equation to have both $D\varrho/Dt = 0$, and $\nabla \cdot (\overline{\rho}\vec{u}) = 0$.

Altogether, the approximate momentum, mass and internal energy conservation equations given by (1.72), (1.73) and (1.76) constitute the anelastic equations for an ideal, adiabatic gas. These are listed together in Table 1.1.

1.12.5 Boussinesq gases

Although the density of the atmosphere decreases to zero going from the ground to outer space, the Boussinesq approximation can still be used to examine internal waves in the atmosphere provided that the air density changes little over the vertical range in which their dynamics are being examined. An upper bound for this range is given by the density scale height, H_ϱ. For example, $H_\varrho \simeq 7.4$ km in the troposphere, so the Boussinesq approximation may reasonably be employed for internal waves with vertical wavelengths and vertical propagation distance no more than a few kilometres.

Like the Boussinesq approximation for a liquid, the momentum equations are approximated by replacing the total density ϱ with the characteristic density ϱ_0 in front of the material derivative to give (1.64). However, the fluctuation density, ρ, is not a convenient variable in the evolution equations for a gas because density is not a conserved quantity; it decreases as it moves upward into lower pressure. What is conserved is the potential temperature, ϑ.

We therefore derive the approximate momentum equations by assuming that both the background density, $\overline{\rho} \simeq \varrho_0$, and background potential temperature, $\overline{\theta} \simeq \vartheta_0$, can be treated as approximately constant in the corresponding anelastic equations (1.72). Multiplying through by ϱ_0, we have

$$\varrho_0 \left(\frac{Du}{Dt} - f_0 v \right) = -\frac{\partial p}{\partial x},$$

$$\varrho_0 \left(\frac{Dv}{Dt} + f_0 u \right) = -\frac{\partial p}{\partial y}, \qquad (1.77)$$

$$\varrho_0 \frac{Dw}{Dt} = -\frac{\partial p}{\partial z} + g \frac{\varrho_0}{\vartheta_0} \theta.$$

For a Boussinesq gas, the internal energy equation is the same as that in the anelastic approximation (1.76). However, the mass conservation law is different. Treating the background density as approximately constant, the anelastic approximation to the mass conservation law (1.73) becomes that for an incompressible fluid (1.66). This condition does not preclude the dynamics of air expansion in reduced pressure. What it does is make this change instantaneous.

Equations (1.77), (1.76) and (1.66) constitute the Boussinesq equations for a gas. In fact, these are mathematically equivalent to the equations for a Boussinesq liquid. The vertical momentum equation can be recast in terms of ρ rather than θ using the relation (see Exercises)

$$\rho/\varrho_0 = -\theta/\vartheta_0. \qquad (1.78)$$

The minus sign is consistent with an increase in density corresponding to a decrease in temperature. The corresponding relationship holds between the background density and potential temperature gradients:

$$\frac{1}{\varrho_0}\frac{d\overline{\rho}}{dz} = -\frac{1}{\vartheta_0}\frac{d\overline{\theta}}{dz}.$$ (1.79)

This, together with (1.78), transforms the internal energy equation for a gas (1.76) into the internal energy equation for a liquid (1.67).

The equivalence of the Boussinesq equations for a gas and liquid is an important result. For one thing, it means that one can perform laboratory experiments using salt-stratified water to simulate the dynamics of internal waves in the atmosphere that have small vertical extent compared with the density scale height.

1.13 Conservation of angular momentum

Manipulation of the equations describing the conservation of momentum yields equations describing the conservation of angular momentum. First we derive formulae in two distinct circumstances that neglect Coriolis effects. Rotational effects are then considered in deriving the law of conservation of potential vorticity.

1.13.1 Vorticity in uniform-density fluid

For simplicity, we examine the case of an inviscid, incompressible, uniform-density Boussinesq fluid in which the effects of background rotation are ignored. The momentum equations are therefore given by (1.64) with $f_0 = 0$ and, for convenience, we write them in terms of the total pressure instead of the dynamic pressure. The curl of the resulting equations is

$$\nabla \times \left(\frac{D\vec{u}}{Dt} = -\frac{1}{\varrho_0}\nabla P - g\hat{z} \right).$$ (1.80)

It is straightforward to show that the curl of the right-hand side is identically zero. The curl of the left-hand side can be manipulated to give

$$\frac{D\vec{\zeta}}{Dt} = (\vec{\zeta} \cdot \nabla)\vec{u},$$ (1.81)

in which

$$\vec{\zeta} \equiv \nabla \times \vec{u}$$ (1.82)

is the mathematical definition of vorticity.[1] Explicitly, in three-dimensional Cartesian co-ordinates

$$\vec{\zeta} \equiv (\zeta_x, \zeta_y, \zeta_z) = \left(\frac{\partial w}{\partial y} - \frac{\partial v}{\partial z}, \frac{\partial u}{\partial z} - \frac{\partial w}{\partial x}, \frac{\partial v}{\partial x} - \frac{\partial u}{\partial y} \right). \qquad (1.83)$$

In particular, consider the solid-body rotation of the x–y plane. Each point on the plane moves with angular frequency Ω so that the Cartesian position of a parcel initially at $(x_0, y_0) = (r\cos(\phi_0), r\sin(\phi_0))$, is given by $(x, y) = (r\cos(\Omega t + \phi_0), r\sin(\Omega t + \phi_0))$. The Cartesian velocity of any point is therefore given by

$$(u, v) = \frac{d}{dt}(x, y) = (-\Omega y, \Omega x).$$

Its associated vorticity is the z-component of the curl of $(u, v, 0)$:

$$\zeta \equiv \frac{\partial v}{\partial x} - \frac{\partial u}{\partial y} = 2\Omega. \qquad (1.84)$$

So the vorticity of a rotating body is twice its angular frequency.

Equation (1.81) shows that vorticity following the fluid motion changes due to vortex stretching and vortex tilting, as illustrated in Figure 1.18. Vortex stretching in the x-direction, for example, is represented by $\zeta_x(\partial u/\partial x)$ and changes to ζ_x due to vortex tilting are represented by $\zeta_y(\partial u/\partial y) + \zeta_z(\partial u/\partial z)$.

If the vorticity everywhere in the fluid is identically zero at some time, (1.81) shows that the vorticity will remain zero for all time. That is, the fluid can be treated

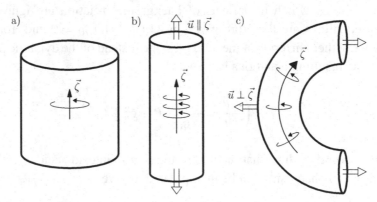

Fig. 1.18. For a fluid parcel with associated vorticity, $\vec{\zeta}$, as shown in a), the vorticity increases due to stretching if the velocity divergence is aligned with the vorticity, as shown in b), and vorticity is generated in directions perpendicular to $\vec{\zeta}$ due to tilting if the shear is oriented perpendicular to $\vec{\zeta}$, as shown in c).

---

[1] Some texts use the symbol ω to denote vorticity. Here we employ ζ in order to avoid confusion with our use of ω to represent wave frequency.

as 'irrotational'. This result will be useful when we consider interfacial waves in layered fluids in Chapter 2.

A mathematical consequence of a fluid being irrotational is that its velocity field can be represented by gradients of a scalar function, ϕ, called the 'velocity potential'. Explicitly,

$$\vec{u} \equiv \nabla \phi. \tag{1.85}$$

It follows immediately from vector identities that the irrotational flow condition $\vec{\zeta} \equiv \nabla \times \vec{u} = \vec{0}$ is automatically satisfied.

1.13.2 Vorticity in stratified fluid

Vorticity can be created spontaneously within the domain if the fluid has non-uniform density. As before we will assume the fluid is Boussinesq and we will ignore the effects of background rotation. We will also assume the motion is two dimensional, taking place only in the x–z plane. We take the curl of the momentum equations (1.64) with $f_0 = 0$ and consider only the y-component of vorticity denoted simply by ζ. Thus we find

$$\frac{D\zeta}{Dt} = \frac{g}{\varrho_0}\frac{\partial \rho}{\partial x}. \tag{1.86}$$

The y-component of the $\vec{\zeta} \cdot \nabla \vec{u}$ term does not appear because v is taken to be zero for flow confined to the x–z plane.

Equation (1.86) shows that vorticity can be created if there is a horizontal density gradient. If the density decreases from left to right, the fluid will rotate in a counter-clockwise direction. For motion in the x–z plane, this corresponds to the component of vorticity in the y-direction being negative, as illustrated in Figure 1.19.

Note that the sign of the vorticity and the corresponding direction of rotation depend upon the plane in which the movement is observed. In the x–y plane, counter-clockwise rotation corresponds to a positive component of vorticity in the z-direction.

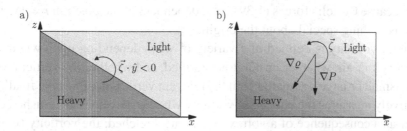

Fig. 1.19. Baroclinic generation of vorticity in a) a two-layer fluid and b) a continuously stratified fluid.

More generally, three-dimensional density variations modify the vorticity equation (1.81) so that it becomes

$$\frac{D\vec{\zeta}}{Dt} = (\vec{\zeta} \cdot \nabla)\vec{u} + \frac{1}{\varrho^2}\left(\nabla\varrho \times \nabla P\right). \tag{1.87}$$

The cross term on the right-hand side of this equation is called the 'baroclinic torque'. If the pressure and density gradients are aligned, then density variations do not contribute to the generation of vorticity, in which case the flow is said to be 'barotropic'. However, if the two vectors are not aligned, then vorticity can be created: heavy fluid to one side sinks while light fluid to the other side rises. Such flows are said to be 'baroclinic', meaning 'pressure breaking'.

1.13.3 Potential vorticity

Potential vorticity ('PV' for short) is conserved in a fluid, stratified or not, in which background rotation is significant and isopycnals (constant density surfaces in the ocean) or isentropes (layers of constant potential temperature in an adiabatic atmosphere) deflect vertically by a small amount compared with the scale of horizontal variations. The assumption of small deflections has the effect of restricting the evolution of vortex lines in the fluid so that they remain vertical, subject only to stretching and compression, as in Figure 1.18b. Vortex tilting, shown in Figure 1.18c, cannot occur. This is the case for large-scale geophysical motions in which rotation, stratification and the small vertical-to-horizontal aspect ratio of the domain all contribute to restricting the vorticity vector to the vertical.

Potential vorticity conservation is analogous to the law for conservation of angular momentum but it includes the effects, respectively, of spin-up and spin-down of fluid columns that are stretched and compressed. As rotating fluid is drawn inwards and its area decreases, it must spin up analogous to the way in which figure skaters spin more quickly by drawing their arms inwards. Unlike the skater analogy, however, there is no central axis of rotation in our circumstance. The vorticity changes anywhere the fluid column changes height, not just at the centre of the rotating fluid. This is because Coriolis forces (1.39) are independent of the position $\vec{r} = (x, y)$ and so there is nothing special about the origin.

Potential vorticity is defined in a variety of ways depending upon whether the fluid is water or air, whether or not it is stratified, and whether the vertical deflections are small or large compared with the relevant vertical depth-scale. In all cases it effectively measures the change of vorticity when vortex lines are crushed closer together as a consequence of a vortex tube being stretched: the vorticity times the area of the fluid element perpendicular to the vorticity vector is conserved. Practically, however, the potential vorticity is represented as the ratio of vorticity over

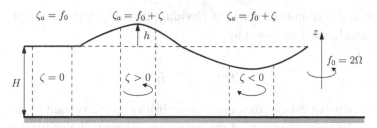

Fig. 1.20. Schematic showing vortex stretching and compression due to a surface wave. The absolute vorticity, ζ_a, is that viewed in a non-rotating frame of reference. In the rotating frame only the relative vorticity, ζ, is observed. Throughout the motion, the potential vorticity q, given by (1.88), remains constant.

a distance parallel to the vortex lines, which is a measure of how much stretching has occurred.

For example, consider the circumstance illustrated in Figure 1.20 of a one-layer, rotating shallow water fluid of mean depth, H. In the undisturbed fluid, a cylindrical column of fluid of infinitesimal horizontal area A is stationary in the rotating frame, but in the absolute frame has absolute vorticity $\zeta_a = f_0 = 2\Omega$ corresponding to that of solid-body rotation, as in (1.84).

Now suppose the surface above this column is displaced by a vertical distance, h. The cylinder of fluid is stretched to a length $H + h$ and so, for its volume to remain the same, its horizontal cross-sectional area must be $A' = AH/(H + h)$. Since the area times the absolute vorticity remains constant during the motion, the fluid column must have gained relative vorticity, ζ, so that $(f_0 + \zeta)A' = f_0 A$. Substituting our expression for A', we find that the relative vorticity ζ is given implicitly by the conservation relation $(f_0 + \zeta)/(H + h) = f_0/H$.

This result leads us to define the potential vorticity, q, for uniform, shallow water fluid on the f-plane to be

$$q \equiv \frac{f_0 + \zeta}{H + h}, \tag{1.88}$$

which is a conserved quantity following the fluid motion: $Dq/Dt = 0$. Sometimes q, given by (1.88), is called the 'barotropic potential vorticity' to emphasize that it is defined for fluid in which the effects of stratification, if present, are negligible.

If h is everywhere much smaller than H, one can take a Taylor series to expand the denominator in (1.88) and instead define the barotropic potential vorticity by

$$q_{BT} \simeq \zeta - f_0 \frac{h}{H}, \tag{1.89}$$

which, unlike q, has units of vorticity. Dividing through by the constant f_0, q_{BT} is defined in nondimensional form by

$$\tilde{q}_{BT} \simeq \frac{\zeta}{f_0} - \frac{h}{H}. \tag{1.90}$$

In these last two expressions the contribution from relative vorticity appears as the first term of the right-hand side of the definition, whereas the contribution due to vortex stretching is the second term.

1.14 Conservation of mechanical energy

In Section 1.7 we used thermodynamics to derive equations arising from the conservation of internal energy. The total energy of a fluid parcel is a combination of the internal energy, together with the mechanical energy associated with its motion (its kinetic energy) and the energy stored by moving against gravity (its potential energy). The usual definition of potential energy is not very useful for describing the energy content of internal gravity waves. We will see that a more practical quantity is the available potential energy. For waves in uniform-density and layered fluids, the conservation of mechanical energy is better cast in terms of the Bernoulli equation.

1.14.1 Kinetic and potential energy

The laws of conservation of kinetic and potential energy can be expressed in a variety of forms of which those relevant to the propagation of internal gravity waves are presented here. Throughout we will assume that the fluid is incompressible.

The flux form of the kinetic energy equation is found by taking the dot-product of $\vec{u}$ with the corresponding momentum equations (1.54). Manipulation of the result and using the mass conservation condition for incompressible fluid (1.30), we find

$$\frac{\partial E_K}{\partial t} = -\nabla \cdot [(E_K + P)\vec{u}] - g\varrho w, \tag{1.91}$$

in which

$$E_K \equiv \frac{1}{2}\varrho|\vec{u}|^2 \tag{1.92}$$

is the kinetic energy density. Equation (1.91) states that kinetic energy changes at a fixed location because of the divergence of the flux of kinetic energy and pressure,

$$\vec{\mathcal{F}}_K \equiv (E_K + P)\vec{u}, \tag{1.93}$$

and because of the work done against gravity to lift fluid upwards, $g\varrho w$.

We define the potential energy density to be

$$E_p \equiv \varrho g z. \tag{1.94}$$

Taking the material derivative of E_p and assuming the fluid is incompressible, we find $DE_p/Dt = g\varrho w$. In flux form, this is written

$$\frac{\partial E_p}{\partial t} = -\nabla \cdot [E_p \vec{u}] + g\varrho w. \tag{1.95}$$

Thus we see that for ascending fluid the kinetic energy loss, $-g\varrho w$ in (1.91), is converted into a gain of potential energy, $+g\varrho w$ in (1.95). Likewise, potential energy is lost and kinetic energy is gained where the fluid descends.

Combining (1.91) and (1.95) we have the flux form of equations expressing the conservation of total energy, $E_T \equiv E_K + E_p$:

$$\frac{\partial E_T}{\partial t} = -\nabla \cdot [(E_T + P)\vec{u}]. \tag{1.96}$$

These energy exchanges are illustrated schematically in Figure 1.21.

One consequence of (1.96) is that energy within a fixed volume of space changes only due to changes in the flux, $\vec{\mathcal{F}}_T = E_T + P$, of energy and pressure across the boundaries of the domain. In particular, if the domain is surrounded by rigid walls through which the normal velocity is zero, then as a direct consequence of the divergence theorem, $\frac{\partial}{\partial t} \iiint E_T \, dV = \frac{\partial}{\partial t} \iint -\vec{\mathcal{F}}_T \cdot \hat{n} \, dS = 0$. Hence the total energy within the domain is constant in time. In reality, viscosity acts as an additional sink of kinetic energy through which mechanical energy is converted into internal (heat) energy.

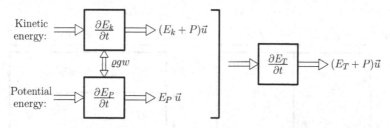

Fig. 1.21. Schematic showing that kinetic and potential energies change, either due to the divergence of their respective fluxes, or due to exchange from one form to the other through the baroclinic conversion term, $\varrho g w$. The total energy only changes due to the divergence of the total energy flux.

1.14.2 Available potential energy

The most straightforward way to compute the total energy in a domain is to integrate the sum of E_K and E_p. However, this is often unilluminating because the integral of E_p is typically much larger than that of E_K. In (1.94), E_p represents the energy required to bring all the fluid against gravity to a level $z = 0$. But this is impossible. In an incompressible fluid a fluid parcel carried from $z = z_0$ to $z = 0$ must be replaced by another fluid parcel carried from $z = 0$ to $z = z_0$, in order to ensure the volume remains unchanged.

A more practical definition measures the change of potential energy when fluid parcels are rearranged to a minimum energy state that keeps the volume fixed, as illustrated in Figure 1.22. The minimum energy state is one with no horizontal changes in density and with constant or monotonically decreasing density with height. Note that no mixing occurs upon rearranging the parcels because this requires energy. In particular, for a Boussinesq liquid this means the statistical distribution of parcels of some density is unchanged upon rearrangement. If half the fluid initially has density ρ_0, then half has that density after being rearranged to its minimum energy state.

The difference in potential energy between the initial and minimum energy states is the 'available potential energy', E_P, because it represents the potential energy available to be converted to kinetic energy. Here we use an upper-case 'P' in the subscript of E to distinguish the available potential energy from the potential energy, E_p. Explicitly, for a liquid $E_P = \iiint g(\varrho - \varrho_f) z \, dV$, in which $\varrho(x,y,z,t)$ and $\varrho_f(z)$ respectively represent the density distribution of the disturbed and minimum energy states. Some of this energy may then go into mixing or it may be dissipated into heat through viscosity.

In the context of internal gravity waves, consider a tank filled with a stratified salt-water solution having background ambient density profile $\bar{\rho}(z)$. Unperturbed, the total potential energy per unit horizontal cross-sectional area is given by integrating (1.94): $\int_0^H g \bar{\rho} z \, dz$. If the fluid is perturbed by internal gravity waves, the total

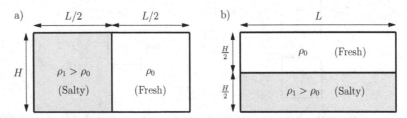

Fig. 1.22. Available potential energy is the difference in potential energy of a system, such as that shown in a) with dense fluid to the right and light fluid to the left, with a system in its minimum energy state, as shown in b).

potential energy is $\int_0^H g(\bar{\rho} + \rho)z\,dz$. Thus the available potential energy per unit cross-sectional area is the difference of the two integrals:

$$E_P = \int_0^H g\rho z\,dz. \tag{1.97}$$

1.14.3 The Bernoulli equation

In the special case of an inviscid fluid whose flow is barotropic (meaning that the fluid's density depends only upon pressure), energy is conserved following the motion of the fluid along streamlines. This result follows from manipulation of the momentum equations in the form (1.35). Using the identity

$$\frac{1}{\varrho}\frac{\partial P}{\partial x} = \frac{\partial}{\partial x}\int \frac{1}{\varrho}\,dP \tag{1.98}$$

and neglecting Coriolis forces, we find

$$\frac{\partial \vec{u}}{\partial t} + \nabla B = \vec{u} \times \vec{\zeta}. \tag{1.99}$$

Here we have defined the 'Bernoulli function'

$$B \equiv \frac{1}{2}|\vec{u}|^2 + \int \frac{1}{\varrho}\,dP + gz. \tag{1.100}$$

In the special case of a uniform liquid having constant density ϱ_0, the Bernoulli function can be represented by $\varrho_0 B = \varrho_0|\vec{u}|^2/2 + P + \varrho_0 gz$ or, using hydrostatic balance, by

$$\varrho_0 B = \frac{1}{2}\varrho_0|\vec{u}|^2 + p, \tag{1.101}$$

in which p is the dynamic pressure.

If the flow is steady, then the time derivative in (1.99) vanishes and we find that B, the energy per unit mass, must be constant following the motion of a fluid parcel along a streamline. This is because the gradient of B points in the same direction as $\vec{u} \times \vec{\zeta}$, which is the direction perpendicular to surfaces formed by streamlines.

If the flow is both steady and irrotational, then $\vec{\zeta} = 0$ and so B is a constant everywhere in the domain. In particular, for a uniform-density fluid, (1.101) shows that the dynamic pressure decreases if the velocity, and hence kinetic energy, increases. This is illustrated in Figure 1.23.

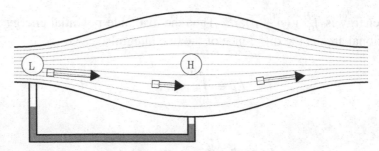

Fig. 1.23. Schematic illustrating that pressure is lower where the flow is relatively faster in steady, uniform-density fluid. The dense fluid (shaded dark in the U-tube below the expanding flow on the left-hand side of the schematic) is sucked up a tube with its opening under the relatively fast flow to the left.

If the flow is unsteady and irrotational, then we can introduce the velocity potential ϕ, defined in (1.85), to determine that the time-dependent Bernoulli function

$$\mathcal{B} \equiv \frac{\partial \phi}{\partial t} + \frac{1}{2}|\vec{u}|^2 + \int \frac{1}{\varrho}\,dP + gz \qquad (1.102)$$

is constant along streamlines. In particular, in uniform density liquid we have

$$\frac{\partial \phi}{\partial t} + \frac{1}{2}|\vec{u}|^2 + \frac{1}{\varrho_0}P + gz = \text{constant} \qquad (1.103)$$

along a streamline.

Although the Bernoulli equation proves useful in describing the dynamics of interfacial waves between homogeneous layers of fluid, it cannot be used to describe internal waves in continuously stratified fluid because, as we will see, such motions are neither barotropic nor irrotational.

1.15 Wave theory

Here we present the mathematical methods for representing and analysing plane waves, which are perfectly sinusoidal and infinite in extent, and wavepackets, which are localized in space. The motion of plane waves is determined by their dispersion relation, which relates the frequency to wavenumber. From this, one can determine the phase and group velocities. Conversely, from the dispersion relations, one can derive a differential equation that predicts how the waves evolve in space and time. This is done here for small-amplitude wavepackets through the development of Schrödinger's equation. Finally, we present formulae for determining cross-correlations between wave fields.

1.15.1 Representation of plane waves

Any small-amplitude disturbance can be represented by a superposition of perfectly sinusoidal waves and its evolution can be described by tracking the individual progress of each sinusoid. If the disturbance is a single sinusoidal wave that extends to infinity, it is called a 'plane wave' or sometimes a 'monochromatic wave', a metaphor referring to light waves of one frequency and, hence, one colour.

Mathematically a plane wave can be prescribed, for example, by

$$\eta(\vec{x},t) = A_0 \cos(\vec{k} \cdot \vec{x} - \omega t), \tag{1.104}$$

$$\eta(\vec{x},t) = A_0 \sin(\vec{k} \cdot \vec{x} - \omega t), \tag{1.105}$$

$$\eta(\vec{x},t) = \mathcal{A}_0 \exp[\imath(\vec{k} \cdot \vec{x} - \omega t)], \tag{1.106}$$

$$\eta(\vec{x},t) = \frac{1}{2}\mathcal{A}_0 \exp[\imath(\vec{k} \cdot \vec{x} - \omega t)] + \text{cc}. \tag{1.107}$$

Here η could represent displacement or fluctuations of velocity components, density, pressure, etc.

In all these forms, $\vec{x}$ can be a vector in one, two or three dimensions depending upon the geometry of the waves under consideration. The quantities ω and $\vec{k}$ represent the frequency and wavenumber vector, respectively. In the first two expressions A_0 is the wave amplitude. In the last two expressions, $\mathcal{A}_0$ is a complex number which is a measure of both amplitude and phase. The abbreviation cc in the last expression denotes the 'complex conjugate' of the preceding term.

Frequency is measured in units of radians per unit time (e.g., s^{-1} for radians per second). In terms of the wave period T, the (angular) frequency is $\omega = 2\pi/T$. For a frequency, ν, measured in cycles per unit time (e.g., cps for cycles per second), we would replace ωt with $2\pi \nu t$ to describe the time-variation of η. To avoid writing 2π explicitly in the arguments to cos and sin, it is typical to write frequency in terms of radians per time.

In three dimensions $\vec{k} = (k_x, k_y, k_z)$ in which k_x, k_y and k_z are the components of the wavenumber vector in the x-, y- and z-directions, respectively. The wavenumber, with units of radians per distance, measures the spatial extent of a periodic wave. For example, $k_x \equiv 2\pi/\lambda_x$, in which λ_x is the wavelength (the distance between two successive crests) in the x-direction. Generally, the wavelength $\lambda = 2\pi/|\vec{k}|$ is the shortest distance between successive crests. Hence the magnitude of the wavenumber $|\vec{k}|$ is small for long waves and is large for short waves.

Typically, a positive frequency $\omega > 0$ indicates that the waves advance forward in time and the orientation of the wavenumber vector indicates the direction in which the waves propagate through space. In particular, for $\omega > 0$ waves move rightwards in the x-direction if $k_x > 0$ and move leftwards if $k_x < 0$. However, there is some ambiguity in the description of waves: one could instead describe leftward-moving waves by setting $\omega < 0$ and $k_x > 0$. Effectively this corresponds to observing rightward-propagating waves moving backwards in time. By convention, we will assume $\omega \geq 0$ when possible and use the wavenumber vector to denote the propagation direction. We will see that in some cases we cannot do this. For example, with deep interfacial waves in a two-layer fluid (see Section 2.3) we must fix the sign of k_x to be positive and describe rightward- and leftward-propagating waves by $\omega > 0$ and $\omega < 0$, respectively.

The term 'amplititude' is sometimes used ambiguously to mean either the value of the field at some position $\vec{x}$ and time t or, specifically, the maximum value of the field. Here we use 'amplitude' to mean the latter. In (1.104) and (1.105), A_0 is the real-valued amplitude of η which, for example, could represent maximum displacement or maximum fluctuation pressure. The two representations differ in the phase of the waves they describe. In (1.104) the wave peaks at the origin at time $t = 0$, whereas the wave given by (1.105) has zero value at that position, reaching a peak a quarter-cycle later. This is illustrated in Figure 1.24a.

In (1.106) $\mathcal{A}_0$ is complex-valued and it is understood that the actual amplitude (the value of η at any position $\vec{x}$ and time t) is the real part of η. Thus if $\mathcal{A}_0 = A_{0r} + \imath A_{0i}$, the actual structure of η is $A_{0r} \cos(\vec{k} \cdot \vec{x} - \omega t) - A_{0i} \sin(\vec{k} \cdot \vec{x} - \omega t)$. It is more illuminating, however, to write $\mathcal{A}_0$ in polar form as

$$\mathcal{A}_0 = A_0 \exp(\imath \phi_0), \tag{1.108}$$

in which the magnitude $A_0 \equiv |\mathcal{A}_0| = (A_{0r}{}^2 + A_{0i}{}^2)^{1/2}$ is the amplitude, and the argument

$$\phi_0 \equiv \tan^{-1}(A_{0i}/A_{0r}) \tag{1.109}$$

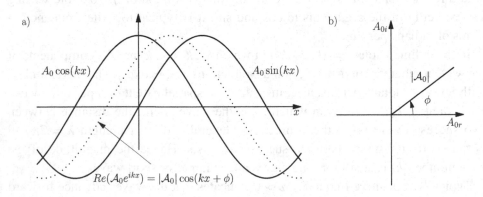

Fig. 1.24. a) Sketch of waves at time $t = 0$ represented by cosine, sine and complex exponential functions. b) Representation of complex amplitude $\mathcal{A}_0 = A_{0r} + \imath A_{0i}$ in complex Cartesian and polar co-ordinates.

is the phase. Strictly speaking, if we treat (A_{0r}, A_{0i}) as a vector in the complex plane, then ϕ_0 is the angle formed between this vector and the positive x-axis, as shown in Figure 1.24b. Using (1.108), the structure of η given by the real-part of the right-hand side of (1.106) is

$$\eta = A_0 \cos(\vec{k} \cdot \vec{x} - \omega t + \phi_0). \qquad (1.110)$$

In particular, if $\mathcal{A}_0 = A_0$ is real and positive, then (1.109) shows that $\phi_0 = 0$ (or an integer multiple of 2π). If $\mathcal{A}_0 = \iota A_0$ with A_0 real and positive, then $\phi_0 = \pi/2$ and (1.110) becomes (1.105).

Although using (1.106) to describe waves seems unintuitive at first, its form is preferable to (1.104) and (1.105) because it makes no explicit assumption about the phase of the wave. The form (1.107) is more cumbersome but provides a useful compromise between describing the wave field explicitly as a real function, while still using a notation that does not explicitly prescribe the phase of the waves. Summing a complex number with its complex conjugate gives twice the real part of the number. Thus the definition of (1.107) is identical to taking the real part of (1.106).

Instead of (1.107), some texts write $\eta = \mathcal{A}_0 \exp[\iota (\vec{k} \cdot \vec{x} - \omega t)] + $ cc. In this notation $|\mathcal{A}_0|$ is the 'quarter peak-to-peak amplitude'. Here, we consistently define η by either (1.106) or (1.107), in which case $|\mathcal{A}_0|$ is the more conventional 'half peak-to-peak amplitude'.

The complex representation also has the advantage that linear operators commute with the process of taking real and imaginary parts while not changing the form of the function upon which $\vec{x}$ and t depend. For example, the x-derivative of the real part of $\exp(\iota kx)$ is $d \cos(kx)/dx = -k \sin(kx)$. This equals the real part of the x-derivative of $\exp(\iota kx)$: $\Re\{\iota k \exp(\iota kx)\} = -k \sin(kx)$.

We will see that this is useful in solving coupled linear partial differential equations in which the relative phases of the different fields in the equations are unknown. This is particularly useful when computing the polarization relations, which interrelate fields associated with waves. For example, the vertical velocity field, w, is related to the vertical displacement field, ξ, by $w = \partial_t \xi$. Assuming $\xi = \mathcal{A}_\xi \exp[\iota (kx - \omega t)]$ and $w = \mathcal{A}_w \exp[\iota (kx - \omega t)]$, the differential relation immediately gives $\mathcal{A}_w = -\iota \omega \mathcal{A}_\xi$. Because $-\iota = \exp(-\iota \pi/2)$, we see that w lags in phase by $90°$ from ξ and the amplitude of the vertical velocity field is $|\mathcal{A}_\xi| \omega$.

The operation of taking the product of wave fields (as would be done, for example, to compute cross-correlations or to derive equations for moderately large amplitude waves) is nonlinear and so the representation of waves by (1.106) is inappropriate and can lead to incorrect computations (see Section 1.15.8). Instead, it is useful to represent their structure by (1.107).

1.15.2 The dispersion relation

For all waves, the frequency depends upon the wavenumber through what is called
a dispersion relation. This describes how a wavepacket spreads, and hence disperses, if waves of different wavelength travel at different speeds. If the waves
are sufficiently large amplitude that nonlinear effects cannot be neglected, then ω
is additionally a function of amplitude. In most of the discussion below we will
consider only small-amplitude waves.

Waves that travel at the same speed for all wavelengths are 'nondispersive'.
Examples of such waves include light waves, sound waves and long waves on
shallow water. The evolution of these waves is prescribed by the well-known 'wave
equation':

$$\frac{\partial^2 \eta}{\partial t^2} = c^2 \nabla^2 \eta. \tag{1.111}$$

Here c is a constant and $\eta(\vec{x}, t)$ represents the amplitude of the wave at position $\vec{x}$
and time t. The symbol $\nabla^2 \equiv \nabla \cdot \nabla$ is the Laplacian operator, sometimes denoted by
the symbol Δ. In three-dimensional Cartesian co-ordinates $\nabla^2 \equiv \partial_x{}^2 + \partial_y{}^2 + \partial_z{}^2$.
Straightforward substitution of (1.106) into (1.111), which amounts to taking a
Fourier transform, gives the dispersion relation for nondispersive waves

$$\omega^2 = c^2 |\vec{k}|^2. \tag{1.112}$$

We will see that short interfacial waves and internal waves in continuously stratified fluid are dispersive. Their dispersion relation is not given by (1.112), but more
generally by

$$\omega = \omega(\vec{k}). \tag{1.113}$$

For example, surface waves in deep water have the dispersion relation

$$\omega^2 = g|\vec{k}|, \tag{1.114}$$

in which $\vec{k} = (k_x, k_y)$.

As in (1.112) and (1.114), the dispersion relation for waves in otherwise stationary
fluid is often given in terms of the squared frequency ω^2. This indicates that two
types of waves are captured by the dispersion relations: those that propagate both
forwards and backwards in time.

1.15.3 Phase velocity

By definition, the speed at which wave crests move is called the phase speed. Of
course, there is nothing special about the wave crests: the motion of the trough or
of any point of constant phase suffices in the definition of phase speed. The phase
velocity describes the direction as well as speed of propagation.

For a one-dimensional wave having structure in the x-direction alone, the phase speed is $c_p \equiv \omega/k = \lambda/T$, meaning the crest moves one wavelength λ in the time of one wave period T.

For waves having structure in two or three dimensions, the phase velocity can be represented in one of two ways. In the first definition, which is standard, we imagine we are sitting on a wave crest moving with the wave in a direction perpendicular to the along-crest direction. In this perspective, the phase velocity is

$$\vec{c}_p \equiv \frac{\omega}{|\vec{k}|}\hat{k} = \frac{\omega}{|\vec{k}|^2}\vec{k}, \tag{1.115}$$

in which $\hat{k} \equiv \vec{k}/|\vec{k}|$ is the unit vector pointing in the direction of $\vec{k}$, as illustrated in Figure 1.25a. In particular, the x-component of the phase speed for a two-dimensional wave is

$$c_{px} = \omega \frac{k_x}{k_x{}^2 + k_y{}^2}.$$

The phase speed is just the magnitude $c_p \equiv |\vec{c}_p| = \omega/|\vec{k}|$.

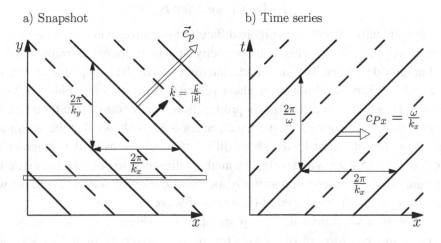

Fig. 1.25. a) Snapshot of plane waves with crests illustrated by solid lines and troughs by dashed lines. The wavelengths in the x and y directions are indicated by double-headed arrows; the unit wavenumber vector $\hat{k}$, indicated by the single arrow, is perpendicular to the crests; the phase velocity $\vec{c}_p$, indicated by the double-tailed arrow, is parallel to $\hat{k}$. b) A time series constructed by examining how the flow evolves within the horizontal window indicated by the thin box in a). The x-component of the phase speed, $\vec{c}_{px} = \omega/k_x$, is not necessarily equal to the x-component of $\vec{c}_p$.

If instead we imagine we are sitting at a fixed position with wave crests moving past, the speeds of the crests in the x, y, and z directions are, respectively

$$c_{Px} = \frac{\omega}{k_x}, \ c_{Py} = \frac{\omega}{k_y}, \ \text{and} \ c_{Pz} = \frac{\omega}{k_z}. \tag{1.116}$$

Here we have used an upper-case 'P' in the subscript to distinguish this definition from that in (1.115). The definitions in (1.116) are useful when analysing time series data either from observations or numerics, as illustrated in Figure 1.25b. It does not make sense to compose the vector $\vec{c}_P$ from the components in (1.116) because $|\vec{c}_P|$ does not equal the appropriate value for the phase speed, $c_p = \omega/|\vec{k}|$. This is only the case for one-dimensional waves.

Using the dispersion relation, the phase speed can be expressed explicitly in terms of the wavenumber alone. For example, using (1.112), one-dimensional shallow water waves have phase speed $c_p = \pm c$, a constant for all wavenumbers. The plus and minus signs correspond to rightward and leftward propagating waves, respectively. For one-dimensional deep water waves with a dispersion relation given by (1.114), the phase speed is $c_p = \pm(g/k)^{1/2}$. So long waves (small k) travel at faster speeds.

1.15.4 Group velocity

More dynamically important yet more difficult to perceive than the phase velocity is the group velocity, $\vec{c}_g$. This is the velocity at which energy is transported by small-amplitude waves. The magnitude and direction of the group velocity are not necessarily the same as those of the phase velocity, as illustrated in Figure 1.26. So one should not look at the direction of propagation of wave crests to infer where the energy is being transported. In particular, Section 3.3.4 shows that the group and phase velocities of internal waves have different directions as well as magnitudes.

In Section 1.15.6 we will develop the mathematics used to describe the evolution of the amplitude envelope of a spatially localized packet of waves. There we will show that $c_g = |\vec{c}_g|$ is the speed of the group of waves.

Elementary texts derive the group speed of one-dimensional waves by considering a superposition of two waves having wavenumbers k_0 and $k_0 + k_\Delta$ and respective frequencies $\omega_0 \equiv \omega(k_0)$ and $\omega_0 + \omega_\Delta$. Assuming $|k_\Delta| \ll |k|$, the difference in frequencies is found from the second term of the Taylor-series expansion of ω about k_0:

$$\omega_\Delta \simeq \omega'(k_0) k_\Delta, \tag{1.117}$$

in which the prime denotes the k derivative of ω.

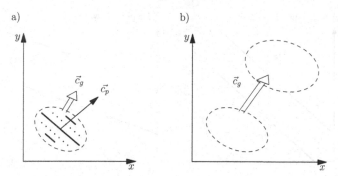

Fig. 1.26. a) The vector $\vec{c}_p$ indicates the speed and direction at which crests (solid lines) and troughs (dotted lines) propagate within a wavepacket (enclosed by dashed lines). This is not necessarily the same as the group velocity, $\vec{c}_g$, which indicates the speed and direction at which the wavepacket as a whole moves, as illustrated in b).

Taking the amplitudes to be the same and arbitrarily matching their initial phase at $x = 0$, the disturbance amplitude of the two waves is given by

$$\eta = A_0 \cos(k_0 x - \omega_0 t) + A_0 \cos[(k_0 + k_\Delta)x - (\omega_0 + \omega_\Delta)t]$$

$$\simeq A_0 \left\{ 1 + \cos[k_\Delta(x - \omega'(k_0)t)] \right\} \cos(k_0 x - \omega_0 t). \qquad (1.118)$$

In deriving the second expression, we have used (1.117) and a double-angle trigonometric identity to write the cosine of two angles as the sum of a product of cosines and sines. The term involving a product of sines has been neglected under the assumption that k_Δ is small.

Equation 1.118 shows that the superposition of waves acts like a plane wave of wavenumber k_0 and frequency ω_0, but whose amplitude changes in space and time as

$$A(x,t) = A_0 \left\{ 1 + \cos[k_\Delta(x - \omega'(k_0)t)] \right\}. \qquad (1.119)$$

This is the so-called 'amplitude envelope' of the wave. Thus we have shown that the superposition of the two waves gives a disturbance in which crests move at the phase speed $c_p = \omega_0/k_0$, but for which the peak of the amplitude envelope moves at speed $c_g = \omega'(k_0)$, the group speed.

More generally, the group velocity determines how a wavepacket travels in one, two or three dimensions. Mathematically, it is given by

$$\vec{c}_g \equiv \nabla_{\vec{k}} \omega, \qquad (1.120)$$

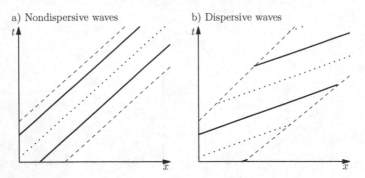

Fig. 1.27. Time series showing the phase lines (solid and dotted) moving within a wavepacket (dashed lines) for waves that are a) nondispersive and b) dispersive with $c_g < c_p$.

in which $\nabla_{\vec{k}}$ is the gradient operator that takes derivatives with respect to each component of the wavenumber vector, $\vec{k}$. For example, the x-component of the group velocity is $c_{gx} = \partial\omega/\partial k_x$.

For nondispersive waves, the group and phase velocity are identical: the wavepackets move in the same direction and at the same speed as the wave crests, as shown in Figure 1.27a. This is not the case for dispersive waves. For example, the group velocity of one-dimensional, rightward-propagating deep water waves is $c_g = (g/k)^{1/2}/2 = c_p/2$. In this case, the group moves in the same direction as the wave crests but at half the speed. Crests advance from the back to the front of a wavepacket during its propagation, as shown in Figure 1.27b.

1.15.5 Representation of wavepackets

A 'wavepacket' is a localized wavy disturbance. Typically this is a superposition of waves having frequencies in a range about a central value, and phases are set so that the amplitude of the wavepacket drops to negligibly small values away from its centre. Such wavepackets are said to be 'quasi-monochromatic' because they behave similarly to (monochromatic) plane waves with single frequency.

For example, a one-dimensional, small-amplitude wavepacket as it evolves over time from a known initial condition can be written as the superposition of plane waves by

$$\eta(x,t) = \int_{-\infty}^{\infty} \hat{\eta}(k)e^{i(kx-\omega t)}\, dk. \tag{1.121}$$

Here, $\hat{\eta}(k)$ is the amplitude (per unit wavenumber) of waves with wavenumber k. This can be determined explicitly from the initial conditions by the inverse Fourier transform

$$\hat{\eta}(k) = \frac{1}{2\pi} \int_{-\infty}^{\infty} \eta(x,0)e^{-\imath kx} \, dx. \tag{1.122}$$

Alternately, the spatial structure of the waves as established from known variations of the amplitude at a boundary ($x = 0$, say) may be represented by

$$\eta(x,t) = \int_{-\infty}^{\infty} \hat{\eta}(\omega)e^{\imath(kx-\omega t)} \, d\omega, \tag{1.123}$$

in which

$$\hat{\eta}(\omega) = \frac{1}{2\pi} \int_{-\infty}^{\infty} \eta(0,t)e^{\imath \omega t} \, dt. \tag{1.124}$$

In these equations, we have defined the transforms so that a constant factor, in this case $1/2\pi$, appears only in the formulae for the inverse transforms (1.122) and (1.124). To exploit symmetry, mathematicians sometimes define the transform pairs with constant factors $(2\pi)^{-1/2}$ leading each integral. We do not do this here so that we can associate $\hat{\eta}$ more directly with the half peak-to-peak amplitude of the waves.

In some idealized studies, a wavepacket is conveniently represented by the product of a plane wave with a smooth, non-negative function which is the 'amplitude envelope'. For example, a one-dimensional 'Gaussian wavepacket' with peak value at the origin is represented by

$$\eta(x,0) = \mathcal{A}_0 \exp\left[-\frac{x^2}{2\sigma^2}\right]e^{\imath k_0 x}. \tag{1.125}$$

This is sketched in Figure 1.28. Here σ measures the width of the wavepacket envelope consisting of waves with wavenumber k_0. If the amplitude $\mathcal{A}_0 = A_0$ is a

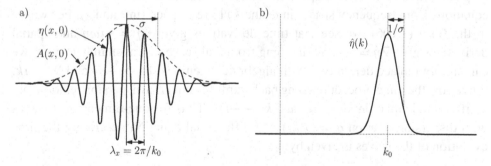

a)

b)

Fig. 1.28. a) A Gaussian wavepacket with initial amplitude envelope $A(x,0) = |\mathcal{A}| \exp(-x^2/2\sigma^2)$ containing waves with wavenumber k_0. b) The Fourier transform of this wavepacket is peaked about wavenumber $k = k_0$. The width of the peak narrows as σ increases but the area under the curve remains constant.

real number, then the actual structure of η, given by the real part of the right-hand side of (1.125), is a Gaussian times a cosine function.

This wavepacket can be thought of as a superposition of waves having non-negligible amplitude only for a range of wavenumbers near k_0. This is revealed by the inverse Fourier transform of $\eta(x,0)$ through (1.122):

$$\hat{\eta}(k) = (\sqrt{2\pi}\sigma)A_0 \exp\left[-\frac{1}{2}\sigma^2(k-k_0)^2\right]. \tag{1.126}$$

The range of wavenumbers with significant amplitude is measured by $1/\sigma$, as shown in Figure 1.28b. In the limit $\sigma k_0 \to \infty$, the wavepacket is a plane wave in which only waves with wavenumber equal to k_0 have significant amplitude. The corresponding Fourier transform (1.126) becomes a Dirac delta function: $\hat{\eta}(k) = \mathcal{A}_0 \delta(k-k_0)$. Substituting this into (1.121) gives $\eta = \mathcal{A}_0 \exp[\iota(k_0 x - \omega(k_0)t)]$, as expected.

For $\sigma k_0 \ll 1$, the wavepacket more closely resembles a 'wave-pulse' than a group of waves. Such waves are not quasi-monochromatic and are usually not considered in the theory of wavepacket propagation.

1.15.6 Plane wave and wavepacket evolution equations

Knowing the dispersion relation, we can formulate a differential equation that describes the evolution of the waves. More generally, one can determine how the amplitude envelope of a quasi-monochromatic wavepacket translates and disperses in time. The equation that describes the latter process for dispersive waves is known as 'Schrödinger's equation'. Although best known for its application in quantum mechanics, generally it is a formula that describes the evolution of dispersive wavepackets.

We begin by showing how to derive a differential equation for the evolution of plane waves given the dispersion relation. This amounts to transforming the equations from frequency space (in ω and $\vec{k}$) to real space (in t and $\vec{x}$). For waves in the form (1.106), we see that time derivatives give $\partial_t \eta = -\iota \omega \eta$ and spatial derivatives give $\nabla \eta = \iota \vec{k} \eta$. So, in going from real space to wavenumber space, we can simply replace derivatives with algebraic quantities: $\partial_t \to -\iota \omega$ and $\nabla \to \iota \vec{k}$. We follow the same procedure going backwards to convert a dispersion relation into a differential equation: $\omega \to \iota \partial_t$ and $\vec{k} \to -\iota \nabla$. Thus for one-dimensional waves with dispersion relation $\omega = \omega(k_x)$, the differential equation describing the time evolution of the waves is given by

$$\left[\iota\frac{\partial}{\partial t}\right]\eta = \left[\omega\left(-\iota\frac{\partial}{\partial x}\right)\right]\eta, \tag{1.127}$$

in which $\eta(x,t)$ represents the structure of the waves in real space.

For example, the one-dimensional analogue of (1.112) gives the dispersion relation of nondispersive waves: $\omega = \pm c k_x$. The upper sign corresponds to rightward-propagating waves and the lower sign to leftward-propagating waves. Then (1.127) gives the differential equation describing their evolution: $\iota \partial_t \eta = \pm c(-\iota \partial_x)\eta$. This is simplified to give

$$\frac{\partial \eta}{\partial t} \pm c \frac{\partial \eta}{\partial x} = 0. \tag{1.128}$$

This advection equation is the one-dimensional analogue of (1.111), separately describing waves propagating rightwards with speed $+c$ and leftwards with speed $-c$.

Next we derive an equation for the evolution of a wavepacket in an unbounded domain. We suppose the structure in terms of η can be written as the Fourier transform

$$\eta(x,t) = \int_{-\infty}^{\infty} \hat{\eta}(k) e^{\iota(kx - \omega t)} \, dk. \tag{1.129}$$

To ensure η is real, we insist that $\hat{\eta}(-k) = \hat{\eta}^\star(k)$, in which the star denotes the complex conjugate.

Initially, the structure of η can be represented by

$$\eta(x,0) = \mathcal{A}(x,0) \, e^{\iota k_0 x}, \tag{1.130}$$

in which $\mathcal{A}$ is the possibly complex-valued amplitude envelope containing waves with wavenumber k_0. For example, $\mathcal{A}(x,0) = \mathcal{A}_0 \exp[-x^2/2\sigma^2]$ for the Gaussian wavepacket in (1.125). Here it is understood that η is given by the real part of the expression on the right-hand side of (1.130).

The initial condition (1.130) together with (1.129) implicitly defines $\hat{\eta}$ through

$$\mathcal{A}(x,0) = \int_{-\infty}^{\infty} \hat{\eta}(k) e^{\iota(k - k_0)x} \, dk. \tag{1.131}$$

As before, $\hat{\eta}$ can be written explicitly in terms of $\mathcal{A}(x,0)$ by the inverse Fourier transform except with k replaced by $k - k_0$.

For example, Figure 1.28 shows a Gaussian wavepacket and its Fourier transform. The initial amplitude is $A_0 = |\mathcal{A}(0,0)|$ and its width is σ. For quasi-monochromatic waves, we assume $\sigma k_0 \gg 1$. In this case the Fourier transform of (1.130) is sharply peaked about wavenumbers with $k \simeq k_0$.

If the waves are quasi-monochromatic initially, we may assume they remain quasi-monochromatic as the wavepacket evolves. Only the amplitude envelope, $\mathcal{A}(x,t)$, will change, albeit slowly compared with the wave period. We therefore wish to develop an equation describing the evolution of $\mathcal{A}(x,t)$ from which the structure $\eta(x,t)$ of the waves can then be determined from

$$\eta(x,t) = \mathcal{A}(x,t) e^{\iota[k_0 x - \omega(k_0)t]}. \tag{1.132}$$

Substituting (1.132) into (1.129) and multiplying both sides by $\exp\{-\iota[k_0 x - \omega(k_0)t]\}$ gives an integral equation for $\mathcal{A}$:

$$\mathcal{A}(x,t) = \int_{-\infty}^{\infty} \hat{\eta}(k) e^{\iota[(k-k_0)x - (\omega(k) - \omega(k_0))t]} \, dk. \tag{1.133}$$

Because the Fourier transform, $\hat{\eta}(k)$, of the wavepacket is sharply peaked about wavenumbers $k \simeq k_0$, the integrand of (1.133) is non-negligible only in a small range about k_0. Thus we need only be concerned with the behaviour of the dispersion relation near $k \simeq k_0$, which can be found through a Taylor-series expansion:

$$\omega(k) \simeq \omega(k_0) + \omega'(k_0)(k - k_0) + \frac{1}{2}\omega''(k_0)(k - k_0)^2. \tag{1.134}$$

Here, the primes denote k-derivatives of ω, and we have chosen to truncate the Taylor series in ω at second order in $(k - k_0)$.

Substituting (1.134) into (1.133), taking x- and t-derivatives, and comparing terms we arrive at the following approximate equation for $\mathcal{A}$:

$$\frac{\partial \mathcal{A}}{\partial t} \simeq -\omega' \frac{\partial \mathcal{A}}{\partial x} + \iota \frac{1}{2} \omega'' \frac{\partial^2 \mathcal{A}}{\partial x^2}. \tag{1.135}$$

The first term on the right-hand side of (1.135) indicates that the wavepacket translates at the group speed $c_g = \omega'(k_0)$. Typically, (1.135) is further simplified by a change of co-ordinates to a frame translating at speed c_g, which has the effect of eliminating this term. Explicitly defining $X = x - c_g t$, (1.135) becomes

$$\frac{\partial \mathcal{A}}{\partial t} \simeq +\iota \frac{1}{2} \omega'' \frac{\partial^2 \mathcal{A}}{\partial X^2}. \tag{1.136}$$

This is Schrödinger's equation.

The right-hand side of (1.136), or equivalently the last term on the right-hand side of (1.135), represents the leading-order effect of dispersion. This describes how the wavepacket may change shape as it propagates because waves with different wavenumbers propagate at different speeds if they are dispersive. Typically, dispersion has the effect of widening the wavepacket while its maximum amplitude gradually decreases.

The discussion in this section has been confined to small-amplitude waves. Consequently, equation (1.136) is sometimes called the linear Schrödinger equation, emphasizing that the equation is linear and also that it captures only the linear dispersion of small-amplitude waves. The extension of this equation to include moderately large-amplitude effects results in the nonlinear Schrödinger equation. Its derivation and interpretation are described in Section 4.2.3.

1.15.7 Wave modes

The complex-exponential representation of waves given by (1.106) is appropriate particularly for propagating waves in an unbounded medium. If physical boundaries or ambient conditions (as in a wave duct) confine the waves to a finite-sized region, it is often more useful to describe the waves in terms of modes. Wave modes are sometimes called 'standing waves', meaning that as the crests move up and down the positions of the nodes between them remain stationary.

One-dimensional, horizontally propagating wave modes can be thought of as the superposition of two propagating waves both having the same frequency and spatial structure, but moving in opposite directions with locked-in phase at the boundaries. Thus, for example, the superposition of rightward- and leftward-propagating waves represented by (1.104) in one dimension is the wave mode

$$A\cos(kx - \omega t) + A\cos(-kx - \omega t) = 2A\cos(\omega t)\cos(kx). \tag{1.137}$$

This represents a stationary wave with spatial structure given by $\cos(kx)$ but whose amplitude changes in time according to $2A\cos(\omega t)$. The variations in time and space have explicitly been separated.

A one-dimensional wave mode is generally represented by

$$\eta(x,t) = \eta(x)\cos(\omega t + \phi_0), \tag{1.138}$$

in which η is a real function describing the spatial structure and amplitude of the wave and ϕ_0 is an arbitrary but constant phase factor. One could describe the displacement of waves by $A\cos(kx)$, for example, in which A represents the maximum vertical displacement of the surface. The wavenumber k varies continuously if the domain is infinitely large, but it can hold only a discrete set of values if the domain is constrained by requiring $0 \leq x \leq L$. In this case, the allowable values of k are $k_n = n\pi/L$ for $n = 1, 2, \ldots$ The zeroth mode of cosine-shaped waves is constant and so is neglected.

The dispersion relation $\omega \equiv \omega(k)$ is the same whether or not the domain is bounded. Because $k = k_n$ can hold only a discrete set of values in a bounded domain the frequency $\omega_n \equiv \omega(k_n)$ is likewise discrete-valued.

With the co-ordinate system set up with the origin at one of the boundaries, cosine functions are appropriate if the slope of the field is required to be zero at the boundary. These are called Neumann boundary conditions, and are used, for

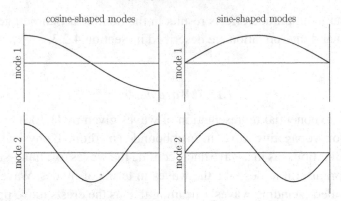

Fig. 1.29. Lowest wave-like modes of one-dimensional waves in a finite-sized domain. Their structure, $\eta(x)$, is represented by cosine functions if the side-wall boundary condition requires zero slope (zero Neumann boundary conditions). It is represented by sine functions if the boundary condition requires zero value (zero Dirichlet boundary conditions).

example, to describe the vertical displacement of small-amplitude waves in a box. Sine functions are appropriate for Dirichlet boundary conditions, which require the value of the field to be zero at the boundary. These would be used, for example, to describe the horizontal velocity field of laterally bounded waves.

The left-hand column of Figure 1.29 shows mode 1 and 2 for laterally bounded cosine-shaped waves and the right-hand column shows these modes for sine-shaped waves. The lowest mode has a wavelength that is twice the extent of the domain while exactly one wavelength of mode-2 waves fills the domain.

Any disturbance in a bounded domain can be written as a sum of modes. If the disturbance must have zero slope conditions at the boundaries, it is written as a superposition of cosine functions using the Fourier cosine series. If the disturbance is necessarily zero at the boundaries, then the Fourier sine series is used.

In our discussion of internal waves in Section 3.5, we will see that in a domain with sloping sides the description of waves as a superposition of modes is not so practical and the use of attractors gives a more useful description.

1.15.8 Cross-correlations

Cross-correlations help to diagnose the transport properties of waves. For example, the vertical transport of mass by waves is characterized by computing the cross-correlation between the density field, ρ, and the vertical velocity field, w: if this correlation is zero, the waves do not transport mass; if positive, mass is transported upwards; if negative, mass is transported downwards.

For spatially periodic one-dimensional waves, the cross-correlation between fields η and ξ is denoted by $\langle \eta \xi \rangle$ in which

$$\langle \eta \xi \rangle \equiv \frac{1}{\lambda} \int_0^\lambda \eta(x) \xi(x)\, dx. \tag{1.139}$$

Here $\lambda = 2\pi/k$ is the horizontal wavelength corresponding to wavenumber k. For example, if $\eta = A_\eta \cos(kx)$ and $\xi = A_\xi \cos(kx)$, the correlation of the two fields is $\langle \eta \xi \rangle = A_\eta A_\xi /2$.

Although (1.139) is given as a spatial average, the cross-correlation can be defined as an average over one period rather than one wavelength. For plane waves defined in space and time, it turns out that it does not matter whether the averaging is done over a wavelength or over a period; the result is the same. Putting $\eta = A_\eta \cos(kx - \omega t)$ and $\xi = A_\xi \cos(kx - \omega t)$ in (1.139) still gives the correlation $\langle \eta \xi \rangle = A_\eta A_\xi /2$.

If η and ξ represent the same field, (1.139) is called the autocorrelation. Its square root is the root-mean-square average from which one can determine the amplitude of the field by multiplying by $2^{1/2}$.

Some shortcuts are possible in the computation of (1.139). For example, suppose η and ξ are the real parts of $\mathcal{A}_\eta \exp(\iota kx)$ and $\mathcal{A}_\xi \exp(\iota kx)$, respectively. Then

$$\langle \eta \xi \rangle = \frac{1}{\lambda} \int_0^\lambda \left[\frac{1}{2} \left(\mathcal{A}_\eta e^{\iota kx} + \text{cc} \right) \right] \left[\frac{1}{2} \left(\mathcal{A}_\xi e^{\iota kx} + \text{cc} \right) \right] dx$$

$$= \frac{1}{4\lambda} \int_0^\lambda \mathcal{A}_\eta \mathcal{A}_\xi e^{2\iota kx} + \mathcal{A}_\eta \mathcal{A}_\xi^\star + \mathcal{A}_\eta^\star \mathcal{A}_\xi + \mathcal{A}_\eta^\star \mathcal{A}_\xi^\star e^{-2\iota kx}\, dx. \tag{1.140}$$

Here cc and the star superscript denote the complex conjugate. Note that we need to represent the fields explicitly as real quantities through (1.107) because taking the product of fields is a nonlinear operation: the real part of their product is not the product of the real parts.

The first and last expressions in the integrand are periodic in x with period λ and so their integral is zero. The middle two terms are constant (in fact, one is the complex conjugate of the other) and so these can be pulled outside the integral. Thus, without resorting to applying trigonometric identities in the integrand, we immediately arrive at a succinct formula for the correlation of two fields associated with a plane wave:

$$\langle \eta \xi \rangle = \frac{1}{4} \left(\mathcal{A}_\eta \mathcal{A}_\xi^\star + \text{cc} \right) = \frac{1}{2} \Re \left(\mathcal{A}_\eta \mathcal{A}_\xi^\star \right) = \frac{1}{2} \Re \left(\mathcal{A}_\eta^\star \mathcal{A}_\xi \right), \tag{1.141}$$

in which $\Re(\mathcal{Z})$ takes the real part of $\mathcal{Z}$.

One immediate consequence of (1.141) is that the cross-correlation is zero between two fields that are out of phase by $90°$. In particular, if $\eta = \Re\{A_\eta \exp(\iota kx)\} \propto \cos(kx)$ and $\xi = \Re\{-\iota A_\xi \exp(\iota kx)\} \propto \sin(kx)$, then $\langle \eta \xi \rangle = 0$.

Exercises

1.1   In an Eulerian frame, the concentration of dye that advects at speed U_0 while diffusing is given by

$$C(x,t) = \frac{C_0}{\sqrt{t}} \exp\left[-\frac{(x-U_0 t)^2}{4\kappa t}\right],$$

in which κ is the diffusivity and C_0 is a constant.

(a) Show that the concentration in the Lagrangian frame is

$$C_L(t) = \frac{C_0}{\sqrt{t}} \exp\left[-\frac{x^2}{4\kappa t}\right].$$

(b) Compute dC_L/dt and show that this equals the material derivative $DC/Dt = \partial_t C + U_0 \partial_x C$.

1.2   Use equation (1.33) and the divergence theorem to show that a conserved quantity s is constant in a volume V if the flux of s across its surface, ∂V, is zero.

1.3   (a) Using the ideal gas law and energy conservation, derive and solve the differential equation (1.11), which relates the pressure to the density of an ideal gas undergoing adiabatic compression and expansion.

(b) A parcel of dry air at sea-level pressure, 101.3 kPa, has temperature 20°C. What is its density?

(c) This parcel is transported adiabatically upwards 8850 m to the top of Mount Everest where the atmospheric pressure is 30 kPa. What is the temperature and density of the parcel at this altitude?

1.4   Show that the potential density of a gas is inversely proportional to its potential temperature.

1.5   (a) Assuming the mass flux of saline fluid is proportional to its density gradient, use the divergence theorem to derive the diffusion equation for a stationary fluid:

$$\frac{\partial \varrho}{\partial t} = \kappa \nabla^2 \varrho. \tag{E1.1}$$

(b) Show that the internal energy equation for an incompressible and diffusive fluid in motion is given by the advection–diffusion equation

$$\frac{D\varrho}{Dt} = \kappa \nabla^2 \varrho. \tag{E1.2}$$

1.6 Consider the situation illustrated in Figure 1.14, in which a puck moves across a frictionless table that rotates counter-clockwise at frequency Ω. The puck moves at constant speed c across the diameter of the table starting at distance R_0 from its centre.

(a) Determine the position, $\vec{x}_a(t)$, of the puck in time with respect to a bystander in the absolute frame of reference. For this problem, put the origin at the centre of the table and align the x-axis with the direction of propagation of the puck.

(b) Suppose an observer in the rotating frame initially sits at the same location as the puck. Determine the position, $\vec{x}(t)$, of the observer over time according to the bystander in the absolute frame.

(c) As the observer rotates with the table, the co-ordinate system both translates and rotates with respect to the absolute frame of reference. Using your results in a) and b), first determine the position of the puck with respect to the observer on the table assuming the co-ordinate system does not rotate, then determine the position $\vec{x}_r(t)$ of the puck in the rotating frame.

(d) Taking two time derivatives of your result in c) gives the apparent acceleration of the puck as seen in the rotating frame of reference. Rewrite this in terms of your expressions for position, $\vec{x}_r$, and velocity, $\vec{x}_r'$. Which terms arise from centripetal forces and which from Coriolis forces?

1.7 (a) Use (1.6) and (1.47) to derive a differential equation for density in an isothermal gas and solving it, show that the density in an isothermal atmosphere is given by (1.50).

(b) Again using (1.47) and your result in a), determine the formula for the pressure of an isothermal gas under the assumption that the pressure vanishes as $z \to \infty$.

(c) Supposing the constant temperature is $T_0 = 288\,\text{K}$, estimate the atmospheric pressure at sea level. Compare this with sea-level pressure at 15°C.

(d) At what depth in the ocean is the difference between the local and surface pressure equal to atmospheric pressure?

1.8 (a) Using the equation for hydrostatic balance and the relationships between density, temperature and pressure for a dry adiabatic gas, show that the temperature decreases with height at the 'adiabatic lapse rate' of $\Gamma = g/C_p \simeq 9.8\,\text{K/km}$.

(b) Suppose the temperature decreases with height as $T(z) = T_0 - \Gamma z$, in which $T_0 = 288\,\text{K}$. Derive the differential equation for pressure and solve it assuming the pressure at $z = 0$ is P_{00}. How does the resulting pressure at $z = 10\,\text{km}$ differ from that in an isothermal atmosphere?

(c) What assumptions were made in deriving the formula for Γ in a) that preclude the observed temperature increase in the stratosphere?

1.9   Show that the statement for background hydrostatic balance written in terms of the background potential temperature, $\bar{\theta}$, and the background Exner pressure, $\bar{\pi}$, is

$$\bar{\theta}\frac{d\bar{\pi}}{dz} = -\Gamma,$$

in which Γ is the adiabatic lapse-rate.

1.10 (a) The density of an isothermal atmosphere is given by (1.50). Assuming the atmosphere is stationary, recast this in the nondimensional form given generically by (1.61) and so give explicit expressions for $\bar{\epsilon}$ and $\bar{r}(z)$ in terms of H and H_ϱ.

(b) Assuming $0 \leq z \leq H \ll H_\varrho$, find the leading two terms in the Taylor-series expansion for $\bar{\rho}$ and recast this in non-dimensional form.

1.11 Derive the equation describing sound wave propagation. Hint: represent the total density by $\varrho = \varrho_0 + \rho(\vec{x}, t)$, in which ϱ_0 is the (constant) density in the absence of waves, and the magnitude of the fluctuation density $\|\rho\|$ is much less than ϱ_0. Then show that the assumption of fast and small fluctuations means the momentum equation can be suitably approximated by $\varrho_0\vec{u}_t = -\nabla P$. Likewise the continuity equation, in the form of (1.57), is approximately $P_t = -c_s{}^2\varrho_0\nabla \cdot \vec{u}$. Combine these results to get (1.58).

1.12 Show that the Boussinesq equations of motion given by (1.64), (1.66) and (1.67) are invariant upon reflection in z if $\bar{\rho}$ is linear in z. Hint: make the substitutions $z \rightarrow -z$, $w \rightarrow -w$, $g \rightarrow -g$ and show the equations are unchanged.

1.13 (a) Show that if s is an arbitrary scalar function, then for incompressible fluid

$$\frac{Ds}{Dt} = \frac{\partial s}{\partial t} + \nabla \cdot (\vec{u}s).$$

(b) Derive the kinetic energy conservation law for an incompressible fluid, (1.91), by taking the dot-product of $\vec{u}$ with the momentum equations, (1.54).

1.14 Derive the approximate formula for the fluctuation potential temperature in terms of the fluctuation density and pressure, as given by (1.70).

1.15 Derive the approximate formula (1.71) relating the components of force in the vertical momentum equation to the potential temperature.

1.16 The anelastic form of the momentum equations is sometimes written in terms of the potential temperature, $\vartheta = \bar{\theta} + \theta$, and the Exner pressure, $\Pi = \bar{\pi} + \pi$, in which Π is given by (1.15).

(a) Write $\bar{\pi}$ and $\bar{\theta}$ in terms of the background density and pressure and so show that hydrostatic balance requires

$$\bar{\theta}\frac{d\bar{\pi}}{dz} = -\frac{g}{C_p},$$

in which C_p is the specific heat at constant pressure.

(b) Ignoring Coriolis forces, show that the following are equivalent to the anelastic form of the momentum equations, (1.72):

$$\frac{D\vec{u}}{Dt} = -\nabla(C_p\bar{\theta}\pi) - C_p\frac{d\bar{\pi}}{dz}\theta\hat{z}. \tag{E1.3}$$

1.17 (a) In the absence of perturbations, use (1.18) to show that the background density and potential temperature gradients in the Boussinesq approximation are related by (1.79).

(b) Now including perturbations, use (1.18) and the approximation $(1 + x)^{-1} \simeq 1 - x$ to show that the fluctuation density and potential temperature are related by (1.78).

1.18 The density of a stationary liquid changes in x–z space according to $\varrho(x,z) = \varrho_0(1 - \alpha_x x - \alpha_z z)$, in which α_x and α_z are positive constants. Compute the corresponding hydrostatic pressure field and find the baroclinic torque $(\nabla\varrho \times \nabla P)/\varrho^2$. For what α_x and α_z is this term zero?

1.19 Compute the available potential energy per unit width for the system shown in Figure 1.22.

1.20 Compute the available potential energy per unit width for a fluid with density distribution given by $\varrho = \rho_1 - (\rho_1 - \rho_0)x/L$ for $0 \le x \le L$ and $0 \le z \le H$.

1.21 Prove the identity (1.98) assuming the fluid is barotropic so that density can be written as a function of pressure alone.

1.22 Hold two sheets of paper from their top edge so that they are suspended two finger-widths apart. Blowing between the two sheets, do you suspect the pages with fly apart or will they be drawn towards each other? Explain your answer.

1.23 Suppose a one-dimensional wave in an unbounded domain has a dispersion relation $\omega^2 = a + bk^2$, in which a and b are real constants. Derive its partial differential equation by following the steps below.

(a) Suppose the structure, $\eta(x,t)$, of the wave can be represented by the Fourier transform (1.121). Take two time and space derivatives of η carrying the derivatives through the integral.

(b) Form a linear superposition of η, η_{tt} and η_{xx} so that, together with the dispersion relation above, the resulting integrand is exactly zero. Thus write the partial differential equation that prescribes the wave motion.

1.24 Suppose the initial structure of a wavepacket is given by $\eta(x,0) = \eta_0(x)$. Using the Fourier representation of one-dimensional waves given by (1.121), show that rightward-propagating nondispersive wavepackets with a dispersion relation $\omega = ck$ have structure $\eta(x,t) = \eta_0(x - ct)$ at any time t. (Do not explicitly use the inverse Fourier transform in the solution of this problem.)

1.25 Use geometric arguments to derive the formulae (1.116) for the phase speed, c_P, observed in a fixed frame from the usual definition (1.115) of the phase speed moving with the wave crests.

1.26 Follow the steps below to show that a small-amplitude, one-dimensional wavepacket moves at the group velocity, $c_g \equiv \omega'(k)$.

(a) Starting with the wavepacket prescribed by (1.121), transform to a frame of reference moving with the group velocity and so show that η is given by the generalized Fourier integral

$$\eta(x,t) = \int_{-\infty}^{\infty} \tilde{\eta}(k) e^{-\imath \Omega(k) t} \, dk,$$

in which $\Omega(k) \equiv \omega - c_g k$ is the extrinsic (Doppler-shifted) frequency.

(b) Find the structure of η for large t in this moving frame. Do this by using the method of stationary phase which states that the integral is dominated by those wavenumbers k_c for which $d\Omega(k_c)/dk = 0$. (For larger or smaller k, the exponential in the integrand oscillates rapidly with tiny changes in k and sums to zero if $\tilde{\eta}$ varies slowly.) Taylor-series expand Ω about k_c keeping the first two non-zero terms and substitute into the integral in a).

(c) Find the asymptotic approximation of this generalized Fourier integral for large t (integration by parts will do), and keep the leading-order terms. In the case $\omega''(k_c) > 0$, show that

$$\eta(x,t) \simeq \tilde{\eta}(k_c)\sqrt{\frac{2\pi}{t\omega''(k_c)}} \exp\left[\imath\left(k_c x - \omega(k_c)t - \frac{\pi}{4}\right)\right]. \qquad (E1.4)$$

Thus at long times, the only waves that appear at position $x = c_g t$ are those with wavenumber k_c, which have a group velocity $c_g = \omega'(k_c)$.

1.27 An isolated square-wave is given by

$$\eta(x) = \begin{cases} -A & -\sigma < x < 0 \\ A & 0 < x < \sigma \\ 0 & \text{otherwise,} \end{cases}$$

in which $\sigma > 0$.

(a) Find the Fourier transform of η.
(b) Find the Fourier cosine series of η assuming the function is bounded so that $|x| \leq L$, with $L \geq \sigma$.
(c) Compare the amplitudes of the lowest two modes for your answer in b) for $L = \sigma$ and $L = 2\sigma$.

1.28 From the infinite series resulting from the Taylor expansion of the dispersion relation $\omega = \omega(k)$ about k_0, derive a general formula for the linear Schrödinger equation written as a time derivative, A_t, on the left-hand side of the equations and with an infinite sequence of successive orders of x-derivatives of A appearing on the right-hand side.

1.29 Here is a maths trick that is useful when computing correlations between waves and which avoids the necessity for using double-angle formulae in the computation of integrals of cosine-squared and sine-squared.

(a) Sketch the graphs of $\cos^2\phi$ and $\sin^2\phi$ for $0 \leq \phi \leq 2\pi$ and so convince yourself by inspection that the areas under the graphs are equal.
(b) Based on your observation in (a) and the fact that $\cos^2\phi + \sin^2\phi = 1$, show that $\langle\cos^2\phi\rangle = \langle\sin^2\phi\rangle = 1/2$.

1.30 In the following problems $\eta = A_\eta \exp[\imath(kx - \omega t)]$ and $\xi = A_\xi \exp[\imath(kx - \omega t)]$, in which A_η and A_ξ are complex amplitudes and it is understood the actual fields are given by the real part of the expressions.

(a) Compute the correlation $\langle\eta\xi\rangle$ between η and ξ by keeping x fixed and integrating in t. Show that you get the same result as (1.141), which was determined by integrating in x.
(b) Suppose $A_\xi = A_\eta e^{\imath\phi}$ in which ϕ is a phase factor. Using (1.141), show that $\langle\eta\xi\rangle = 0$ if $\phi = \pm\pi/2$.

2

Interfacial waves

2.1 Introduction

In this chapter we consider waves at the interface between two fluids of different density and waves in multi-layer fluids. We derive the fully nonlinear equations for waves of arbitrary amplitude and then determine approximate formulae appropriate for small-amplitude waves. These are expressed through coupled linear partial differential equations for which analytic solutions may be found through standard methods. Approximate solutions for nonlinear waves are considered in Chapter 4.

A 'one-layer fluid' has uniform density and is bounded above by a free surface that may oscillate up and down, for example, due to gravitational forces. Waves in a one-layer fluid are specifically referred to here as 'surface waves'. Waves on the ocean surface can be treated as existing in a one-layer fluid if the density variation with depth in the ocean negligibly affects their dynamics.

A 'two-layer fluid' describes a fluid of one density underlying a second fluid of smaller density. 'Interfacial waves' propagate at the interface between the two fluids. In one sense, surface waves on the ocean are interfacial waves in that they propagate at the interface between water and air. However, in this book we distinguish surface waves from interfacial waves by requiring for the latter that the density difference between the upper and lower layer fluids is a small fraction of the density of either layer. Thus undular displacements of an interface between fresh and salt water are considered for waves in a two-layer fluid. Likewise, waves can exist at an atmospheric inversion or at a thermocline in lakes and oceans. These can be treated as interfacial waves provided their horizontal wavelength is much longer than the inversion or thermocline depth.

Although surface waves are not internal gravity waves, below we nonetheless review some of their properties because this discussion helps lay the groundwork for the theory of interfacial internal gravity waves. We then go on to consider the effects of background shear and rotation upon interfacial waves.

2.2 Surface waves

Surface wave theory is described in detail in numerous textbooks; only its relevant features are presented here. For the most part, we consider waves at a water–air interface in which Coriolis forces, viscosity and surface tension (capillary) effects are neglected, and the fluid is assumed to be incompressible. As a one-layer fluid, the density of water is assumed to be uniform and we assume the ambient is stationary, meaning that any motion in the fluid is due only to wave-induced perturbations. Likewise, we assume there is no mean wind, so the surface stress is zero. For mathematical convenience, we further suppose the waves are one-dimensional having structure only in the vertical and in one horizontal direction.

2.2.1 Finite-amplitude equations

The most significant dynamical difference between surface and (internal) interfacial waves is that the perturbation of air by water waves has negligible influence upon the propagation of the waves themselves. Thus the theory of surface waves typically ignores the presence of air altogether, treating the free surface as a region of constant pressure.

We describe the waves in a Cartesian co-ordinate system with x horizontal and z vertical, as shown in Figure 2.1. The corresponding velocity components are u and w, respectively. The domain of the wave is given by all values x and z lying below the surface, which is prescribed by the vertical displacement, $\eta(x,t)$, from the unperturbed surface.

A straightforward approach to determine the motion of the fluid interior would solve the momentum equations (1.64) for incompressible fluid. Rather than do this, it is typical to invoke angular momentum conservation to describe the motion of surface waves. In the absence of waves the vorticity field of the stationary ambient is everywhere zero. Vorticity cannot be introduced at the boundaries because the fluid is inviscid and there are no surface stresses. Furthermore, the density of the fluid is constant and so vorticity cannot be introduced by baroclinic torques within the

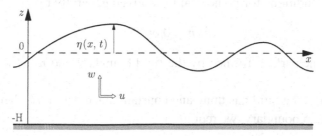

Fig. 2.1. Set-up for the description of waves in a one-layer fluid.

Interfacial waves

fluid, as in (1.87). Therefore the vorticity within the fluid must be zero everywhere for all time: the fluid is irrotational. As a consequence, we can define the 'velocity potential', ϕ, such that

$$(u, w) = \left(\frac{\partial \phi}{\partial x}, \frac{\partial \phi}{\partial z} \right). \tag{2.1}$$

This guarantees that the y-component of vorticity, $(\nabla \times (u, 0, w)) \cdot \hat{y}$, is zero.

If we further assume the fluid is incompressible, then (1.30) together with (2.1) give

$$\nabla^2 \phi \equiv \frac{\partial^2 \phi}{\partial x^2} + \frac{\partial^2 \phi}{\partial z^2} = 0. \tag{2.2}$$

This elliptical, homogeneous partial differential equation is sometimes called the 'potential equation' or 'Laplace's equation'. Even though it is linear, it can be used to describe the motion of large-amplitude waves. Nonlinear effects enter through the prescription of surface boundary conditions.

The time dependence of ϕ comes from the surface boundary condition which specifies that the pressure is the same everywhere on the surface. For mathematical convenience, we will take this reference pressure to be zero. Using the time-dependent form of the Bernoulli equation (1.102) with $\varrho = \varrho_0$, and using (2.1) this condition becomes

$$\left[\frac{\partial \phi}{\partial t} + \frac{1}{2} |\nabla \phi|^2 \right] \bigg|_{z=\eta} = -g\eta. \tag{2.3}$$

The surface displacement is itself related to the velocity potential by matching the vertical velocity at the surface:

$$\frac{\partial \phi}{\partial z} \bigg|_{z=\eta} = \frac{D\eta}{Dt}, \tag{2.4}$$

in which the material derivative is $D/Dt = \partial_t + u \, \partial_x$. The z-derivative term is neglected because η is a function of x and t alone.

In addition to the surface boundary condition, if the fluid is contained within a basin we must also specify conditions at the rigid bottom and side boundaries. These are the conditions for no normal flow, given generally by

$$\vec{u} \cdot \hat{n} = 0 \text{ on } \mathcal{S}, \tag{2.5}$$

in which $\mathcal{S}$ is the surface formed by the rigid boundary and $\hat{n}$ is the unit normal vector to $\mathcal{S}$.

In particular, if the fluid has finite and constant mean depth, H, then for no flow across the bottom boundary, we require

$$\phi_z(x, -H, t) = 0, \tag{2.6}$$

in which the subscript denotes a partial derivative with respect to z. If instead the fluid is unbounded below, then the condition on ϕ specifies that it must remain bounded as $z \to -\infty$. If the fluid is bounded by vertical side-walls at $x = 0$ and $x = L$ then the no normal flow conditions are

$$\phi_x(0, z, t) = \phi_x(L, z, t) = 0. \tag{2.7}$$

Both (2.6) and (2.7) are zero Neumann boundary conditions on ϕ. They require that the component of $\nabla\phi$ normal to the boundary is zero. No restriction is placed upon the motion parallel to the boundary. We will see that the fluid is free to slide along the boundary, which is a consequence of neglecting viscosity. Thus (2.6) and (2.7) also constitute 'free-slip' boundary conditions.

2.2.2 Small-amplitude approximation

An exact analytic solution of the linear partial differential equation (2.2) with the nonlinear surface boundary conditions (2.3) and (2.4) is generally impossible, although approximate solutions can be found in special limits. If the amplitude of the waves is small compared with their horizontal scale, then the boundary conditions become linear partial differential equations. Furthermore, if the domain is unbounded or if the bottom of the domain is horizontal then explicit analytic solutions can be found.

Assuming η is small, a function f of η can be expressed as a Taylor series about $\eta = 0$:

$$f(\eta) \simeq f(0) + \eta f'(0) + \dots \tag{2.8}$$

We substitute this into (2.3) and (2.4) keeping only those terms that are linear in ϕ and η: because η, and hence ϕ, are small any terms involving products of these functions are much smaller than the linear terms and can be neglected.

Thus the equations of motion for waves in fluid of finite depth H give the boundary value problem

$$
\begin{aligned}
\phi_{xx} + \phi_{zz} &= 0 && -H < z < 0 \\
\phi_z &= 0 && z = -H \\
\phi_z &= \eta_t && z = 0 \\
\phi_t + g\eta &= 0 && z = 0
\end{aligned}
\tag{2.9}
$$

Assuming the domain has infinite horizontal extent, the equations are simplified by seeking plane wave solutions that are periodic in time with frequency ω and are horizontally periodic in space with horizontal wavenumber k. Although this may seem restrictive, in effect we are performing a Fourier transform and it is understood that disturbances with arbitrary spatial structure can be represented as a superposition of plane waves using (1.121).

Because the domain is bounded above and below we cannot assume a plane wave structure in z. Instead we suppose the z-dependent part of the wave structure is separable from the x- and t-dependent parts. Thus we assume the velocity potential can be written as

$$\phi(x,z,t) = \hat{\phi}(z)e^{\iota(kx-\omega t)}, \tag{2.10}$$

and the surface displacement is

$$\eta(x,t) = Ae^{\iota(kx-\omega t)}. \tag{2.11}$$

Here $\hat{\phi}(z)$ may be a complex function of depth z and it is understood that the actual velocity potential of the wave is given by the real part of (2.10). This is where the complex representation comes in handy. If we wrote $\phi = \hat{\phi}\cos(kx - \omega t)$ and $\eta = A\cos(kx - \omega t)$, we would be making an assumption about the relative phase of the two fields. In fact, we will see that they are 90° out of phase.

We could have assigned a complex-valued amplitude $\mathcal{A}$ to η so that the actual maximum displacement of the waves would be $|\mathcal{A}|$. However, without loss of generality we set up the co-ordinate system so that a crest lies above the origin at time $t = 0$. Hence the constant amplitude A is assumed to be positive and real valued.

Substituting (2.10) and (2.11) into the system of equations (2.9), taking derivatives and cancelling exponential terms in z gives the ordinary differential equation

$$\hat{\phi}'' - k^2\hat{\phi} = 0, \tag{2.12}$$

with boundary conditions

$$\hat{\phi}'(-H) = 0$$
$$\hat{\phi}'(0) = -\iota\omega A \tag{2.13}$$
$$-\iota\omega\hat{\phi}(0) + gA = 0,$$

in which primes denote ordinary z-derivatives.

Equation (2.12) has exponential solutions. But, in light of the boundary condition at $z = -H$, it is more convenient to write the solutions in terms of the hyperbolic sine and cosine functions, sinh and cosh, respectively. Explicitly, applying the first two boundary conditions in (2.13) we have

$$\hat{\phi}(z) = -\iota\frac{A\omega}{k\sinh(kH)}\cosh k(z+H). \tag{2.14}$$

The presence of the complex number ι in this relation means that the velocity potential is $90°$ out of phase with the surface displacement field. Explicitly, assuming A is real, the actual surface displacement is the real part of (2.11),

$$\eta(x,t) = A\cos(kx - \omega t), \tag{2.15}$$

and the actual velocity potential is

$$\phi(x,z,t) = \frac{A\omega}{k\sinh kH}\cosh k(z+H)\sin(kx - \omega t). \tag{2.16}$$

2.2.3 Dispersion relation, phase and group speeds

The last boundary condition of (2.13) gives the dispersion relation of the waves. Using (2.14), we find

$$\omega^2 = gk\tanh(kH), \tag{2.17}$$

which is plotted in Figure 2.2.

For prescribed k, the frequency given by (2.17) is an eigenvalue of the boundary value problem (2.12) with (2.13). Whether k is positive or negative, the right-hand side of (2.17) is positive. Typically, one would take the positive square root to define the frequency and let the directionality of the waves be set by the sign of k. However, for surface and interfacial waves, particularly in the deep water limit, it is practical to assume k is positive and let the sign of ω establish the propagation direction.

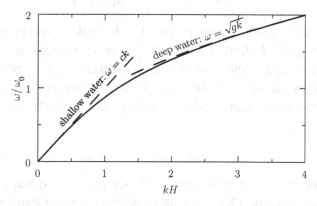

Fig. 2.2. Dispersion relation for small-amplitude surface waves in fluid of mean depth, H. The dashed lines correspond to the dispersion relation in the shallow and deep water limits. The wavenumber is normalized by H and the frequency is normalized by the characteristic value $\omega_0 = (g/H)^{1/2}$.

The deep water limit is expressed mathematically by $kH \gg 1$. Putting this in (2.17) gives the dispersion relation of deep water waves:

$$\omega^2 = gk, \tag{2.18}$$

in which the assumption $k > 0$ has now been made explicit. Waves that are very long compared with the fluid depth are called 'shallow water waves'. Their dispersion relation is determined by taking the limit $kH \ll 1$ in (2.17) to give

$$\omega^2 = c^2 k^2, \text{ with } c \equiv \sqrt{gH}. \tag{2.19}$$

Thus shallow water waves are nondispersive.

The dispersion relations for finite-depth, shallow water and deep water waves are listed in Table 2.1 along with other relevant properties of small-amplitude waves in these three circumstances. These dispersion curves are also plotted in Figure 2.2. Evidently, waves can be treated as shallow if $k \lesssim 1/2H$, and they can be treated as deep if $k \gtrsim 2/H$. The transition region occurs for wavelengths approximately between $3H$ and $12H$.

Using the one-dimensional version of (1.115) and (1.120), the dispersion relation predicts that rightward-propagating wave crests move at phase speed

$$c_p \equiv \frac{\omega}{k} = \sqrt{\frac{g}{k} \tanh(kH)}, \tag{2.20}$$

and that wavepackets move at group speed

$$c_g \equiv \frac{\partial \omega}{\partial k} = \frac{1}{2} c_p \left(1 + \frac{2kH}{\sinh 2kH} \right). \tag{2.21}$$

In particular, in the shallow water limit (2.19) shows the waves are nondispersive with $c_p = c_g = \sqrt{gH}$. In finite-depth water the group velocity is slower than c_p having magnitude $c_p/2$ in the limit of deep water waves, $kH \to \infty$. So the crests of these waves advance to the front of a wavepacket whereas the crests of shallow water waves move at the same speed as the wavepacket, as illustrated in Figure 1.27.

2.2.4 Polarization relations

Once η has been prescribed in the form (2.11) for given A and k and with ω given by the dispersion relation (2.17), all other fields of interest can be determined. The expressions for these fields in terms of A and k are known as the 'polarization relations'. Just as ω is an eigenvalue of the boundary value problem, the polarization relations represent the corresponding eigenfunctions.

Table 2.1. *Table of properties for small-amplitude surface waves on water of finite depth, shallow depth (kH → 0) and great depth (kH → ∞). Values are given for waves in the (x, z) plane with no background rotation. Properties include the dispersion relation and the relationship of group speed c_g to phase speed c_p. Polarization relations are given for fields relative to a given surface displacement $\eta = A\cos\varphi$, in which $\varphi \equiv kx - \omega t$. These fields are normalized by characteristic values: A for displacement fields (η, ξ_x and ξ_z), Aω for velocity fields (u and w), and $A\varrho_0 g$ for the pressure field (p). Correlations between pairs of fields are given representing averages of energy (E), energy flux ($\mathcal{F}_E$), the Stokes drift (u_S) and the time-averaged momentum flux ($\mathcal{F}_M$).*

	Finite-depth waves	Shallow water waves	Deep water waves
	$\omega = (gk\tanh kH)^{1/2}$	$\omega = (gH)^{1/2}k$	$\omega = (gk)^{1/2}$
	$c_g = \dfrac{1}{2}c_p\left(1 + \dfrac{2kH}{\sinh 2kH}\right)$	$c_g = c_p$	$c_g = \dfrac{1}{2}c_p$
$\dfrac{1}{A}\eta$	$\cos\varphi$	$\cos\varphi$	$\cos\varphi$
$\dfrac{1}{A\omega/k}\phi$	$\dfrac{\cosh k(z+H)}{\sinh kH}\sin\varphi$	$\dfrac{1}{kH}\sin\varphi$	$e^{kz}\sin\varphi$
$\dfrac{1}{A\omega}u$	$\dfrac{\cosh k(z+H)}{\sinh kH}\cos\varphi$	$\dfrac{1}{kH}\cos\varphi$	$e^{kz}\cos\varphi$
$\dfrac{1}{A\omega}w$	$\dfrac{\sinh k(z+H)}{\sinh kH}\sin\varphi$	$\left(1+\dfrac{z}{H}\right)\sin\varphi$	$e^{kz}\sin\varphi$
$\dfrac{1}{A}\xi_x$	$-\dfrac{\cosh k(z+H)}{\sinh kH}\sin\varphi$	$-\dfrac{1}{kH}\sin\varphi$	$-e^{kz}\sin\varphi$
$\dfrac{1}{A}\xi_z$	$\dfrac{\sinh k(z+H)}{\sinh kH}\cos\varphi$	$\left(1+\dfrac{z}{H}\right)\cos\varphi$	$e^{kz}\cos\varphi$
$\dfrac{1}{A\varrho_0 g}p$	$\dfrac{\cosh k(z+H)}{\cosh kH}\cos\varphi$	$\cos\varphi$	$e^{kz}\cos\varphi$
$\langle E\rangle$	$\dfrac{1}{2}\varrho_0 gA^2$	$\dfrac{1}{2}\varrho_0 gA^2$	$\dfrac{1}{2}\varrho_0 gA^2$
$\langle \mathcal{F}_E\rangle$	$\varrho_0 g\dfrac{\omega}{2k}\left(1+\dfrac{2kH}{\sinh 2kH}\right)A^2$	$\varrho_0 g\dfrac{\omega}{k}A^2$	$\varrho_0 g\dfrac{\omega}{2k}A^2$
u_S	$\dfrac{1}{2}\omega k\dfrac{\cosh 2k(z+H)}{\sinh^2 kH}A^2$	$\dfrac{1}{2}\dfrac{\omega}{kH^2}A^2$	$\omega k e^{2kz}A^2$
$\langle \mathcal{F}_M\rangle_t$	$\dfrac{1}{2}\varrho_0 g\left(1+\dfrac{2kH}{\sinh 2kH}\right)A^2$	$\varrho_0 gA^2$	$\dfrac{1}{2}\varrho_0 gA^2$

From (2.10) with (2.14) the velocity potential is

$$\phi(x,z,t) = \frac{-\iota A\omega}{k\sinh kH}\cosh k(z+H)\,e^{\iota(kx-\omega t)},\tag{2.22}$$

the real part of which gives the actual field represented by (2.16).

Taking derivatives, the velocity fields are

$$u(x,z,t) = \frac{A\omega}{\sinh(kH)}\cosh k(z+H)\,e^{\iota(kx-\omega t)}\tag{2.23}$$

and

$$w(x,z,t) = \frac{-\iota A\omega}{\sinh(kH)}\sinh k(z+H)\,e^{\iota(kx-\omega t)}.\tag{2.24}$$

The actual velocities are listed in Table 2.1. As expected, $w=0$ at $z=-H$, consistent with the condition for no normal flow through the bottom boundary.

Equation (2.23) shows that the horizontal velocity is in phase with the surface displacement field. For rightward-propagating waves, the fluid moves from left to right beneath a wave crest and from right to left beneath a trough, as shown in Figure 2.3a. The vertical velocity field, given by (2.24), is out of phase with η and u. Its maximum value occurs at the inflection point leading the wave crest.

We compute the horizontal and vertical displacements of fluid parcels, ξ_x and ξ_z, by solving $\partial\xi_x/\partial t = u$ and $\partial\xi_z/\partial t = w$, respectively. This amounts to dividing

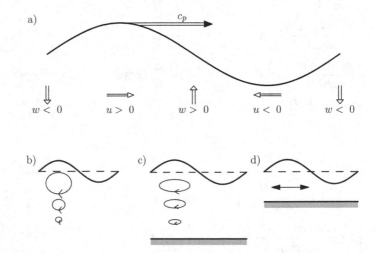

Fig. 2.3. a) Velocity below surface at different phases of a rightward-propagating wave. Also shown are parcel trajectories below b) deep water waves, c) finite-depth waves and d) shallow water waves.

(2.23) and (2.24) through by $-\iota\omega$. Thus

$$\xi_x(x,z,t) = \frac{\iota A}{\sinh(kH)} \cosh k(z+H) \, e^{\iota(kx-\omega t)} \tag{2.25}$$

and

$$\xi_z(x,z,t) = \frac{A}{\sinh(kH)} \sinh k(z+H) \, e^{\iota(kx-\omega t)}. \tag{2.26}$$

As expected, at $z = 0$ we have $\xi_z(x,0,t) = Ae^{\iota(kx-\omega t)} = \eta$.

The dynamic pressure field, p, may be determined from the linearized form of the Bernoulli equation (1.99), in which the hydrostatic background is removed using $P = P_0 - \varrho_0 g z + p(\vec{x},t)$. Thus we have

$$\varrho_0 \frac{\partial \phi}{\partial t} + p = 0. \tag{2.27}$$

This formula can also be derived from either of the horizontal or vertical momentum equations in (1.64). Thus we find

$$p(x,z,t) = \frac{A\varrho_0\omega^2}{k \sinh kH} \cosh k(z+H) \, e^{\iota(kx-\omega t)} = \varrho_0 c_p u(x,z,t). \tag{2.28}$$

The formulae presented here have been given in general for finite-depth waves. The polarization relations for shallow water waves may be determined from these by taking the limit $kH \to 0$, in which case $\sinh(kH) \to kH$, $\cosh k(z+H) \to 1 + (k(z+H))^2/2$ and $\sinh k(z+H) \to k(z+H)$. Thus, for example, the velocity potential given by (2.16) becomes

$$\phi \simeq \frac{A\omega}{k^2 H} \left\{ 1 + \frac{1}{2}[k(z+H)]^2 \right\} \sin(kx - \omega t) \simeq \frac{A\omega}{k^2 H} \sin(kx - \omega t).$$

The last expression is the leading-order approximation, but the middle expression is necessary to evaluate the leading-order vertical velocity field $w = \partial\phi/\partial z$.

Similarly, in the deep water limit one can use the approximations $\sinh(kH) \to \exp(kH)/2$ and $\cosh k(z+H), \sinh k(z+H) \to \exp[k(z+H)]/2$. From (2.16), the velocity potential for deep water waves is

$$\phi \simeq \frac{A\omega}{k} e^{kz} \sin(kx - \omega t).$$

The resulting polarization relations given in terms of the vertical displacement amplitude are listed in Table 2.1.

2.2.5 Fluid parcel motion

We can reconstruct the approximate path followed by an individual fluid parcel during the passage of a small-amplitude wave using (2.25) and (2.26). Supposing the parcel is near (x_0, z_0), its actual position $(x(t), z(t))$ at time t is given by

$$x(t) \simeq x_0 - \frac{A}{\sinh kH} \cosh k(z_0 + H) \sin(kx_0 - \omega t) \qquad (2.29)$$

and

$$z(t) \simeq z_0 + \frac{A}{\sinh kH} \sinh k(z_0 + H) \cos(kx_0 - \omega t). \qquad (2.30)$$

Here, consistent with our small-amplitude wave assumption, we have supposed that the displacements from the original position are small so that we can approximate $x \simeq x_0$ and $z \simeq z_0$ on the right-hand side of (2.25) and (2.26) after we take the real part of these expressions. These restrictions will be eased somewhat in the discussion concerning the Stokes drift in Section 2.2.8.

Eliminating t in these equations shows that the fluid parcel follows an elliptical path given by

$$\frac{(x - x_0)^2}{\cosh^2 k(z_0 + H)} + \frac{(z - z_0)^2}{\sinh^2 k(z_0 + H)} = \frac{A^2}{\sinh^2 kH}. \qquad (2.31)$$

The ratio of the minor to major axis varies from 0 to 1 as kH increases from 0 to ∞. Thus the parcel oscillates back and forth horizontally in shallow water and moves in circular orbits beneath waves in deep water. The extent of the excursion from (x_0, z_0) is proportional to the wave amplitude and, except for shallow water waves, depends upon the distance from the surface. The paths of fluid parcels in deep, finite-depth and shallow water are sketched in Figure 2.3.

2.2.6 Energy transport

The kinetic energy equation is generally given by (1.91). However, this simplifies for the case of small-amplitude surface waves. Adding u times the x-momentum equation to w times the vertical momentum equation in (1.64), and using the incompressibility condition in the form (1.30), we find

$$\frac{\partial}{\partial t}\left(\frac{1}{2}\varrho_0[u^2 + w^2]\right) = -\frac{\partial(up)}{\partial x} - \frac{\partial(wp)}{\partial z}.$$

Integrating this between the bottom and surface, the second term on the right-hand side vanishes and so we have the flux-form of the conservation of kinetic energy

$$\frac{\partial E_K}{\partial t} = -\frac{\partial \mathcal{F}_K}{\partial x}, \qquad (2.32)$$

in which

$$E_K(x,t) \equiv \int_{-H}^{0} \frac{1}{2}\varrho_0(u^2 + w^2)\,dz \qquad (2.33)$$

and

$$\mathcal{F}_K(x,t) \equiv \int_{-H}^{0} up\,dz. \qquad (2.34)$$

In these expressions we have assumed the wave amplitude is small enough that the upper bound of the z-integral can be set to zero.

Averaging (2.33) either over one wavelength or over one period in time determines the mean kinetic energy (per unit horizontal area), $\langle E_K \rangle$. Using (2.23) and (2.24), and averaging using (1.141) we find

$$\langle E_K \rangle = \frac{1}{4}\varrho_0 \frac{\omega^2 A^2}{k \tanh kH} = \frac{1}{4}\varrho_0 g A^2. \qquad (2.35)$$

Surprisingly, with use of the dispersion relation (2.17), we see that the mean kinetic energy is independent of both k and H.

The total potential energy of a column of fluid is found by integrating $\varrho_0 gz$ from the bottom at $z = -H$ to the surface at $z = \eta$. The available potential energy, E_P, associated with surface waves is the difference between the total potential energies in the presence and absence of waves. Thus, by analogy with (1.97)

$$E_P(x,t) \equiv \int_{0}^{\eta} \varrho_0 gz\,dz = \frac{1}{2}\varrho_0 g \eta^2. \qquad (2.36)$$

Averaging over one wavelength gives

$$\langle E_P \rangle = \frac{1}{4}\varrho_0 g A^2. \qquad (2.37)$$

This is the same result as (2.35). Indeed, it is generally true of conservative, small-amplitude waves that the mean kinetic and available potential energy are equal. This is referred to as the 'equipartition of energy'.

Therefore, the mean total energy associated with fluid parcels moving due to waves is

$$\langle E_T \rangle \equiv \langle E_K \rangle + \langle E_P \rangle = \frac{1}{2}\varrho_0 g A^2. \qquad (2.38)$$

The horizontal transport of mass by small-amplitude waves is $\varrho_0 u$, which is zero when averaged over one period. Therefore the mean energy flux is given by the flux of kinetic energy alone. So the energy conservation law for small-amplitude waves at a fixed point in space and time is

$$\frac{\partial E_T}{\partial t} = -\frac{\partial \mathcal{F}_E}{\partial x}, \qquad (2.39)$$

in which $\mathcal{F}_E \equiv \mathcal{F}_K$. Using the relation $p = \varrho_0 c_p u$, we find

$$\langle \mathcal{F}_E \rangle = \varrho_0 c_p \int_{-H}^{0} \langle u^2 \rangle \, dz = \frac{1}{4} c_p \left(1 + \frac{2kH}{\sinh 2kH} \right) \varrho_0 g A^2. \qquad (2.40)$$

Together with (2.21) we recognize that

$$\langle \mathcal{F}_E \rangle = c_g \langle E_T \rangle. \qquad (2.41)$$

This holds not only for surface waves but for all small-amplitude waves. It is the statement that energy is transported by waves at the group velocity.

Although we have referred to $\langle E_T \rangle$ as 'energy', in fact it is the energy per unit horizontal area. In order to assess the energy, in joules, associated with one wavelength one could multiply by the wavelength and spanwise (along-crest) extent. Equivalently we could assess how much energy crosses a spanwise length in a given time. This is given by multiplying the energy flux (2.41) by the spanwise extent and the time, λ/c_g, which is how long the wavepacket takes to transport energy a distance of one wavelength. The flux $\langle \mathcal{F}_E \rangle$ times the spanwise extent gives the mean power associated with the waves.

For example, deep water waves with a wavelength of $\lambda = 10$ m and an amplitude of 0.1 m have energy per unit spanwise extent of 50 J/m over one wavelength. The frequency of the waves is 2.5 s^{-1} (a period of $T \simeq 2.5$ s) so the group speed is $c_g \simeq 2.0$ m/s. Therefore, using (2.41), the power delivered by the waves per unit spanwise extent is 10 W/m. Note that the energy associated with one wavelength is not computed from the flux by multiplying by T, the time for a crest to travel one wavelength. It is found by multiplying by $\lambda/c_g = 5$ s, the time for the group to travel one wavelength.

Some care must be taken in the interpretation of the average energy and its flux. It is not obvious that we can substitute (2.38) and (2.41) into the energy conservation law (2.39) because, strictly speaking, the averaging operation cannot be taken inside the derivatives. For example, if we average both sides of (2.39) over one period in time, we find

$$\frac{1}{T} \Delta E = -\frac{\partial}{\partial x} \langle \mathcal{F}_E \rangle_t, \qquad (2.42)$$

in which $\langle \rangle_t$ denotes time-averaging. Although we can take the averaging operator inside the x-derivative, when acting on the left-hand side of (2.39), the result gives the difference in energy, ΔE, between two periods.

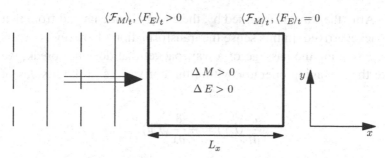

Fig. 2.4. Schematic showing energy and momentum deposition resulting from a wave breaking within a control volume of length L_x.

Of course, if the disturbance is a plane wave $\Delta E = 0$. Likewise, the right-hand side of (2.42) is zero because the average energy flux is constant in x. So as a plane wave passes by, the energy at a point does not change over one period because as much energy is transported towards the point as moves away from it. The energy of the fluid itself changes only if the divergence in the flux is non-zero. This can happen in one of two ways.

One way is through wave breaking. For example, consider the situation depicted in Figure 2.4. A plane wave passes into the left-hand side of a control volume (at $x = 0$) with an associated mean energy flux $\langle \mathcal{F}_E \rangle_t$. Over a distance L_x the waves break within the volume so that no waves exit on the right-hand side: $\langle \mathcal{F}_E \rangle |_{x=L_x} = 0$. Integrating (2.42) over the length of the control volume, we see that the power per unit spanwise extent delivered to the fluid in this volume by breaking waves is

$$L_x(\Delta E)/T = \langle \mathcal{F}_E \rangle_t |_{x=0}. \tag{2.43}$$

Therefore, in this scenario we interpret $\langle \mathcal{F}_E \rangle$ as the power per unit spanwise extent that would be extracted from the waves when they dissipate. The power would go into the mean flow, lost to dissipation or put into the motion of a solid object, for example, if breaking against the hull of a ship. Whatever the cause, the transformation of energy is irreversible.

A second way to interpret (2.42) is through reversible energy changes resulting from the passage of a quasi-monochromatic wavepacket, as illustrated for example in Figure 1.28a. The mean energy flux associated with the waves is given by (2.40) except that now the amplitude is $A = A(X, T)$, a slowly varying function of space and time. $A(X, T)$ is the amplitude envelope of the wavepacket. Therefore, the right-hand side of (2.42) is non-zero. As a wavepacket approaches a point, the energy at that point increases in time because the amplitude of the waves increases. As the wavepacket moves away from the point, the energy decreases as the amplitude decreases. After the wave has passed by, the energy is unchanged from that before the wavepacket arrived. In this sense the transformation of energy is reversible.

When considering the passage of a wavepacket that does not break, we can in effect take the averaging operator within the derivatives on both sides of (2.39) to give

$$\frac{\partial}{\partial t} \langle E_T \rangle = -\frac{\partial}{\partial x} \langle \mathcal{F}_E \rangle, \tag{2.44}$$

in which averaging can be performed over one wavelength in space or one period in time. It is understood that $\langle E_T \rangle$ and $\langle \mathcal{F}_E \rangle$ are slowly varying functions prescribed through the structure of the amplitude envelope.

Below we will see that the same two interpretations can be ascribed to the corresponding law for momentum conservation. The interpretation in which momentum is irreversibly deposited to the mean flow due to wave breaking is discussed in Section 2.2.7. The interpretation in which the mean flow reversibly accelerates then decelerates during the passage of a wavepacket is discussed in Section 2.2.9. Therein the procedure of averaging over wavepackets is considered with more mathematical rigour.

2.2.7 Momentum deposition

The fully nonlinear horizontal momentum equation in the Boussinesq approximation is given by (1.64). Setting $f_0 = 0$ (ignoring background rotation) and using the incompressibility condition in the form (1.30), this is written in flux-form by

$$\frac{\partial}{\partial t}(\varrho_0 u) = -\frac{\partial}{\partial x}(\varrho_0 u^2 + p) - \frac{\partial}{\partial z}(\varrho_0 uw). \tag{2.45}$$

Vertically integrating both sides of this equation between $z = -H$ and $z = 0$ gives

$$\frac{\partial M}{\partial t} = -\frac{\partial}{\partial x}(\mathcal{F}_M + p) - \varrho_0(uw|_{z=0}), \tag{2.46}$$

in which

$$M \equiv \int_{-H}^{0} \varrho_0 u \, dz \tag{2.47}$$

and

$$\mathcal{F}_M \equiv \int_{-H}^{0} \varrho_0 u^2 \, dz. \tag{2.48}$$

To help interpret (2.46), we average the right-hand side of (2.46) over one period in time to give

$$\frac{1}{T}\Delta M = -\frac{\partial}{\partial x} \langle \mathcal{F}_M \rangle_t, \tag{2.49}$$

in which we have denoted time-averaging with the symbol $\langle \rangle_t$.

In evaluating the right-hand side of (2.49), the pressure field averages to zero. Likewise the product of u with w vanishes upon averaging because these fields are $90°$ out of phase. The remaining quantity on the right-hand side of (2.49) describes the divergence of the time-averaged flux $\langle \mathcal{F}_M \rangle_t$, which represents the transport of momentum by waves per unit horizontal area. Explicitly,

$$\langle \mathcal{F}_M \rangle_t = \frac{1}{2}\left(1 + \frac{2kH}{\sinh 2kH}\right) \varrho_0 g A^2 = \varrho_0 g \frac{c_g}{c_p} A^2. \qquad (2.50)$$

The left-hand side of (2.49) denotes the change in momentum over one period that results from the divergence in the momentum flux. Of course, if we substitute (2.50) into (2.49), the right-hand side becomes zero because for plane waves $\langle \mathcal{F}_M \rangle_t$ is independent of x. This is consistent with the left-hand side being zero because M is the same from one period to the next. Therefore (2.49) states that no net momentum is imparted to the fluid as a consequence of the passage of plane waves, a circumstance familiar to surfers who lie on their boards far from shore waiting to catch a 'good' wave. Although waves pass beneath them during their wait, the surfers themselves hardly approach the shore.

The averaged momentum equation in the form (2.49) is useful when considering the deposition of transported momentum by breaking waves. Consider, for example, the situation illustrated in Figure 2.4 in which rightward-propagating waves break within a control volume of length L_x. The average momentum flux (per unit spanwise length) on the left side of the control volume is given in terms of the incident waves by (2.50). Within the box the waves break and no waves exit through the right-hand side. So the flux of momentum due to waves on the right side of the control volume is zero. Because the momentum flux diverges within the volume the momentum of the fluid within the volume must increase. That is to say, wave breaking results in the acceleration of the fluid itself. This is the acceleration felt by surfers who have caught a 'good' wave that propels them rapidly towards the shore as it breaks.

Given this interpretation of the flux-form of the momentum conservation law, we can use (2.49) to estimate the average increase in speed of the fluid as a result of wave breaking over a distance L_x and a time T. Integrating over the length of the control volume, we have

$$\Delta M = \frac{T}{L_x} \langle \mathcal{F}_M \rangle_t \big|_{x=0}. \qquad (2.51)$$

To estimate the resulting near-surface flow, we assume that the acceleration occurs primarily over a depth, $\mathcal{H}$, characteristic of the vertical scale of significant horizontal motion due to waves. Explicitly, we define

$$\mathcal{H} = \tanh(kH)/k. \tag{2.52}$$

For deep water waves, $kH \gg 1$ and $\mathcal{H} \simeq k^{-1}$; for shallow water waves, $kH \ll 1$ and $\mathcal{H} \simeq H$. Consequently, we define the average change in the speed over the characteristic depth to be $\Delta U \equiv \Delta M / (\varrho_0 \mathcal{H})$.

Using (2.49), (2.50) and (2.51) we therefore estimate

$$\Delta U \simeq \frac{T}{L_x} \frac{g}{\mathcal{H}} \frac{c_g}{c_p} A^2 = \frac{\lambda}{L_x} c_g (Ak)^2 \coth^2(kH). \tag{2.53}$$

The last expression has used the definitions $\omega = 2\pi/T$ and $k = 2\pi/\lambda$, as well as the dispersion relation for finite-depth surface waves (2.17). In the deep water limit, (2.53) becomes

$$\Delta U \simeq \frac{1}{2} \frac{\lambda}{L_x} c_p (Ak)^2, \tag{2.54}$$

whereas the shallow water limit gives

$$\Delta U \simeq \frac{\lambda}{L_x} c_p (A/H)^2. \tag{2.55}$$

Assuming the breaking distance is at least as large as the wavelength, in both cases the average change of the flow speed due to small-amplitude waves is small compared to the phase speed, c_p.

By way of example, suppose deep water waves with wavelength $\lambda = 10\,\mathrm{m}$ and amplitude $A = 0.1\,\mathrm{m}$ break over a distance of $L_x = 50\,\mathrm{m}$. Equation (2.54) predicts that a current would develop with speed $0.002\,\mathrm{m/s}$. Of course the speed resulting from large waves breaking near shore can be substantially larger.

In using these formulae, one should keep in mind the approximations underlying their derivations. We have assumed that the background flow is stationary and that the process of propagation, breaking and momentum deposition occurs over a fluid of constant depth. The process of wave breaking at a beach is much more complicated as a result of the varying bottom depth and the formation of rip tides. Consideration of these dynamics is beyond the scope of this book.

2.2.8 The Stokes drift

Although the approximations arising in (2.31) capture the leading-order motion of fluid parcels, a less restrictive calculation shows that fluid parcels do not follow closed elliptical orbits. Instead, over one cycle they are displaced moderately in the direction of the wave's propagation, as illustrated in Figure 2.5.

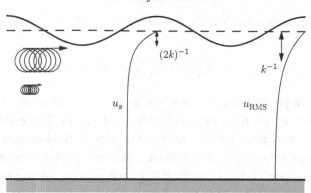

Fig. 2.5. Stokes drift resulting from period-averaging the horizontal flow associated with spiraling fluid parcel paths beneath surface waves. The magnitude of this wave-induced mean flow, u_S, decays more quickly with distance below the surface than the root-mean-square-averaged horizontal flow, u_{RMS}, associated with the waves. For deep water waves with horizontal wavenumber k, u_S decays exponentially over a distance $(2k)^{-1}$, whereas u decays as k^{-1}.

This occurs because the magnitude of the velocity is larger near the surface than at greater depths. Along the upper part of its orbit a fluid parcel moves horizontally with greater speed than when it moves in the opposite direction at the lower part of its orbit.

To determine the approximate magnitude of the net drift, we account for small displacements of the fluid parcel away from its mean position at $\vec{x}_0 = (x_0, z_0)$ by performing a Taylor series expansion. Thus we estimate the horizontal velocity to be

$$u(x, z, t) \simeq u(\vec{x}_0, t) + (x - x_0) \left. \frac{\partial u}{\partial x} \right|_{\vec{x}_0} + (z - z_0) \left. \frac{\partial u}{\partial z} \right|_{\vec{x}_0} \simeq (u + \xi_x \partial_x u + \xi_z \partial_z u)|_{\vec{x}_0},$$

(2.56)

in which ξ_x and ξ_z are the horizontal and vertical displacement fields, respectively, associated with the waves. Each term on the right-hand side is then averaged over one period in time. Using (2.23), (2.25) and (2.26), we find that the first term on the right-hand side is zero, reiterating our statement that the mean velocity is zero if we suppose the fluid parcel resides negligibly far from $\vec{x}_0$. However, accounting for small but finite displacements gives us a non-negligible mean contribution from the remaining two terms.

Dropping the zero subscript on the components of $\vec{x}_0$, we have the formula for the period-averaged net horizontal velocity:

$$u_S(z) = \frac{1}{2} c_p (Ak)^2 \frac{\cosh 2k(z + H)}{\sinh^2(kH)}.$$

(2.57)

In the limit of deep water waves, for which $kH \gg 1$, this is called the 'Stokes drift'. Explicitly,

$$u_S(z) \simeq c_p(Ak)^2 e^{2kz} = (gk^3)^{1/2} A^2 e^{-2k|z|}, \qquad (2.58)$$

in which we have put absolute value signs in the exponent of the last expression to emphasize that u_S decreases exponentially with depth. The e-folding depth is $(2k)^{-1}$, half that associated with the horizontal velocity field of deep water waves.

The Stokes drift is more generally called the 'wave-induced mean flow', referring to the mean advection of fluid resulting from the passage of waves. It is distinct from the irreversibly established mean flow that results from breaking waves.

The formulae (2.57) and (2.58) have been written to emphasize that the speed of the Stokes drift is much smaller than the phase speed of the waves. For example, deep water waves at the point of breaking have relative amplitude $Ak \simeq 0.2$. So u_S is at most about four per cent of the phase speed and generally is much slower for small-amplitude waves. This is why surfers waiting to catch a wave may appear to be stationary as other waves pass beneath them. But over time they will drift, however slowly, in the direction of wave propagation.

From (2.57) we may also compute a formula for the Stokes drift in the shallow water limit:

$$u_S \simeq \frac{1}{2}(g/H^3)^{1/2} A^2. \qquad (2.59)$$

If we imagine H is fixed, we see that this expression is much smaller than the Stokes drift near the surface associated with deep water waves for which $k \gg 1/H$. This is consistent with our view that the drift results from the faster motion near the surface of fluid parcels following elliptical paths. In the shallow water limit, the displacement of fluid parcels does not change with depth and so the drift is negligible.

Although examination of the shallow water circumstance helps with our interpretation of the Stokes drift associated with deep water waves, we will see in what follows that (2.57) does not necessarily represent the actual mean horizontal motion associated with horizontally localized wavepackets in finite-depth fluid. The net horizontal mass transport coinciding with the momentum transport by a wavepacket means there must be a corresponding opposing mass flux at great depth. The dynamics of this return flow can be assumed to have negligible impact upon the near-surface processes only if the fluid is sufficiently deep.

2.2.9 Momentum transport

Here we consider the transport of momentum by quasi-monochromatic wavepackets. Because we will be taking products of fields, we explicitly represent the waves in the real-valued form (1.107). In particular, the horizontal velocity field associated with a quasi-monochromatic deep water wavepacket is

$$u = \frac{1}{2} A_u e^{kz} \exp[\imath (kx - \omega t)] + \text{cc}, \tag{2.60}$$

in which $A_u(X, T)$ is the amplitude envelope.

The arguments to A_u are called 'slow variables'. Whereas u varies quickly as x varies over a distance k^{-1}, significant changes to the amplitude envelope only occur over its width σ, which is assumed to be much larger than k^{-1} as shown, for example, in Figure 1.28a. The leading-order changes to A_u at a fixed point in space occur because the wavepacket moves at the group velocity c_g. We take this into account by including this translation in our definition of the slow space and time variables X and T. Explicitly, we take $X = \epsilon(x - c_g t)$, in which $\epsilon = (\sigma k)^{-1} \ll 1$ is a nondimensional measure of the wavepacket extent, σ. It follows from the structure of Schrödinger's equation (1.136) that the amplitude envelope in the translating frame should evolve in time on the slow scale measured by $T = \epsilon^2 t$.

Likewise, we can represent the vertical velocity and pressure fields in the form of (2.60) in which the amplitude envelopes $A_w(X, T) = -\imath A_u$ and $A_p(X, T) = (\varrho_0 g / \omega) A_u$ are expressed in terms of $A_u(X, T)$ using the polarization relations for plane waves. These are good approximations for a quasi-monochromatic wavepacket.

Substituting the wavepacket representations of u, w and p into the right-hand side of (2.45) gives terms proportional to $\exp[\pm \imath (kx - \omega t)]$, $\exp[\pm 2\imath (kx - \omega t)]$ and $\exp[\pm 0\imath (kx - \omega t)] \equiv 1$. The first results from the pressure term and the last two result from expanding the quadratic u^2 and uw terms as, for example, in the integrand of (1.140). Through the quadratic terms the waves interact with themselves to excite superharmonic waves (of wavenumber $2k$) and mean disturbances (of wavenumber 0). The second of these we associate with terms that accelerate a wave-induced mean flow $\langle u \rangle$.

Thus we find that mean flow induced by self-interaction of waves is given at leading order by (see exercise)

$$\frac{\partial}{\partial t} \langle U \rangle_L = -\frac{\partial}{\partial x} \left\langle \frac{1}{2} |A_u(X, T)|^2 e^{2kz} \right\rangle. \tag{2.61}$$

Note that there is no contribution from the last term on the right-hand side of (2.45) because the polarization relations show that A_u and A_w are 90° out of phase. By pulling out the zero wavenumber contribution, we have in effect averaged the rapidly varying complex exponentials for each term within the derivatives of (2.61), keeping only the slowly varying amplitude envelope.

Interfacial waves

Just as A_u depends upon X and T, so do we expect the wave-induced mean flow $\langle U \rangle_L$ to depend upon these variables. That is, the envelope surrounding the mean flow translates at the group velocity but otherwise its structure evolves slowly in horizontal space and time. Explicitly, from the chain rule of partial derivatives we have $\partial_t \langle U \rangle_L = -\epsilon c_g \partial_X \langle U \rangle_L + \epsilon^2 \partial_T \langle U \rangle_L$. Assuming ϵ is small, the $\epsilon^2 \partial_T$ term can be neglected. Likewise, $\partial_x = \epsilon \partial_X$ on the right-hand side of (2.61).

Thus we arrive at the approximate equation for the evolution of the wave-induced mean flow due to waves:

$$-\epsilon c_g \frac{\partial}{\partial X} \langle U \rangle_L \simeq -\frac{\partial}{\partial X} \left\langle \frac{1}{2} |A_u(X,T)|^2 e^{2kz} \right\rangle.$$

This involves X-derivatives on both sides which can be integrated out. Assuming no pre-existing mean flow, we find

$$\langle U \rangle_L = \frac{1}{2c_g} |A_u|^2 e^{2kz}. \tag{2.62}$$

Finally, using the polarization relations for deep water waves we can rewrite this expression in terms of the vertical displacement amplitude envelope $A_\xi(X,T) \simeq A_u(X,T)/\omega = A_u(X,T)k/c_p$. For deep water waves $c_g = c_p/2$. And so we have

$$\langle U \rangle_L = c_p |A_\xi|^2 k^2 e^{2kz}. \tag{2.63}$$

Notice that (2.63) does not explicitly require the amplitude envelope to have finite extent. We can take the limit $\epsilon \to 0$ in the prescription of the amplitude envelope to get the plane wave limit in which the envelope function $A_\xi(X,T) \to A$ is constant. Thus we see that our expression for the wave-induced mean flow of deep water waves (2.63) is identical to our expression for the Stokes drift (2.58).

This derivation shows that the Stokes drift is a direct consequence of momentum conservation and results from the divergence of the momentum flux per unit mass. By analogy with (2.41), the above derivation shows that $\langle u^2 \rangle = c_{gx} \langle U \rangle_L$, which is the statement that the momentum flux per unit mass is the momentum per unit mass, $\langle U \rangle_L$, times the group velocity. For this reason, the Stokes drift is also sometimes referred to as the 'pseudomomentum' per unit mass. The prefix 'pseudo' is meant to distinguish the wave-induced mean flow from background currents that have momentum even in the absence of waves. But one should not assume that pseudomomentum refers to false momentum. If waves propagate into an otherwise stationary fluid, a net (period-averaged) surface flow with associated momentum really does result.

In Section 2.2.7 we explained that the divergence of momentum flux that occurs when waves break results in the change of momentum to the mean flow. Now it becomes clear that this mean momentum is not created spontaneously. Wave breaking acts to transform the reversible mean flow associated with waves (the Stokes drift) into an irreversible mean flow that moves in the absence of waves.

Rather than compute the divergence of the momentum flux to derive the change in flow speed as in (2.53), let us attempt to estimate this quantity based upon the Stokes drift. We define the average flow over the characteristic depth $\mathcal{H}$, defined by (2.52), to be

$$\langle u_S \rangle = \frac{1}{\mathcal{H}} \int_{-H}^{0} u_S \, dz, \tag{2.64}$$

and we substitute into the integral the general formula for the Stokes drift, (2.57). After some algebraic manipulation and using the dispersion relation we find

$$\langle u_S \rangle = \frac{1}{2} c_p (Ak)^2. \tag{2.65}$$

For deep water waves there is a direct mathematical analogy between the Stokes drift and the mean flow that arises due to wave breaking. Before the waves break, the divergence in the momentum flux which results in the wave-induced mean flow averaged over a depth $\mathcal{H} = k^{-1}$ is given by (2.54) with $L_x = \lambda$. This is exactly the expression given by (2.65). And so we see the result of deep water wave breaking is just to take the average momentum over one wavelength and spread it out over the breaking distance L_x.

For finite-depth waves the connection between the Stokes drift and the momentum deposition arising from wave breaking is complicated by the non-negligible effect of returning undercurrents which we have neglected both in the examination of breaking and in the theory of momentum transport by quasi-monochromatic wavepackets. For example, the method used to derive (2.62) and (2.63) for deep water waves does not give the general formula for the Stokes drift (2.57) if the corresponding approach is taken for finite-depth waves.

Part of the reason for the discrepancy is a subtle mathematical inconsistency that arises when considering wavepackets of finite horizontal extent in finite-depth fluid. By analogy with (2.62), the mean flow over the wavepacket extent is computed to be

$$\langle U \rangle_L = \frac{1}{2c_g} |A_u(X,T)|^2 \frac{\cosh^2 k(z+H)}{\sinh^2 kH}. \tag{2.66}$$

For rightward-propagating finite-depth waves, $\langle U \rangle_L$ is non-negligible and positive from the top to the bottom of the domain. So (2.66) predicts the whole fluid column is advected rightwards by the wave-induced mean flow below the middle of the

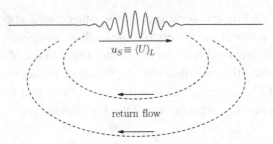

Fig. 2.6. Schematic showing the near-surface flow $\langle u \rangle$ associated with the Stokes drift and the deeper and weaker return flow established so that the total mass transport over the fluid column is zero.

wavepacket, but far ahead and behind it the fluid column is stationary. Thus the motion necessarily requires fluid to pile up in front and to be depleted behind the wavepacket. This would result in a net adverse pressure gradient driving a return flow.

For deep water waves, this return flow can occur well below the depth at which the disturbance fields due to the waves are substantial, as illustrated in Figure 2.6. But for finite-depth waves, the return flow must interact non-negligibly with the waves. Because the calculation of A_u and c_g in (2.66) neglects the presence of a return flow, the computation of $\langle U \rangle_L$ is physically inconsistent.

In comparison, the computation of the Stokes drift for finite-depth waves (2.57) does not suffer from this inconsistency because it assumes the disturbance field is a plane wave and so the fluid column drifts rightwards at the same rate everywhere in the horizontal.

2.2.10 Shallow water equations

The polarization relations for shallow water waves given in Table 2.1 show that u and p are constant with depth while w decreases linearly with depth. Here we take advantage of this structure to derive a system of partial differential equations particularly suitable for shallow water waves. This will be useful when, in Section 4.4, we consider finite-amplitude effects for long interfacial waves.

In shallow water, vertical accelerations are negligibly small so that the dynamic pressure balances the forces occurring due to the displacements of the surface. Thus we have

$$p = \varrho_0 g \eta, \tag{2.67}$$

which states that the fluctuation as well as background fields are in hydrostatic balance. This is apparent from the polarization relations for η and p given in the middle column of Table 2.1. Putting (2.67) into the linearized form of the horizontal momentum equation and neglecting Coriolis terms gives

$$\frac{\partial u}{\partial t} + g\frac{\partial \eta}{\partial x} = 0. \tag{2.68}$$

The vertical displacement field is related to the velocity field through the condition for incompressibility in the form (1.30). Taking the two-dimensional form of this equation in the x–z plane, we integrate from the bottom at $z = -H$ to the free surface at $z = \eta$. The horizontal velocity field is approximately constant with depth below shallow water waves, so the integral of $\partial_x u$ becomes $H\partial_x u$. The vertical velocity field is zero at the bottom and $w = \partial_t \eta$ at the top. Thus we have

$$\frac{\partial \eta}{\partial t} + H\frac{\partial u}{\partial x} = 0. \tag{2.69}$$

Eliminating u between (2.69) and (2.68) gives the well-known one-dimensional form of the wave equation:

$$\frac{\partial^2 \eta}{\partial t^2} = c^2\frac{\partial^2 \eta}{\partial x^2}, \tag{2.70}$$

in which $c^2 = gH$. This equation describes small-amplitude one-dimensional shallow water waves as well as the propagation of other nondispersive waves such as sound and electromagnetic (light) waves. As expected, the dispersion relation for waves satisfying (2.70) is given by (2.19).

2.3 Interfacial waves in a two-layer fluid

Instead of surface waves, which exist at a water–air interface, we now consider waves that propagate along the interface between two layers of fluid having comparable density. For example, these could represent undulations of the thermocline or an atmospheric inversion. If the height over which the density change between the layers is small compared with the wavelength of the undulations, thermoclines and inversions can be represented approximately by infinitesimally thin interfaces.

Throughout this section we will assume the upper- and lower-layer fluids each have uniform density and the thickness of the interface between them is so small compared with the horizontal wavelength of the interfacial waves that it can be treated as infinitesimally thin. Although the two layers are miscible, we will neglect heat and salt diffusion so that the density discontinuity at the interface persists for all time. To begin with we will not make any assumptions about the magnitude of the density jump across the interface.

For large-scale atmospheric motions, buoyancy is usually characterized in terms of the potential temperature. However, since the change in potential temperature and density is small across an inversion, the fluid can be treated as Boussinesq. Therefore, the dynamics of interfacial waves in the atmosphere and ocean are both well represented using density as a measure of buoyancy (see Section 1.12.5). We further assume that the fluid is not rotating, which is justifiable for relatively short period waves. Rotational effects will be considered in Section 2.7.

2.3.1 Equations of motion

Consider a horizontally unbounded two-layer fluid which can be unbounded vertically or which can have horizontal upper and lower boundaries, as illustrated in Figure 2.7. In the absence of waves, this fluid is characterized by the ambient density profile

$$\bar{\rho}(z) = \begin{cases} \rho_1 & z \geq \eta \\ \rho_2 & z < \eta, \end{cases} \tag{2.71}$$

in which $\rho_1 < \rho_2$ for a stably stratified fluid.

The equations prescribing the motion of interfacial waves require the fluid to be irrotational and incompressible both above and below the interface. Defining velocity potentials ϕ_1 and ϕ_2 respectively above and below the interface, we therefore have two partial differential equations describing the interior motion in each layer:

$$\begin{aligned} \nabla^2 \phi_1 &= 0 \quad z > \eta \\ \nabla^2 \phi_2 &= 0 \quad z < \eta. \end{aligned} \tag{2.72}$$

These are coupled by the nonlinear interface conditions that require continuity of pressure via Bernoulli's formula (1.103) evaluated at the interface,

$$\left\{ \rho_1 \left[\frac{\partial \phi_1}{\partial t} + \frac{1}{2} |\nabla \phi_1|^2 + gz \right] = \rho_2 \left[\frac{\partial \phi_2}{\partial t} + \frac{1}{2} |\nabla \phi_2|^2 + gz \right] \right\} \Big|_{z=\eta}, \tag{2.73}$$

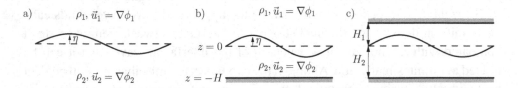

Fig. 2.7. Variables associated with interfacial waves in a) infinitely deep fluid, b) semi-infinite fluid, and c) finite-depth fluid.

and vertical velocity,

$$\frac{\partial \phi_1}{\partial z}\bigg|_{z=\eta} = \frac{\partial \phi_2}{\partial z}\bigg|_{z=\eta} = \frac{D\eta}{Dt}. \qquad (2.74)$$

The system of equations is completed by specifying upper and lower boundary conditions. If the fluid is unbounded we insist that the velocity potential vanishes far from $z = 0$. Otherwise we require that there is no normal flow across the rigid boundaries above and/or below the interface. In particular, if the rigid boundaries are horizontal we require the vertical velocity, $\partial \phi / \partial z$, to be zero at the upper and/or lower boundary.

In the consideration of small-amplitude waves, the equations are linearized first by changing the boundary conditions on the partial differential equations for ϕ_1 and ϕ_2 at the interface:

$$\nabla^2 \phi_1 = 0 \quad z > 0$$
$$\nabla^2 \phi_2 = 0 \quad z < 0. \qquad (2.75)$$

The linearized interface conditions, appropriate for small-amplitude waves, become

$$\rho_1 \frac{\partial \phi_1}{\partial t}\bigg|_{z=0} = \rho_2 \frac{\partial \phi_2}{\partial t}\bigg|_{z=0} + \rho_2 g' \eta \qquad (2.76)$$

and

$$\frac{\partial \phi_1}{\partial z}\bigg|_{z=0} = \frac{\partial \phi_2}{\partial z}\bigg|_{z=0} = \frac{\partial \eta}{\partial t}. \qquad (2.77)$$

In (2.76) we have defined g' to be

$$g' \equiv g(\rho_2 - \rho_1)/\rho_2. \qquad (2.78)$$

In the limit in which the upper layer has a density much smaller than the lower layer (as for a water–air interface), $g' \to g$, in which case the motion is governed by the dynamics of the lower layer alone as prescribed by (2.9). If the density difference between the layers is small, then $g' \ll g$. Hence, it is referred to as the 'reduced gravity'. In the atmosphere, the reduced gravity can be written in terms of the potential temperature difference between the two layers by $g' \equiv g(\theta_1 - \theta_2)/\theta_2$.

Assuming the upper and lower boundaries are horizontal, if not infinitely far away, the boundary conditions are independent of x and y. Therefore the partial differential equations (2.75) with interface conditions (2.76) and (2.77) may be solved analytically using Fourier transforms in horizontal space and in time. For simplicity, we orient the co-ordinate system so as to focus upon waves moving in the x-direction alone.

Explicitly, we assume the displacement of the interface is represented by a plane wave as in (2.11):

$$\eta(x,t) = Ae^{i(kx-\omega t)}. \tag{2.79}$$

The velocity potentials in the upper and lower layers are likewise represented respectively by

$$\phi_1(x,z,t) = \hat{\phi}_1(z)e^{i(kx-\omega t)} \text{ and } \phi_2(x,z,t) = \hat{\phi}_2(z)e^{i(kx-\omega t)}. \tag{2.80}$$

Substituting these expressions into the equations for small-amplitude waves (2.75) together with the linearized interface and boundary conditions gives a coupled set of ordinary differential equations whose eigenvalues give the dispersion relation for interfacial waves and the corresponding eigenfunctions give the polarization relations that connect the structure of the flow above and below the interface.

We will proceed first by examining small-amplitude interfacial waves in an unbounded fluid.

2.3.2 Interfacial waves in infinite-depth fluid

Using (2.75) and (2.80) and assuming the fluid is unbounded above and below, the corresponding velocity potentials in the upper and lower layers are

$$\phi_1 = \mathcal{A}_1 e^{-kz} e^{i(kx-\omega t)}, \quad z > 0, \tag{2.81}$$

and

$$\phi_2 = \mathcal{A}_2 e^{kz} e^{i(kx-\omega t)}, \quad z < 0. \tag{2.82}$$

Here we have defined the exponentials in z with the implicit assumption that k is positive. In this way both ϕ_1 and ϕ_2 are bounded in the vertical.

The interface conditions prescribe how the complex amplitudes $\mathcal{A}_1$ and $\mathcal{A}_2$ depend upon the surface displacement amplitude A and how they determine the dispersion relation.

Explicitly, by substituting (2.79), (2.81) and (2.82) into the interface conditions (2.76) and (2.77), we arrive at an algebraic eigenvalue problem involving a coupled set of equations in the amplitudes A, $\mathcal{A}_1$ and $\mathcal{A}_2$:

$$\begin{pmatrix} -i\rho_1\omega & i\rho_2\omega & \rho_2 g' \\ -k & -k & 0 \\ -k & 0 & i\omega \end{pmatrix} \begin{pmatrix} \mathcal{A}_1 \\ \mathcal{A}_2 \\ A \end{pmatrix} = \vec{0}. \tag{2.83}$$

The eigenvalue found by setting the matrix determinant to zero gives the dispersion relation for interfacial waves in an infinitely deep two-layer fluid:

$$\omega^2 = gk\frac{\rho_2 - \rho_1}{\rho_2 + \rho_1} = g'k\frac{\rho_2}{\rho_2 + \rho_1}. \tag{2.84}$$

As expected, in the limit $\rho_1 \ll \rho_2$ we recover the dispersion relation (2.18) for deep waves at a water–air interface. In the Boussinesq limit, in which $\rho_1 \to \rho_2^-$ (the '$-$' superscript is included to emphasize that the limit approaches ρ_2 from below), we have

$$\omega^2 \simeq \frac{1}{2}g'k. \tag{2.85}$$

Therefore, if the density difference between the layers is small, the waves propagate with significantly smaller frequencies and phase speeds than surface waves having comparable wavenumber k. As in the case of deep water waves, the group speed, c_g, is half the phase speed, c_p.

Given ω, the corresponding eigenvector of (2.83) gives

$$\mathcal{A}_1 = -\mathcal{A}_2 = \iota\frac{\omega}{k}A. \tag{2.86}$$

Explicitly, we have found the polarization relations for the velocity potentials in terms of the interface displacement. Putting (2.86) into (2.81) and (2.82) gives

$$\phi_1(x,z,t) = \iota\frac{\omega}{k}Ae^{-kz}e^{\iota(kx-\omega t)} \quad\Rightarrow\quad -\frac{\omega}{k}Ae^{-kz}\sin(kx-\omega t)$$
$$\phi_2(x,z,t) = -\iota\frac{\omega}{k}Ae^{kz}e^{\iota(kx-\omega t)} \quad\Rightarrow\quad \frac{\omega}{k}Ae^{kz}\sin(kx-\omega t), \tag{2.87}$$

in which the real part has been taken to get the rightmost expressions.

From (2.87) and defining the velocities in the upper and lower layers as in (2.1), we can go on to determine the horizontal and vertical velocities in each layer. As with surface waves, the horizontal velocity field in the lower layer is in phase with the vertical displacement field. However, the upper layer horizontal velocity is equal and opposite to that in the lower layer. As sketched in Figure 2.8, this means that the interface acts like a vortex sheet having infinite shear that periodically changes sign between wave crests and troughs.

The existence of vorticity does not contradict our original assumption that each layer is irrotational. The vorticity predicted by the model is confined to the interface and does not extend into the interior of each layer. Vorticity is generated at the interface through the action of baroclinic torques where the density and pressure gradients are misaligned.

In some circumstances the shear can be so great as to cause the interface to wrap up into vortices. Linear theory does not account for such dynamics although

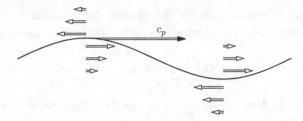

Fig. 2.8. Horizontal velocity field associated with a rightward-propagating interfacial wave.

it can be used to predict the initial development of the instabilities, as discussed in Section 2.6. For the purposes of understanding the propagation of interfacial waves, we can assume that instabilities at the interface do not significantly affect the wave dynamics provided the waves are long compared to the distance over which the interface thickens due to any potential mixing.

The mean kinetic energy associated with small-amplitude interfacial waves is given by the sum of the kinetic energies associated with the motions in each layer:

$$\langle E_K \rangle = \int_0^\infty \frac{1}{2} \rho_1 \left\langle \phi_{1x}^2 + \phi_{1z}^2 \right\rangle dz + \int_{-\infty}^0 \frac{1}{2} \rho_2 \left\langle \phi_{2x}^2 + \phi_{2z}^2 \right\rangle dz. \qquad (2.88)$$

The range of integration in each layer is bounded at $z = 0$ because integrating to $z = \eta$ introduces terms that are negligibly small.

As in Section 2.2.6, we may derive a formula for the available potential energy associated with interfacial waves. Alternately, because this is a conservative system we know that the kinetic and available potential energy are equipartitioned. Thus the total energy is

$$\langle E_T \rangle = 2 \langle E_K \rangle = \frac{1}{2} \rho_2 g' A^2. \qquad (2.89)$$

This reduces to (2.38) in the limit $\rho_1 \ll \rho_2$ with the characteristic density ϱ_0 taken to be the lower-layer density ρ_2. If g' is small, (2.89) shows that a fixed energy input provided on sufficiently slow time-scales can create interfacial waves with much larger amplitudes than surface waves generated by the same energy input. It is left as an exercise to find an explicit expression for the mean energy flux $\langle \mathcal{F}_E \rangle$ and to show that $\langle \mathcal{F}_E \rangle = c_g \langle E_T \rangle$.

2.3.3 Interfacial waves in finite-depth fluid

Now consider a two-layer fluid with a semi-infinite upper-layer fluid and a finite-depth lower-layer fluid, as illustrated in Figure 2.7b. This is sometimes referred to as a 'one-and-a-half-layer fluid'. The co-ordinate system is set up with $z = 0$ being the depth of the undisturbed interface, and the lower layer is bounded below by a free-slip horizontal plane at $z = -H$.

The boundary value problem is the same as before except now the vertical velocity of the lower-layer fluid at the bottom boundary is constrained to be zero:

$$\frac{\partial \phi_2}{\partial z}\bigg|_{z=-H} = 0. \tag{2.90}$$

Again we suppose the waves have small amplitude prescribed by (2.79). In the upper-layer fluid ϕ_1 decreases exponentially with height, as in (2.81). To satisfy the bottom boundary condition, the motion of the lower-layer fluid is prescribed by the velocity potential

$$\phi_2 = \mathcal{A}_2 \cosh[k(z+H)]e^{i(kx-\omega t)} \quad -H < z < 0. \tag{2.91}$$

Applying the linearized form of the interface conditions (2.76) and (2.77), the resulting eigenvalue problem ultimately gives the dispersion relation

$$\omega^2 = g'k\frac{\rho_2}{\rho_1 + \rho_2 \coth kH}. \tag{2.92}$$

This is plotted for different values of ρ_2/ρ_1 in Figure 2.9.

As expected, in the limit $\rho_1 \to 0$, (2.92) reduces to the dispersion relation for surface waves (2.17). For short wavelength waves, $kH \gg 1$ so $\coth(kH) \simeq 1$. Hence (2.92) reduces to (2.84), which is the dispersion relation for interfacial waves in infinitely deep fluid.

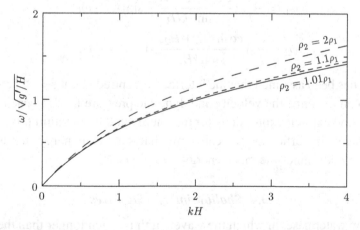

Fig. 2.9. Dispersion relation for interfacial waves in a one-and-a-half-layer fluid as given by (2.92) with $\rho_2 = 2\rho_1$ (long-dashed line), $\rho_2 = 1.1\rho_1$ (short-dashed line) and $\rho_2 = 1.01\rho_1$ (solid line). The frequency is normalized by $(g'/H)^{1/2}$ and the wavenumber is nondimensionalized by the depth H of the finite-depth layer.

For a Boussinesq fluid in which ρ_2 is only moderately larger than ρ_1, the dispersion relation takes the simpler form

$$\omega^2 = \frac{1}{2} g' k \left(1 - e^{-2kH} \right).$$ (2.93)

This is well represented by the solid line in Figure 2.9.

Finally, we consider the circumstance in which the depths H_1 and H_2 of the upper- and lower-layer fluids, respectively, are sufficiently small compared with the wavelength that neither can be treated as infinitely large. This situation is illustrated in Figure 2.7c. The primary application of this circumstance is for interfacial waves at thermoclines in lakes and oceans in which the upper boundary is the surface. As will be shown in Section 2.4, the displacement of the surface due to interfacial waves is typically so small that a rigid surface boundary condition can justifiably be assumed. We further assume that the density difference between the upper- and lower-layer fluid is small so that we can adopt the Boussinesq approximation.

Following the approach laid out above, we find the dispersion relation is given by

$$\omega^2 = g' k \frac{1}{\coth kH_1 + \coth kH_2}.$$ (2.94)

It is straightforward to check that this reduces to (2.93) in the limit as $H_1 \to \infty$ with $H_2 \equiv H$.

With the interfacial displacement given explicitly as the real function $\eta = A\cos(kx - \omega t)$, the polarization relations give the velocity potential in the upper and lower layers respectively as

$$\phi_1 = -\frac{\omega}{k} A \frac{\cosh k(z - H_1)}{\sinh kH_1} \sin(kx - \omega t)$$

$$\phi_2 = \frac{\omega}{k} A \frac{\cosh k(z + H_2)}{\sinh kH_2} \sin(kx - \omega t),$$ (2.95)

in which it has been assumed that the interface is centred about $z = 0$. From (2.95), one can go on to define the velocity and dynamic pressure fields in each layer and from these find explicit expressions for the energy and momentum flux associated with the waves. In particular, one can show that the mean energy flux is equal to the group velocity times the mean energy: $\langle \mathcal{F}_E \rangle = c_g \langle E_T \rangle$.

2.3.4 Shallow interfacial waves

The shallow water case, in which the wavelength is much longer than the depth of one or both layers of fluid, is of particular interest because it is representative of a variety of geophysical circumstances. For example, it describes disturbances on an atmospheric inversion with wavelengths of the order of kilometres and longer.

Similarly, it describes undulations of the thermocline that have a long horizontal scale compared to the mixed-layer depth, but a short horizontal scale compared with the depth of the ocean or lake.

First we consider the shallow-water limit for a one-and-a-half-layer Boussinesq fluid. Taking $kH \ll 1$ in the dispersion relation (2.93) gives

$$\omega^2 = (g'H) k^2. \tag{2.96}$$

As for shallow waves in a one-layer fluid, this corresponds to the dispersion relation of nondispersive waves. All sufficiently long waves move at the same speed:

$$c = \sqrt{g'H}. \tag{2.97}$$

A heuristic that is typically invoked for a wave in a one-and-a-half-layer fluid is to assume the shallow layer with depth H plays the role of the shallow layer in a one-layer fluid except that g is replaced with g' in the corresponding one-layer shallow water formulae. For example, from (2.38) the mean energy per unit horizontal area of shallow interfacial waves is $\langle E_T \rangle = \varrho_0 g' A^2 / 2$. From (2.40), taking $kH \ll 1$, the mean energy flux is $\langle \mathcal{F}_E \rangle = c_g \varrho_0 g' A^2 / 2$, in which $c_g = c$, given by (2.97).

For long waves in a two-layer Boussinesq fluid, the dispersion relation is found by taking $kH_1 \ll 1$ and $kH_2 \ll 1$ in (2.94). As expected, the waves are nondispersive, their frequency satisfying $\omega = \pm ck$. In this case, however, the speed of the waves is given by

$$c = \sqrt{g' \frac{H_1 H_2}{H_1 + H_2}}. \tag{2.98}$$

That is, the waves move as though in a one-layer fluid with reduced gravity g' and depth

$$\bar{H} \equiv \frac{H_1 H_2}{H_1 + H_2}, \tag{2.99}$$

the harmonic mean of H_1 and H_2.

In the limit $H_1 \gg H_2$, (2.98) becomes (2.97) with $H = H_2$. Similarly, for $H_1 \ll H_2$ we get the same result but with $H = H_1$. That is to say, small-amplitude interfacial waves in a shallow upper layer behave the same as interfacial waves in a shallow lower layer. This is expected in the Boussinesq approximation, whose equations are invariant upon reflection in z.

The polarization relations of shallow interfacial waves can be determined from those found for finite-depth waves by taking the long wave limit. In particular, the leading order structure of the velocity potential in the upper and lower layer, respectively, of a shallow, two-layer Boussinesq fluid is given by

$$\phi_1 = -\frac{c}{kH_1}A\sin(kx - \omega t)$$

$$\phi_2 = \frac{c}{kH_2}A\sin(kx - \omega t). \qquad (2.100)$$

Here, A is the amplitude of the displacement of the interface centred about $z = 0$.

To evaluate the vertical velocity field, one needs to retain the order θ^2 terms in the expansion of $\cosh(\theta)$ before taking z-derivatives of the velocity potential. Thus the velocity field in the upper layer is

$$(u_1, w_1) = -A\frac{c}{H_1}\left(\cos(kx - \omega t),\; k(z - H_1)\sin(kx - \omega t)\right), \qquad (2.101)$$

and in the lower layer is

$$(u_2, w_2) = A\frac{c}{H_2}\left(\cos(kx - \omega t),\; k(z + H_2)\sin(kx - \omega t)\right). \qquad (2.102)$$

As expected, the horizontal velocity is independent of depth above and below the interface and the relatively small vertical velocity varies linearly with depth with maximum value $\partial\eta/\partial t = A\omega\sin(kx - \omega t)$ occurring at the interface.

The dynamic pressure fields in the upper and lower layers can be found from the linearized horizontal momentum equation. Explicitly, we find

$$p = \begin{cases} -\rho_1 c^2 A/H_1 \cos(kx - \omega t) & 0 < z < H_1 \\ \rho_2 c^2 A/H_2 \cos(kx - \omega t) & -H_2 < z < 0. \end{cases} \qquad (2.103)$$

Using (2.98), this can be written more succinctly in terms of the interfacial displacement so that $p = -\rho_1 g'(\bar{H}/H_1)\eta$ above the interface and $p = \rho_2 g'(\bar{H}/H_2)\eta$ below the interface.

These relations allow us to recast the description of shallow water waves in terms of coupled partial differential equations for η and the horizontal velocity, u, defined equivalently by

$$u \equiv -\frac{H_1}{\bar{H}}u_1 = \frac{H_2}{\bar{H}}u_2. \qquad (2.104)$$

Analogous to (2.68) and (2.69), the horizontal momentum equation becomes

$$\frac{\partial u}{\partial t} + g'\frac{\partial\eta}{\partial x} = 0, \qquad (2.105)$$

and the vertically integrated continuity equation for an incompressible fluid becomes

$$\frac{\partial\eta}{\partial t} + \bar{H}\frac{\partial u}{\partial x} = 0. \qquad (2.106)$$

The form of the equations is similar to that of the shallow water equations for a one-layer fluid except that gravity has been replaced by the reduced gravity g' (2.78), the fluid depth has been replaced by the harmonic mean depth $\bar{H}$ (2.99), and it is understood that u is the depth-weighted horizontal velocity in each layer, given by (2.104).

In the special case of waves in a semi-infinite fluid with a shallow lower layer, their motion is represented by (2.105) and (2.106), in which $\bar{H} = H_2$, the depth of the shallow layer. From (2.104), $u \equiv u_2$ and the velocity u_1 in the upper layer is negligibly small.

2.4 Interfacial waves in multi-layer fluids

In Section 2.3 we examined interfacial waves in a two-layer fluid. Although the algebra can be daunting, it is theoretically straightforward to extend these concepts to waves in a fluid with any number of interfaces.

Here we restrict our examination to small-amplitude waves in an unbounded three-layer fluid having two interfaces. This illustrates how a single system can support more than one class of waves and it provides validation after the fact that the rigid-lid approximation is appropriate for interfacial waves in an ocean or lake with a free surface.

We consider a fluid with an infinitely deep upper layer of density ρ_0, a finite-depth middle layer of density ρ_1 and an infinitely deep lower layer of density ρ_2. The co-ordinate system is established so that the undisturbed upper and lower interfaces are situated at $z = H/2$ and $-H/2$, respectively, as shown in Figure 2.10. Disturbances in the upper, middle and lower layers are represented by the velocity potentials ϕ_0, ϕ_1 and ϕ_2, respectively, and displacements of the upper and lower interfaces are given by η_1 and η_2, respectively.

We seek periodic solutions in horizontal space and time, with the understanding that any small-amplitude disturbance can be represented by a superposition of these periodic waves. Ignoring motions in the y-direction, we therefore assume

$$\eta_1 = \mathcal{A}_1 e^{i(kx-\omega t)} \quad \text{and} \quad \eta_2 = \mathcal{A}_2 e^{i(kx-\omega t)}. \tag{2.107}$$

Generally, the amplitudes $\mathcal{A}_1$ and $\mathcal{A}_2$ can be complex constants. However, we have the freedom to set up our co-ordinates so that η_1 has a maximum value at $x = 0$ and $t = 0$. This makes $\mathcal{A}_1 = A$, a positive real constant. As such, if $\mathcal{A}_2$ is a real and positive constant, then the crests of the lower interface directly underlie the crests of the upper interface. If $\mathcal{A}_2$ is a real and negative constant, then troughs underlie crests. We will see that the equations of motion require that these are the only two possibilities.

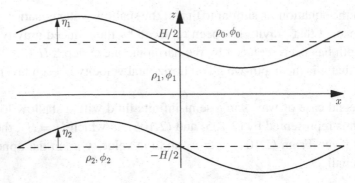

Fig. 2.10. Set-up for the mathematical description of a three-layer fluid that is unbounded above and below and whose middle layer has undisturbed depth H, as indicated by the horizontal dashed lines.

Because the fluid in each layer is irrotational, we have $\nabla^2 \phi_i = 0$ for $i = 0, 1, 2$. Assuming, as we did for η_1 and η_2, that solutions are periodic in horizontal space and time, we can write

$$\phi_i(x,z,t) = \hat{\phi}_i(z)\, e^{\iota(kx - \omega t)}, \quad i = 0, 1, 2. \tag{2.108}$$

The partial differential equations for ϕ_i become the simple ordinary differential equations

$$\hat{\phi}_i'' - k^2 \hat{\phi}_i = 0, \quad i = 0, 1, 2. \tag{2.109}$$

That is, $\hat{\phi}_i$ is given by exponential functions that increase and decrease with height or, equivalently, by hyperbolic sine and hyperbolic cosine functions. The latter are convenient for representing $\hat{\phi}_1$ because they decompose the wave structure into one with odd and even symmetry. Exponentials are useful for representing $\hat{\phi}_0$ and $\hat{\phi}_2$ because the requirement for bounded functions as $z \to \pm\infty$ allows us immediately to reject one of the two possible solutions.

Explicitly, we have

$$\hat{\phi}_0(z) = \mathcal{B}_0 e^{-kz}, \qquad\qquad\qquad z > H/2,$$
$$\hat{\phi}_1(z) = \mathcal{B}_1 \cosh(kz) + C_1 \sinh(kz), \qquad |z| \le H/2, \tag{2.110}$$
$$\hat{\phi}_2(z) = \mathcal{B}_2 e^{kz}, \qquad\qquad\qquad z < -H/2.$$

Here $\mathcal{B}_0$, $\mathcal{B}_1$, C_1 and $\mathcal{B}_2$ are (possibly complex) constants.

The constants are determined by the interface conditions for small-amplitude waves, formulae that are analogous to (2.76) and (2.77):

$$\rho_0 \frac{\partial \phi_0}{\partial t} + \rho_0 g \eta_1 = \rho_1 \frac{\partial \phi_1}{\partial t} + \rho_1 g \eta_1, \qquad z = H/2,$$

$$\frac{\partial \phi_0}{\partial z} = \frac{\partial \phi_1}{\partial z} = \frac{\partial \eta_1}{\partial t}, \qquad z = H/2,$$

$$\rho_1 \frac{\partial \phi_1}{\partial t} + \rho_1 g \eta_2 = \rho_2 \frac{\partial \phi_2}{\partial t} + \rho_2 g \eta_2, \qquad z = -H/2,$$

$$\frac{\partial \phi_1}{\partial z} = \frac{\partial \phi_2}{\partial z} = \frac{\partial \eta_2}{\partial t}, \qquad z = -H/2.$$

(2.111)

Using (2.107), (2.108) and (2.110), we get six equations in six unknown constants $\mathcal{A}_1 (= A)$, $\mathcal{A}_2$, $\mathcal{B}_0$, $\mathcal{B}_1$, $\mathcal{C}_1$ and $\mathcal{B}_2$. This can be recast as a matrix eigenvalue problem in which the eigenvalue is the frequency ω.

After some algebra, we find that the dispersion relation is given by the roots of a quadratic polynomial in ω^2:

$$\omega^4 - gk \frac{\rho_2 - \rho_0}{\rho_2 + \rho_0} (1 + \alpha_{kH}) \omega^2 + (gk)^2 \alpha_{kH} = 0, \qquad (2.112)$$

in which the nondimensional quantity α_{kH} is given by

$$\alpha_{kH} = \left\{ \frac{2\rho_1(\rho_0 + \rho_2)}{(\rho_2 - \rho_1)(\rho_1 - \rho_0)} \left[\frac{1}{1 - \exp(-2kH)} \right] - 1 \right\}^{-1}. \qquad (2.113)$$

The corresponding eigenvectors relate, for example, the amplitude of the lower interface to that of the upper interface:

$$\mathcal{A}_2 = \frac{2\rho_1}{\rho_2 - \rho_1} \left\{ \frac{\rho_1 + \rho_0}{\rho_1 - \rho_0} \left[\frac{\rho_2 + \rho_1}{\rho_2 - \rho_1} \frac{\omega^2}{gk} - 1 \right] e^{kH} + \left(\frac{\omega^2}{gk} - 1 \right) e^{-kH} \right\}^{-1} A. \quad (2.114)$$

Because the coefficient of A on the right-hand side of (2.114) is real we know the waves at the upper and lower interface are either in phase, if the coefficient is positive, or 180° out of phase, if it is negative.

We also have

$$\mathcal{B}_1 = \frac{-\iota\omega}{k \sinh(kH/2)} (A - \mathcal{A}_2), \quad C_1 = \frac{-\iota\omega}{k \cosh(kH/2)} (A + \mathcal{A}_2). \qquad (2.115)$$

So the velocity potential is an even function in z if $\mathcal{A}_2 \simeq -A$ and it is an odd function in z if $\mathcal{A}_2 \simeq A$. Conversely, the vertical velocity field $\partial_z \phi$ is an odd function in the first case and an even function in the second.

For long waves ($kH \ll 1$) the system reduces to that of a two-layer fluid. Indeed, in this limit $\alpha_{kH} \to 0$, and the equation for the dispersion relation simplifies to give

$$\omega^2 = gk \frac{\rho_2 - \rho_0}{\rho_2 + \rho_0}. \tag{2.116}$$

This is analogous to the two-layer dispersion relation, (2.84). Likewise, substitution of (2.116) into (2.114) gives $\mathcal{A}_2 = \mathcal{A}_1 = A$, as expected. The velocity potential has $\mathcal{B}_1 = 0$ so that in the thin middle layer $\hat{\phi}_1 = -2\iota\,(\omega/k)A\sinh(kz)$. The corresponding structure of the vertical velocity field is $\hat{w}_1 = -2\iota\omega A\cosh(kz)$, which is an even function in z.

For finite H, α_{kH} does not vanish. However, the corresponding structure of the waves whose dispersion relation is an extension of (2.116) is maintained in the sense that the upper and lower layer interfaces move up and down in phase. These are called 'sinuous waves'. Reflecting the fact that $\mathcal{A}_2 \simeq A$, so that the vertical velocity field is an approximately even function in z, they are also sometimes referred to as 'even modes'. The dispersion relation of these modes arises as one of the roots of the quadratic polynomial in ω^2 given by (2.113).

The other root yields a second dispersion relation corresponding to waves in which the upper and lower interfaces are 180° out of phase. Because their structure resembles the pinching associated with varicose veins, these modes are sometimes called 'varicose waves'. If the density-jumps across the two interfaces are identical, then the corresponding vertical velocity field is an odd-function of z; there is no vertical motion at $z = 0$ and the flow simultaneously moves either towards or away from the centreline. Because of the structure of w in this limiting case, these waves are also sometimes referred to as 'odd modes'.

Particularly interesting is the case with $\rho_0 \ll \rho_1, \rho_2$, corresponding to a two-layer fluid with a free surface. That is, ρ_1 and ρ_2 could be the density of warm and cool water, respectively, and ρ_0 could be the density of air, which is negligibly small compared with the density of water.

Letting $\rho_0 \to 0$ in the coefficient of ω^2 in (2.112), the resulting equation factors to give the two distinct dispersion relations. One of these, which corresponds to that for even modes, is the same as that for surface waves on deep water:

$$\omega^2 = gk. \tag{2.117}$$

Substituting this into (2.114), we see that the lower-layer interface is in phase with the waves on the free surface but is smaller in amplitude so that

$$\mathcal{A}_2 = Ae^{-kH}. \tag{2.118}$$

The structure of these waves is sketched in Figure 2.11a.

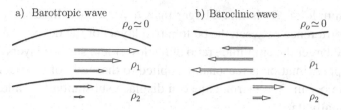

Fig. 2.11. Surface and interface displacements and corresponding horizontal velocity field associated with a) barotropic and b) baroclinic waves in a two-layer fluid with a free surface.

Even if no lower interface was present, surface waves would displace the fluid at a depth H with this amplitude. (For example, see the formula for ξ_z of deep-water waves in Table 2.1.) So the density jump at the lower interface is inconsequential: the fluid moves as if it was a one-layer fluid. Hence these are called 'barotropic waves'. The terminology indicates that pressure gradients are normal to both interfaces and, hence, the local density gradients. Practically, it means that the density change across the lower interface negligibly affects the wave dynamics.

For $\rho_0 \simeq 0$ and for non-zero H, a second non-trivial solution of (2.114) for ω^2 exists which corresponds to that for odd modes. These have a dispersion relation given generally by $\omega^2 = \alpha_{kH} g k$. In the oceanic circumstances of a thermocline below a free surface we may take the Boussinesq approximation, $\rho_2 - \rho_1 \ll \rho_2$, so that the dispersion relation is then given by

$$\omega^2 \simeq \frac{1}{2} g' k (1 - e^{-2kH}), \qquad (2.119)$$

in which $g' = g(\rho_2 - \rho_1)/\rho_2$ is the reduced gravity based on the two lower layers. This is the same result as (2.93), which describes waves at the interface of a finite-depth fluid with a rigid upper boundary.

Because of the relatively small value of g', these waves oscillate more slowly than barotropic waves and their structure differs significantly. From (2.114) we find

$$\mathcal{A}_2 = -A \frac{\rho_1}{\rho_2 - \rho_1} e^{kH}. \qquad (2.120)$$

Because $\mathcal{A}_2$ is opposite in sign to A the waves are 180° out of phase: crests on the surface overlie troughs at the lower interface and vice versa. These waves are sketched in Figure 2.11b. Just as we found in Section 2.3.2, the horizontal velocity above the lower interface is indeed opposite to that below this interface, with vorticity being generated at the interface by baroclinic torques. For this reason, these are sometimes called baroclinic waves to distinguish them from barotropic waves. The term baroclinic emphasizes that the fluid's stratification is essential for the existence of these waves.

The magnitude of $\mathcal{A}_2$ is much larger than A for a baroclinic wave. The tinier the density difference between the bottom two layers or the deeper the lower layer interface, the larger the amplitude ratio of the lower and surface layers. This is why a rigid-lid approximation is reasonably applied to the study of interfacial waves in lakes and the ocean. The thermocline can displace substantially while the surface remains virtually flat.

The energy of both barotropic and baroclinic waves is equipartitioned between kinetic and available potential energy. For Boussinesq barotropic waves, the contribution of the bottom layer is negligible and the energy associated with the wave is approximately the same as that for surface waves in a one-layer fluid. Analogous to (2.38), this is

$$\langle E_T \rangle_{\mathrm{BT}} \simeq \frac{1}{2} \varrho_0 g A^2, \tag{2.121}$$

in which ϱ_0 (equal to ρ_1, say) is a characteristic density of the bottom two layers.

The energy associated with the baroclinic wave is set primarily by disturbances on the lower interface. For short waves compared with H, we have

$$\langle E_T \rangle_{\mathrm{BC}} \simeq \frac{1}{2} \varrho_0 g' |\mathcal{A}_2|^2. \tag{2.122}$$

The presence of the reduced gravity in (2.122) means that the same energy is capable of kicking up a baroclinic wave of much larger amplitude at the interface than the corresponding barotropic wave amplitude at the surface.

2.5 Laterally bounded interfacial waves

So far we have considered waves that are laterally unbounded, which might reasonably represent waves in the atmosphere and open ocean. Additional restrictions are imposed upon interfacial waves when they are confined within a basin such as a lake or a tank in the laboratory, and when their horizontal wavelength is comparable to the horizontal extent of the domain.

If the side-walls are vertical, then the vertical structure and the dispersion relation of the contained waves are the same as those for plane waves. However, whereas k takes on a continuous range of values for plane waves, lateral boundaries restrict the allowable wavenumbers to a discrete spectrum of wave modes.

All possible interfacial disturbances can be represented by a superposition of modes in what essentially forms a Fourier series. Such a series could be used to describe a horizontally propagating wavepacket with negligible amplitude near the side-walls, but this is impractical. It is more convenient to represent such motion by a Fourier transform, which is an integral superposition of a continuous spectrum of waves (1.121). This is what we do in the study of waves in the open

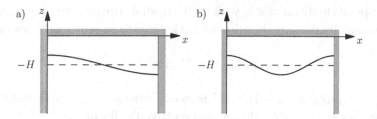

Fig. 2.12. Vertical cross-sections showing a) mode-1 and b) mode-2 interfacial waves in a two-layer fluid with a rigid surface and which is unbounded below.

ocean – we neglect the bounding continents. It is for this reason that the study of laterally-confined disturbances usually focuses upon waves having horizontal wavelength comparable to the horizontal extent of the basin in which they are contained. These are referred to as 'low-order' modes.

Here we consider interfacial waves in the simple two-dimensional geometry of an infinitely deep basin with vertical side-walls at $x = 0$ and L and with a density interface whose mean depth is situated at $z = -H$. This is illustrated in Figure 2.12. The top of the domain at $z = 0$ is treated as a rigid horizontal boundary, consistent with the discussion in Section 2.4.

The governing equations are analogous with those discussed in Section 2.3.3 except that here the amplitude decreases exponentially with depth in the lower layer and the velocity potential in the upper layer depends upon z as a hyperbolic cosine function so as to satisfy the rigid surface condition. The interfacial conditions, (2.73) and (2.74), give a dispersion relation analogous to that in (2.92) except with ρ_2 and ρ_1 interchanged in the fraction on the right-hand side. The Boussinesq form of the dispersion relation is identical to (2.93):

$$\omega^2 = \frac{1}{2} g' k \left(1 - e^{-2kH} \right),$$

in which H is the depth of the upper layer and $g' = g(\rho_2 - \rho_1)/\rho_1$ is the reduced gravity based upon the upper and lower layer densities, ρ_1 and ρ_2, respectively.

For a basin of length L, the side-boundary conditions require that there is no horizontal flow for all z at $x = 0$ and L. Thus we require

$$\frac{\partial \phi_i}{\partial x} = 0 \quad \text{at } x = 0, L \tag{2.123}$$

for $i = 1, 2$. Because u is in phase with η, the zero Neumann boundary condition (2.123) incidentally requires that the slope of the interface is always zero at the side-walls. Thus we require that ϕ_i and η have x-dependence of the form $\cos(k_n x)$, in which $k_n \equiv n\pi/L$ for positive integers n. The $n = 0$ solution is usually ignored since this corresponds to disturbances with no horizontal structure. Explicitly, solutions are represented by the superposition of modes having interfacial displacement

$$\eta = A\cos(k_n x)e^{-\iota\omega_n t} \quad \text{for} \quad 0 \le x \le L, \tag{2.124}$$

in which $\omega_n = \omega(k_n)$ for $n = 1, 2, \dots$ Likewise, insisting that $\partial_z\phi_1 = 0$ at $z = 0$ and $\phi_2 \to 0$ as $z \to -\infty$ requires that ϕ_1 and ϕ_2 have the form

$$\phi_1 = \mathcal{A}_1 \cosh k_n z \, \cos(k_n x) \, e^{-\iota\omega_n t} \qquad 0 \le x \le L, z > -H$$
$$\phi_2 = \mathcal{A}_2 e^{k_n z} \cos(k_n x) \, e^{-\iota\omega_n t} \qquad\qquad 0 \le x \le L, z < -H. \tag{2.125}$$

The actual values of the velocity potentials are determined by taking the real part of these expressions. The amplitudes $\mathcal{A}_1$ and $\mathcal{A}_2$ are determined in terms of the interfacial displacement amplitude A by the polarization relations, the same as those for horizontally unbounded waves.

The lowest two modes are illustrated in Figure 2.12. The lowest mode, for which $k_1 = \pi/L$, has a horizontal wavelength which is twice the domain size so that when the waves crest at one side of the domain they form a trough at the other. This is called a 'mode-1' wave. The mode-2 wave shown in Figure 2.12b has exactly one wavelength spanning the domain so that $k_2 = 2\pi/L$.

The difference between (2.124) and its horizontally unbounded counterpart (2.79) is subtle but carries important dynamical consequences. Not only does $k = k_n$ appear as a discrete variable in (2.124), but the structure of the mode itself has changed: whereas extracting the real part of (2.79) gives $A\cos(kx - \omega t)$ if A is real, the real part of (2.124) gives $A\cos(k_n x)\cos(\omega_n t)$. The former describes a left-to-right moving wave (for positive k and ω); the latter describes a standing wave. The positive sign of k_n has no directional meaning but is simply a measure of the wavelength.

In principle, interfacial modes can be represented by the superposition of leftward- and rightward-propagating waves with locked phase and identical wavelengths (equal to an integral fraction of twice the domain size). This follows directly from the trigonometric identity

$$A\cos(kx)\cos(\omega t) = \frac{1}{2}A\cos(kx - \omega t) + \frac{1}{2}A\cos(-kx - \omega t). \tag{2.126}$$

A significant dynamical difference between propagating and standing waves is that the former transport energy and the latter do not. This is clear from the identity (2.126). Waves with wavenumber k transport energy rightwards and those with wavenumber $-k$ transport energy leftwards by an equal and opposite amount. By superimposing the propagating waves to form a standing wave, the energy flux due to each cancels.

Although the discussion so far has focused upon a semi-infinite fluid with no bottom boundary, these ideas extend straightforwardly to interfacial waves in multi-layer fluids that are vertically bounded or unbounded. The dispersion relation and polarization relations are unchanged but the lateral boundary conditions restrict the allowable wavenumbers to $k_n = n\pi/L$ with $n = 1, 2, \ldots$

2.6 Shear flows

So far we have assumed the fluid is stationary in the absence of waves. However, in many geophysical circumstances currents and winds have changing horizontal speed with height. These are shear flows.

For example, the flow of fresh water into the ocean at an estuary could be approximated by a two-layer fluid in which the upper layer moves with uniform speed over stationary saline fluid. In another example that we have seen, the horizontal velocity field of baroclinic waves is oriented in opposite directions on either side of the interface. In some circumstances the shear at the interface can be so strong as to cause the interface to wrap up into vortices.

Here we will derive the equations for interfacial waves in the presence of shear in multi-layered fluids. In certain circumstances we will see that the resulting dispersion relations allow for a complex-valued frequency whose real part is the frequency in the normal sense. If the imaginary part is non-zero, this means the amplitude of the disturbance grows exponentially in time. The derivation and analysis of equations describing such potentially unstable disturbances comprise what is called 'hydrodynamic stability theory'.

The essence of stability theory for temporally growing disturbances is to assess whether complex ω exists for any (real-valued) k and, if so, to determine for what k the growth rate is largest. Spatial instability, in which ω is real and k is complex, and absolute-convective instability, in which both may be complex, are not considered here.

2.6.1 Derivation of equations

For simplicity, here we neglect Coriolis forces and assume the fluid is laterally unbounded, incompressible and two-dimensional, having structure in the x- and z-directions alone. The fluid is composed of n layers, each with uniform density ρ_i for $i = 1 \ldots n$. That is, the background density $\bar{\rho}(z)$ is piecewise-constant. The ambient flow is horizontal, varying only in the vertical. This is referred to as a 'parallel flow'.

In general, we would like to allow vertical shear within a layer, so we cannot assume the fluid is irrotational. Instead we will work with the momentum equations for an incompressible fluid. The motion of waves results in velocity fluctuations $\vec{u} = (u, w)$ so that the total velocity field is $(\bar{U} + u, w)$. Consistent with linear theory for small-amplitude disturbances, we will assume that the fluctuation quantities are small but that $\bar{U}$ can be large.

Substituting the total velocity in the x-momentum equation of (1.72), setting $f_0 = 0$ and keeping only those terms that are linear in fluctuation quantities (hence, for example, we keep $\bar{U} \partial_x u$ but discard $u \partial_x u$ in the advection terms of the material derivative), we find

$$\frac{\partial u}{\partial t} + \bar{U} \frac{\partial u}{\partial x} + w\bar{U}' = -\frac{1}{\rho} \frac{\partial p}{\partial x}. \tag{2.127}$$

Here the prime denotes an ordinary z-derivative.

Similarly, the z-momentum equation becomes

$$\frac{\partial w}{\partial t} + \bar{U} \frac{\partial w}{\partial x} = -\frac{1}{\bar{\rho}} \frac{\partial p}{\partial z}. \tag{2.128}$$

These equations apply within each layer of the multi-layer fluid. There is no buoyancy term in (2.128) because within a uniform-density liquid this is balanced by the background hydrostatic pressure. Buoyancy effects are felt through the fluctuation pressure gradient which in turn depends upon the vertical displacement of interfaces that bound each layer.

Because the fluid is incompressible, we can represent the velocity fields by derivatives of the streamfunction, ψ. Separating the background and fluctuation parts, we define the fluctuation streamfunction so that

$$(u, w) = \left(-\frac{\partial \psi}{\partial z}, \frac{\partial \psi}{\partial x} \right). \tag{2.129}$$

This can be substituted into (2.127) and (2.128) to give two equations in the unknown functions ψ and p.

We could combine these to derive a single differential equation for, ψ, say. However, the mathematics is less cumbersome if we first Fourier transform the equations in x and t. This can be done because the coefficients of the equations depend only upon z. Substituting

$$\psi = \hat{\psi}(z) \, e^{i(kx - \omega t)} \text{ and } p = \hat{p}(z) \, e^{i(kx - \omega t)}, \tag{2.130}$$

the momentum equations become

$$(\bar{U} - c)\hat{\psi}' - \bar{U}'\hat{\psi} = \frac{1}{\rho}\hat{p} \tag{2.131}$$

and

$$-k^2(\bar{U} - c)\hat{\psi} = -\frac{1}{\bar{\rho}}\hat{p}'. \qquad (2.132)$$

For convenience, we have defined $c \equiv \omega/k$. This is the phase speed of the waves when ω and k are both real-valued.

Assuming that $\bar{\rho}$ is constant in each layer and eliminating $\hat{p}$ from these equations gives Rayleigh's equation

$$\hat{\psi}'' - \left(\frac{\bar{U}''}{\bar{U} - c} + k^2\right)\hat{\psi} = 0. \qquad (2.133)$$

This describes the structure of waves everywhere within a slab of fluid of uniform density. In particular, if there is no background flow, then Rayleigh's equation is identical in form to (2.109). Whereas that formula gives the vertical structure of the velocity potential, (2.133) gives the vertical structure of the streamfunction. Both predict that the waves have amplitudes that change exponentially over a vertical e-folding distance k^{-1}.

The coupling between layers is once again determined by interface conditions that require continuity of mass and pressure. These formulae must be written in terms of the streamfunction and must account for changes in the background flow across the interfaces.

For material on the interface to stay there we require $w = D\eta/Dt$ to be continuous, in which η is the vertical displacement of the interface. Linearizing the material derivative, the statement that the vertical displacement at the lower side of the interface is equal to that at the upper side is given by the condition that

$$\frac{\hat{\psi}}{\bar{U} - c} \qquad (2.134)$$

is continuous across the interface.

Requiring that the pressure does not jump discontinuously across the interface, we must have continuity of the following quantity:

$$\bar{\rho}\left[(\bar{U} - c)\hat{\psi}' - \bar{U}'\hat{\psi} - \frac{g}{\bar{U} - c}\hat{\psi}\right]. \qquad (2.135)$$

This is found through manipulation of the horizontal and vertical momentum equations.

The complete description of small amplitude disturbances in a multi-layer shear flow is given by the ordinary differential equation (2.133) together with interface conditions (2.134) and (2.135), and conditions for boundedness at infinity or no normal flow at rigid upper and lower boundaries.

Table 2.2. *General and special conditions used to match the solutions for the streamfunction amplitude* $\hat{\psi}(z)$ *across interfaces where the background horizontal velocity,* $\bar{U}$, *and/or density,* $\bar{\rho}$, *change. The quantities in the middle and left column must simultaneously hold the same values above and below the interface. The expressions involve* $c \equiv \omega/k$, *which is the phase speed* c_{Px} *one would extract from a horizontal time series.*

	Material continuity	Pressure continuity
General	$\hat{\psi}/(\bar{U}-c)$	$\bar{\rho}\left[(\bar{U}-c)\hat{\psi}' - \bar{U}'\hat{\psi} - g/(\bar{U}-c)\hat{\psi}\right]$
$\bar{U}$ continuous	$\hat{\psi}$	$\bar{\rho}\left[(\bar{U}-c)\hat{\psi}' - \bar{U}'\hat{\psi} - g/(\bar{U}-c)\hat{\psi}\right]$
$\bar{\rho}$ continuous	$\hat{\psi}/(\bar{U}-c)$	$(\bar{U}-c)\hat{\psi}' - \bar{U}'\hat{\psi}$
$\bar{\rho}, \bar{U}, \bar{U}'$ continuous	$\hat{\psi}$	$\hat{\psi}'$
Boussinesq	$\hat{\psi}/(\bar{U}-c)$	$(\bar{U}-c)\hat{\psi}' - \bar{U}'\hat{\psi} - g\bar{\rho}/[\varrho_0(\bar{U}-c)]\hat{\psi}$

   The interface conditions can be simplified under special circumstances, summarized in Table 2.2. In particular, if the background density varies continuously, it follows from (2.134) that (2.135) reduces to

$$(\bar{U} - c)\hat{\psi}' - \bar{U}'\hat{\psi}. \tag{2.136}$$

Otherwise, if the density jump is small compared to the characteristic density, ρ_0, then (2.136) reduces to the condition that the following quantity must vary continuously across the interface:

$$(\bar{U} - c)\hat{\psi}' - \bar{U}'\hat{\psi} - \frac{g\bar{\rho}}{\rho_0}\left[\frac{\hat{\psi}}{\bar{U}-c}\right]. \tag{2.137}$$

This is the Boussinesq form of the pressure continuity condition which states that the density jump is only important in its effect upon buoyancy forces.

2.6.2 Rayleigh waves

Analytic solutions of Rayleigh's equation (2.133) are readily found if the background velocity $\bar{U}$ is constant or if it varies linearly with height, in which case $\bar{U}$ prescribes a uniform shear flow. In this case, the $\bar{U}$ dependence vanishes from (2.133) and so the streamfunction amplitude, $\hat{\psi}$, is given by exponential functions.

   Though sometimes algebraically cumbersome, we can straightforwardly find analytic solutions for $\hat{\psi}$ and the dispersion relation if $\bar{U}$ is piecewise-linear, meaning that it is subdivided into vertical ranges over which $\bar{U}$ is either constant or varies

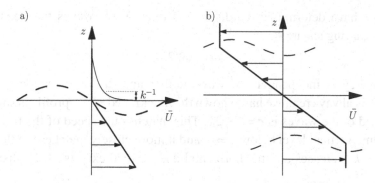

Fig. 2.13. a) Waves associated with a kinked-shear velocity profile. The vertical extent decays exponentially over a distance k^{-1} from the kink and the waves move in the direction of the shear. b) Wave-pairs associated with a shear layer. The waves on the upper flank move rightwards with respect to the upper layer flow; the waves on the lower flank move leftwards with respect to the lower layer flow. The background flow is shown as the solid black line and the waves are illustrated by dashed lines.

linearly with height. In each range we have a pair of exponential solutions which couple from one layer to the next through the interface conditions (2.134) and (2.135).

First we demonstrate this with the simple circumstance of a uniform-density fluid having the following kinked-shear flow profile in an unbounded domain:

$$\bar{U} = \begin{cases} 0 & z \geq 0 \\ -s_0 z & z < 0 \end{cases}, \tag{2.138}$$

in which the shear s_0 below $z = 0$ is constant. This ambient flow field is illustrated in Figure 2.13a.

Requiring bounded solutions to (2.133) gives the streamfunction amplitude in the upper and lower layer:

$$\hat{\psi} = \begin{cases} \mathcal{A}e^{-kz} & z \geq 0 \\ \mathcal{B}e^{kz} & z < 0. \end{cases} \tag{2.139}$$

The interface conditions (2.134) and (2.135) applied at $z = 0$ give the matrix equation

$$\begin{pmatrix} 1 & -1 \\ ck & ck - s_0 \end{pmatrix} \begin{pmatrix} \mathcal{A} \\ \mathcal{B} \end{pmatrix} = \begin{pmatrix} 0 \\ 0 \end{pmatrix}. \tag{2.140}$$

This eigenvalue problem has nontrivial solutions if the determinant of the matrix is zero, that is if $2ck - s_0 = 0$ in which $c = \omega/k$. The structure of the waves is illustrated in Figure 2.13a.

Thus we have determined the dispersion relation for waves that exist due to vertically varying shear:

$$\omega = s_0/2. \tag{2.141}$$

The frequency is independent of wavenumber and the group velocity is zero. Because ω is always real, we have shown that the kinked-shear profile is stable. The phase speed of the waves is $c = s_0/2k$. This matches the speed of the background flow at some vertical level below $z = 0$, and it more closely matches the flow speed at $z = 0$ as k increases and the horizontal and vertical extents of the disturbance decrease.

These waves, driven by shear-induced pressure fluctuations, have no generally accepted name. For convenience here we will describe them as Rayleigh waves, although this terminology should not be confused with that used to describe surface-trapped waves in solids.

2.6.3 Shear layer instability in uniform-density fluid

If $\bar{U}$ describes a flow with uniform shear over a finite depth, it is called a shear layer. Here we will show that a shear layer in uniform density fluid is unstable through resonant coupling between a pair of Rayleigh waves. The process is often referred to as Kelvin–Helmholtz instability, after the two scientists who first examined the phenomenon.

The background flow is prescribed by

$$\bar{U} = \begin{cases} -U_0 & z \geq H \\ -U_0\frac{z}{H} & |z| < H \\ U_0 & z \leq -H \end{cases}. \tag{2.142}$$

This velocity profile is effectively the result of splicing together two kinked-shear flow profiles of the form (2.138), as shown in Figure 2.13b.

Requiring bounded solutions to (2.133), we find

$$\hat{\psi} = \begin{cases} \mathcal{A}e^{-kz} & z \geq H \\ \mathcal{B}_1 \sinh kz + \mathcal{B}_2 \cosh kz & |z| < H \\ Ce^{kz} & z \leq -H \end{cases}. \tag{2.143}$$

Here the solutions in the middle region have been written in terms of hyperbolic functions in order to take advantage of symmetry as in (2.110) for interfacial waves in a three-layer fluid.

Applying the interface conditions at $z = \pm H$ gives four equations in the four unknowns $\mathcal{A}$, $\mathcal{B}_1$, $\mathcal{B}_2$ and C. Solving the eigenvalue problem gives the dispersion

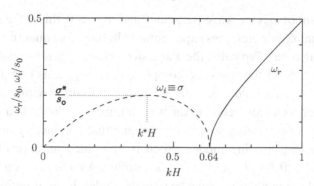

Fig. 2.14. Frequency (ω_r, solid line) and instability growth rate ($\omega_i \equiv \sigma$, dashed line) of horizontally periodic disturbances with wavenumber k superimposed upon an unstratified piecewise-linear shear layer of depth H. For $kH \lesssim 0.64$, $\omega_r = 0$; for $kH \gtrsim 0.64$, $\omega_i = 0$.

relation

$$\omega^2 = \frac{1}{4}s_0^2 \left[(1 - 2kH)^2 - e^{-4kH} \right], \tag{2.144}$$

in which $s_0 \equiv U_0/H$. The corresponding (complex-valued) dispersion relation is plotted in Figure 2.14 for positive frequency when ω is real and for positive growth rate $\sigma = \omega_i$ when $\omega^2 < 0$.

In the limit $kH \to \infty$, the dispersion relation becomes $\omega \to \pm s_0(1 - 2kH)/2 = \pm(s_0/2 - U_0 k)$. The positive root corresponds to the dispersion relation (2.141) for Rayleigh waves with a constant flow, $-U_0$, overlying a uniform shear flow. The negative root corresponds to Rayleigh waves with a shear flow overlying a constant flow with speed U_0. Indeed, for large k the vertical extent of the Rayleigh waves at the upper and lower kink in the shear is so small that one disturbance does not feel the influence of the other.

Thus we see that the dispersion relation (2.144) describes the co-existence of a pair of Rayleigh waves that couple together when their horizontal and vertical extent, k^{-1}, is comparable to H, as illustrated in Figure 2.13b.

Long-wavelength disturbances on either flank are not independent of each other and resonantly couple to form growing modes with zero phase speed. In particular, if $0 < kH \lesssim 0.64$, then the right-hand side of (2.144) is negative, meaning that ω is a pure imaginary number.

The fact that ω is complex-valued, means that the flow is unstable. To see this, suppose in general we have $\omega = \omega_r + \iota\sigma$. Substituting this into the formula for the streamfunction in (2.130) gives

$$\psi = \hat{\psi}(z) e^{\iota(kx - \omega_r t)} e^{\sigma t}. \tag{2.145}$$

Thus the real part of ω is the wave frequency and, if $\sigma > 0$, this is the growth rate of waves whose amplitude increases exponentially in time. The quantity $1/\sigma$ is known as the 'e-folding time'. Typically, the dispersion relation for ω is a polynomial with real coefficients. So if the roots are complex, they appear as complex-conjugate pairs. That is to say, if ω is complex, then an unstable solution with $\sigma > 0$ must exist.

Although there is no restriction to the wavenumber of allowable unstable waves, in reality the fastest growing solution is the one that is typically observed. From (2.144), the fastest growing mode, which is computed numerically, occurs for wavenumber $k^\star \simeq 0.398/H$, and the corresponding growth rate is $\sigma^\star \simeq 0.20 s_0$.

Recall that these predictions have been made under the assumption that the disturbances are small amplitude. So, although the unstable waves grow exponentially, they do not do so without bound. When their amplitude is sufficiently large, nonlinear effects become significant. In the case of the unstable shear layer, the waves grow and develop into coherent vortices sometimes called 'Kelvin–Helmholtz billows'. Eventually they turbulently break down.

Although linear theory cannot predict the long-time evolution of the flow, the wavelength of the most unstable mode sets the horizontal scale of the instability, even as it develops nonlinearly. Thus we expect the unstable shear layer to develop into a train of vortices separated approximately by $2\pi/k^\star \simeq 16H$.

2.6.4 Shear instability of interfacial waves

So far we have examined waves and instability in a fluid with uniform density. If in addition the system consists of layers of fluid with different density, then the system can support interfacial waves as well as Rayleigh waves. Density interfaces modify the interaction between pairs of Rayleigh waves in a Kelvin–Helmholtz unstable flow. They also lead to two new classes of instability. One is called 'Holmboe instability'. This results from the resonant coupling of an interfacial wave with a Rayleigh wave. The other class of instability has no generally accepted terminology, but will be referred to here as Taylor instability. This results from the resonant coupling of two interfacial waves mediated by a shear layer.

2.6.4.1 Kelvin–Helmholtz instability

First we examine how Kelvin–Helmholtz instability, examined in Section 2.6.3 for a uniform density fluid, is affected by the presence of density interfaces. Consider the shear layer given by (2.142) which moves in a three layer fluid with background density

$$\bar{\rho} = \begin{cases} \rho_1 & z \geq H \\ \frac{1}{2}(\rho_1 + \rho_2) & |z| < H \\ \rho_2 & z \leq -H \end{cases} . \tag{2.146}$$

For mathematical simplicity, this is set up so that density interfaces exist at the kinks of the shear profile and the density-jumps across each interface are equal.

The stability problem is solved by assuming the streamfunction amplitude has the form (2.143) but now the effect of discontinuity in the density profiles must be accounted for in the interface conditions (2.134) and (2.135).

After some algebra, the dispersion relation is given by the roots of the following quartic polynomial:

$$\tilde{\omega}^4 - \frac{1}{4}\left[8\tilde{k}^2 - 4(1 - \mathrm{Ri}_b)\tilde{k} + 1 - e^{-4\tilde{k}}\right]\tilde{\omega}^2$$
$$+ \frac{1}{4}\tilde{k}^2\left[(2\tilde{k} - 1 - \mathrm{Ri}_b)^2 - (1 + \mathrm{Ri}_b)^2 e^{-4\tilde{k}}\right], \tag{2.147}$$

in which the frequency and wavenumber have been expressed nondimensionally by $\tilde{\omega} = \omega/s_0$ and $\tilde{k} = kH$.

In (2.147) we have introduced the bulk Richardson number defined by

$$\mathrm{Ri}_b \equiv \frac{g'/H}{s_0^2}, \tag{2.148}$$

in which $s_0 = U_0/H$ is the shear at mid-depth and $g' = g\frac{\rho_2 - \rho_1}{(\rho_2 + \rho_1)/2}$ is the reduced gravity. This is a characteristic measure of the way in which buoyancy forces may retard or overcome shear instability. Taking the limit $\mathrm{Ri}_b \to 0$ in (2.147), we recover the dispersion relation (2.144) as two of the roots.

Although the presence of density-jumps might be expected to stabilize the flow, it turns out that the flow is unstable even for large Ri_b. Larger density-jumps act to reduce the growth rate of the most unstable mode and to increase the corresponding wavenumber. For example, Figure 2.15a plots the frequency and growth rate as a function of wavenumber in the case $\mathrm{Ri}_b = 1$. This should be compared with the corresponding plot for the unstratified shear layer (for which $\mathrm{Ri}_b = 0$) shown in Figure 2.14.

Expecting that both ω_r and ω_i are zero at the conceptual boundary between stable and unstable modes, we find that coupled Rayleigh waves are unstable if their wavenumber k lies within a finite range of values given by

$$\frac{2kH}{1 + \exp(-2kH)} - 1 < \mathrm{Ri}_b < \frac{2kH}{1 - \exp(-2kH)} - 1. \tag{2.149}$$

The so-called 'marginal stability curves' given by (2.149) are plotted in the stability regime diagram shown in Figure 2.15b. In the limit $\mathrm{Ri}_b \to 0$, corresponding to a uniform-density fluid, the range of wavenumbers for which the flow is unstable is given by the values of kH for which the right-hand side of (2.144) is negative. The frequency and growth rate for modes in this case are plotted in Figure 2.14.

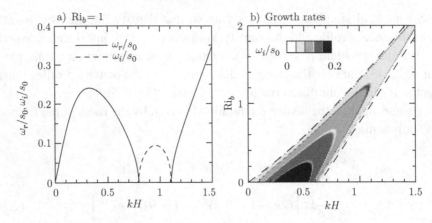

Fig. 2.15. a) Frequency (solid line) and growth rate (dashed line) versus wave-number of modes associated with a piecewise-linear shear layer in a three-layer fluid whose middle-layer density is the average of the upper and lower layers. Values are computed for the bulk Richardson number $\mathrm{Ri}_b \equiv (g\Delta\rho/\rho_0)H/U_0^2 = 1$, in which ρ_0 is the characteristic density taken as the value of $\bar\rho$ at $z = 0$. b) Growth rates as they depend upon the wavenumber and Ri_b. The dashed lines indicate the marginal stability boundaries, given by (2.149). For large Ri_b, the most unstable mode has a nondimensional wavenumber $kH \simeq (\mathrm{Ri}_b + 1)/2$.

Corresponding to each eigenvalue, ω, is the eigenfunction $\hat\psi(z)$, which gives the structure of the Kelvin–Helmholtz modes. The z-dependence of $\hat\psi$ in each layer is given generally by (2.143) in which the coefficients $\mathcal{A}$, $\mathcal{B}_1$, $\mathcal{B}_2$ and C effectively form an eigenvector. Substituting ω for given k and Ri_b into the equations defining the interface conditions, one can explicitly solve for $\mathcal{A}$, $\mathcal{B}_1$ and C in terms of $\mathcal{B}_2$. The result can be normalized by setting $\mathcal{B}_2 = 1$.

Figure 2.16a shows the perturbation streamfunction amplitude computed in this way for the most unstable mode of the shear layer with $\mathrm{Ri}_b = 1$. Note that the function has real and imaginary parts as a consequence of $\omega = \iota 0.094 s_0$ being complex. To help interpret this result, Figure 2.16b shows greyscale contours of the perturbation streamfunction $\psi(x,z) = \Re\{\hat\psi e^{\iota kx}\}$. Here the amplitude and phase are established effectively by setting $t = 0$ and $\mathcal{B}_2 = 1$. At $x = 0$, ψ is identical to the real part of $\hat\psi$; at $x = 3/4\lambda$, $\psi(x,z)$ is identical to the imaginary part of $\hat\psi$. At other values of x the streamfunction is computed from $\hat\psi_r(z)\cos(kx) - \hat\psi_i(z)\sin(kx)$.

The structure of the growing instability is best illustrated by superimposing the perturbation streamfunction upon the background streamfunction $\bar\psi$, defined implicitly through $\bar U = -d\bar\psi/dz$. This is plotted in Figure 2.16d. Contours of the total streamfunction $\psi_T(x,z) = \bar\psi(z) + \epsilon\psi(x,z)$ are shown in Figure 2.16e. Here, the amplitude of the perturbation is established through the choice of ϵ. As time progresses, the amplitude grows exponentially as $\exp(\omega_i t)$. Thus we see that the

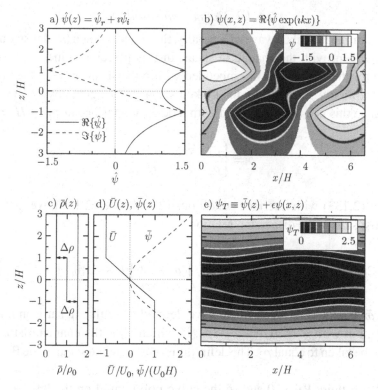

Fig. 2.16. a) Vertical structure of the perturbation streamfunction amplitude for an unstable disturbance in a shear layer with density interfaces at $z = \pm H$. It is computed for the most unstable mode in the case $\mathrm{Ri}_b = 1$, for which $k^* H = 0.96$ and $\omega/s_0 \simeq \iota 0.094$. The eigenfunction is normalized so that $\hat{\psi}(0) = 1$. b) Corresponding spatial structure of the perturbation streamfunction $\psi(x, z) = \Re\{\hat{\psi} \exp(\iota k x)\} = \Re\{\hat{\psi}\} \cos(kx) - \Im\{\hat{\psi}\} \sin(kx)$. Background profiles of c) density and d) velocity (solid line) and streamfunction (dashed line), the latter being defined by $\bar{U} = -d\bar{\psi}/dz$. e) Contours of the total streamfunction $\psi_T = \bar{\psi} + \epsilon \psi$ in which the amplitude of the perturbation has been taken as $\epsilon = 0.1$.

instability distorts the streamlines, pulling them apart near the centre of the shear layer between $x = \lambda/4$ and $3\lambda/4$.

The waves grow by extracting energy from the background shear flow. The growth rate decreases as the density-jump across the interfaces (measured by Ri_b) becomes larger because some of the kinetic energy extracted must go into the available potential energy associated with the disturbance.

As the instability grows to very large amplitude, the linear theory approximations used to derive the form of the disturbance are no longer valid. Nonlinear simulations show that the disturbance saturates at finite amplitude if Ri_b is sufficiently large. In weak stratification however, the contours can wrap up to form a vortex centred in the middle of the shear layers.

2.6.4.2 Holmboe instability

Next we examine Holmboe waves. Generally, these occur when the density interface is not coincident with the change in shear. In this piecewise-linear problem it is necessary for the interface to be embedded within the shear in order for instability to occur. Explicitly, we consider the kinked shear profile given by (2.138) and we now suppose this is a two-layer fluid with a density interface at $z = -H < 0$:

$$\bar{\rho} = \begin{cases} \rho_1 & z \geq -H \\ \rho_2 & z < -H \end{cases}. \tag{2.150}$$

Solving (2.133) with interface conditions (2.134) and (2.135), we find that the waves must satisfy the dispersion relation

$$\tilde{\omega}^3 - \left(2\tilde{k} + \frac{1}{2}\right)\tilde{\omega}^2 + \tilde{k}\left(\tilde{k} + 1 - \frac{1}{2}\mathrm{Ri}_b\right)\tilde{\omega} - \frac{1}{2}\tilde{k}\left(\tilde{k} - \frac{1}{2}\mathrm{Ri}_b\left[1 - e^{-2\tilde{k}}\right]\right) = 0, \tag{2.151}$$

in which $\tilde{\omega} \equiv \omega/s_0$, $\tilde{k} \equiv kH$, and we have defined the bulk Richardson number as in (2.148), but with $g' \equiv g(\rho_2 - \rho_1)/\rho_0$, in which ρ_0 is the characteristic density, which can be taken to equal ρ_2. The definitions of Ri_b are the same in the Boussinesq limit.

In uniform fluid, $\mathrm{Ri}_b = 0$ and so the cubic polynomial on the left-hand side of (2.151) has one root $\tilde{\omega} = 1/2$, corresponding to the dispersion relation for Rayleigh waves (2.141). The remaining double root is $\tilde{\omega} = -\tilde{k}$. For finite Ri_b, this double root introduces a new class of disturbance that evolves due to interfacial waves at $z = -H$ interacting with Rayleigh waves centred near $z = 0$. For a range of k, these double-roots are complex-valued, meaning that the coupling between the two types of waves renders them unstable. The complex dispersion relation is plotted for the case $\mathrm{Ri}_b = 1$ in Figure 2.17a. The thick solid line in the figure also shows the phase speed $c_p = \omega_r/k$.

Unlike shear instability, whose dispersion relation is given by the roots of (2.144), the complex roots of (2.151) have non-zero real as well as imaginary parts. Thus Holmboe waves are not stationary, but as they grow they also propagate in the direction of the shear flow below $z = 0$. This is a distinguishing feature of Holmboe waves.

The values of kH and Ri_b that result in unstable waves are shown in Figure 2.17b. Consistent with our analysis, we see that the growth rate is zero if $\mathrm{Ri}_b = 0$, meaning that there is no density interface. Instability occurs even for very small Ri_b with the most unstable mode having a wavenumber that increases as Ri_b increases. The growth rate is fastest if $\mathrm{Ri}_b \simeq 0.56$, in which case the most unstable mode has a nondimensional wavenumber $kH \simeq 0.89$ and growth rate $\omega_i = 0.14s_0$.

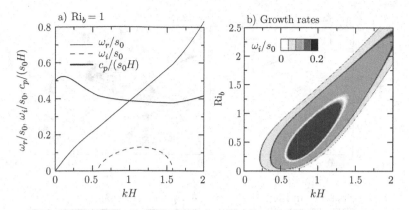

Fig. 2.17. Frequency (ω_r, solid line) and growth rate ($\omega_i \equiv \sigma$, dashed line) of the Holmboe instability resulting from a density interface lying a distance H above a kinked-shear profile with shear s_0 below $z = 0$. The results are plotted for the complex roots of (2.151) with the positive imaginary part for the case with $Ri_b = 1$.

The structure of the most unstable mode in the case $Ri_b = 1$ is shown in Figure 2.18. Unlike the corresponding Kelvin–Helmholtz mode shown in Figure 2.16, the Holmboe mode has a cusped structure that peaks near the kink of the shear profile. As time progresses, this mode grows in amplitude and propagates rightwards.

In many laboratory and geophysical circumstances Holmboe waves are created by a finite-depth shear layer in which the density interface is offset from the midpoint of the shear. In this case Holmboe waves appear on the upper and lower flanks of the shear layer. Their presence is distinguished from Kelvin–Helmholtz instability by the appearance of leftward- and rightward-propagating waves on either flank of the shear layer. Whereas Kelvin–Helmholtz waves have the same speed as the midpoint of the shear and wrap into vortices as they develop nonlinearly, finite-amplitude Holmboe waves form cusps that move in opposite directions above and below the shear layer. The cusps are more pronounced on one flank if the interface is closer to the kink in the shear profile on that flank.

2.6.4.3 Taylor instability

Finally, we examine the circumstance in which two interfacial waves interact through a uniform shear flow. Explicitly, we consider the circumstance in which the background flow is given by

$$\bar{U} = -s_0 z \tag{2.152}$$

for all z, and the three-layer density profile is given by (2.146).

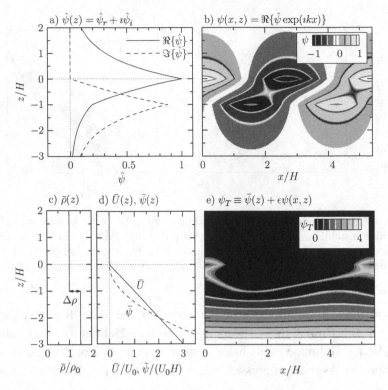

Fig. 2.18. As in Figure 2.16 but showing the structure of the most unstable Holmboe mode in a kinked-shear flow with a density interface at $z = -H$. The functions are computed for the case $\mathrm{Ri}_b = 1$, in which circumstance the most unstable mode has a wavenumber $k^\star H = 1.16$ and complex frequency $\omega/s_0 \simeq 0.443 + \iota 0.129$. a) Vertical structure of the perturbation streamfunction amplitude function $\hat\psi(z)$, b) the corresponding spatial structure of the perturbation streamfunction, background profiles of c) density and d) velocity (solid line) and streamfunction (dashed line), and e) contours of the total streamfunction $\psi_T = \bar\psi + \epsilon\psi$ in which $\epsilon = 0.3$.

Again we take the streamfunction amplitude to have the form (2.143). For mathematical simplicity we assume the fluid is Boussinesq and so apply the interface conditions (2.134) and (2.137). Thus we find the dispersion relation is given by the roots of

$$\tilde\omega^4 - \tilde k\left(2\tilde k + \mathrm{Ri}_b\right)\tilde\omega^2 + \frac{1}{4}\tilde k^2\left[(2\tilde k - \mathrm{Ri}_b)^2 - \mathrm{Ri}_b{}^2\exp(-4\tilde k)\right] = 0. \qquad (2.153)$$

Here $\tilde\omega \equiv \omega/s_0$, $\tilde k \equiv kH$ and the bulk Richardson number is given by (2.148).

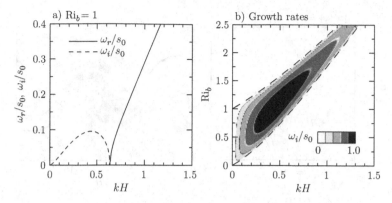

Fig. 2.19. As in Figure 2.15 but showing the dispersion relation associated with a Taylor mode in a uniform shear flow with a three-layer density profile whose middle-layer density is the average of the upper and lower layers. a) Frequency (solid line) and growth rate (dashed line) versus wavenumber of modes computed in the case $\mathrm{Ri}_b = 1$. b) Growth rates as they depend upon the wavenumber and Ri_b. The dashed lines indicate the marginal stability boundaries, given by (2.154).

The roots of (2.153) are complex and the flow unstable if

$$\frac{2\tilde{k}}{1 + \exp(-2\tilde{k})} < \mathrm{Ri}_b < \frac{2\tilde{k}}{1 - \exp(-2\tilde{k})}. \qquad (2.154)$$

These stability boundaries are indicated by the dashed lines in Figure 2.19b. In a shear flow with no density interfaces ($\mathrm{Ri}_b = 0$), no instability occurs. Like Kelvin–Helmholtz modes, the frequency (and hence phase speed) of the disturbances is zero if the mode is unstable. Here, however, the fastest growing modes generally have a much smaller wavenumber and correspondingly larger horizontal and vertical extents.

The structure of the most unstable mode in the case $\mathrm{Ri}_b = 1$ is shown in Figure 2.20. This is qualitatively similar to the structure of Kelvin–Helmholtz modes but it must be kept in mind that there is no kink in the shear flow and so the dynamics driving the instability are fundamentally different. The presence of shear allows interfacial waves in a three-layer fluid to grow in amplitude through extracting kinetic energy from the background shear.

2.7 Interfacial waves influenced by rotation

The Earth's rotation is important for waves having periods longer than many hours and typically on the order of days. Such waves also tend to be of broad horizontal extent – so wide that shallow water theory may be applied.

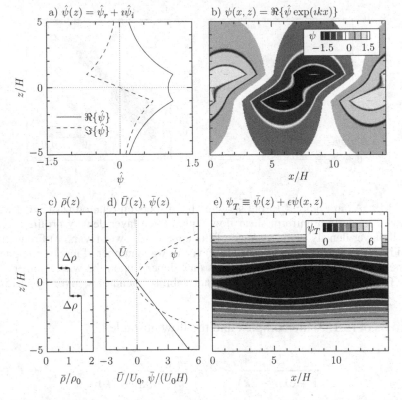

Fig. 2.20. As in Figure 2.16 but showing the structure of the most unstable Taylor mode in a uniform shear flow with density interfaces at $z = \pm H$. The functions are computed for the case $\mathrm{Ri}_b = 1$, in which circumstance the most unstable mode has a wavenumber $k^\star H = 0.44$ and growth rate $\omega_i / s_0 \simeq 0.096$. a) Vertical structure of the perturbation streamfunction amplitude function $\hat{\psi}(z)$, b) the corresponding spatial structure of the perturbation streamfunction, background profiles of c) density and d) velocity (solid line) and streamfunction (dashed line), and e) contours of the total streamfunction $\psi_T = \bar{\psi} + \epsilon \psi$ in which $\epsilon = 0.3$.

For this reason, here we present the theory for waves in a two-layer shallow water fluid in which buoyancy is felt through the reduced gravity $g' = g(\rho_2 - \rho_1)/\rho_2$, and the equivalent depth is the harmonic mean of the upper and lower layer depths (2.99): $\bar{H} = H_1 H_2 / (H_1 + H_2)$.

A shallow water two-layer fluid describes interfacial waves that are long compared with the depth of the upper and lower layer fluids, as would be the case for long oceanic waves at the thermocline.

A shallow water one-and-a-half-layer fluid, recovered in the limit $\bar{H} \to H$, represents waves moving along an interface that separates a relatively thin slab of fluid from an effectively infinitely deep fluid layer. This might describe long-wavelength disturbances at an atmospheric inversion or at the thermocline in the ocean. In the former case, for the Boussinesq approximation to hold, the waves should be long, but not so long that their wavelength is comparable to the density scale height of the atmosphere. In the latter case, the waves should not be so long that their wavelength is comparable to the total ocean depth.

The one-layer case corresponds to the limits $g' \to g$ and $\bar{H} \to H$. This represents the barotropic motion of waves with wavelength long compared with the total fluid depth as would be appropriate, for example, to describe tides in the mid-ocean, which have wavelengths much longer than the characteristic depth of 4 km.

2.7.1 Small-amplitude wave equations

We account for rotational effects by adding the horizontal components of the Coriolis force (1.39) to the linearized form of the shallow water equations. As in the derivation of (2.105) and (2.106), we define the x-component of the velocity field by the depth-weighted average (2.104). In problems involving rotation, we also need to include the y-component of the velocity field, which likewise is given in terms of the depth-weighted average. Explicitly, we define

$$(u, v) \equiv -\frac{H_1}{\bar{H}}(u_1, v_1) = \frac{H_2}{\bar{H}}(u_2, v_2). \tag{2.155}$$

Including the effects of rotation, the x-component of the horizontal momentum equation (2.68) becomes

$$\frac{\partial u}{\partial t} - f_0 v = -g' \frac{\partial \eta}{\partial x}. \tag{2.156}$$

The corresponding y-component of the momentum equation is

$$\frac{\partial v}{\partial t} + f_0 u = -g' \frac{\partial \eta}{\partial y}. \tag{2.157}$$

The system is closed by including the linearized form of the continuity equation (2.69) extended to x–y space:

$$\frac{\partial \eta}{\partial t} + \bar{H} \left(\frac{\partial u}{\partial x} + \frac{\partial v}{\partial y} \right) = 0. \tag{2.158}$$

The Coriolis parameter, $f_0 \equiv 2\Omega_e \sin \phi_0$, is a constant determined by the motion of the waves about a fixed latitude ϕ_0. In this approximation, the waves are said to propagate on the f-plane and f_0 is called the inertial frequency.

For motions on Earth that are influenced by Coriolis forces, it is typical to align the x- and y-axes with lines of latitude and longitude, respectively, in which case k_x and k_y are the zonal and meridional components of the wavenumber vector, $\vec{k}$, and u and v are the zonal and meridional components of the velocity. However, on the f-plane there is no dynamical distinction between north–south and east–west motion and so it is not necessary to employ such terminology. In the discussion of motions on the f-plane, we avoid reference to planetary co-ordinates in order to emphasize that the orientation of the x- and y-axes is arbitrary.

2.7.2 Inertial waves

Inertial waves are shallow water waves that move under the influence of Coriolis forces on the f-plane. The terminology comes from the fact that the inertial frequency f_0 is non-negligible in its influence upon the frequency and structure of the waves. Particularly for surface waves on the f-plane, they are often called 'Poincaré waves'. Here we use the former terminology to describe both surface and interfacial waves.

The three coupled partial differential equations (2.156), (2.157) and (2.158) may be solved in a variety of ways. One option is to combine them to form a single partial differential equation in one variable. In terms of η, we find

$$\frac{\partial^2 \eta}{\partial t^2} = c^2 \nabla_h^2 \eta - f_0^2 \eta, \qquad (2.159)$$

in which $\nabla_h \equiv (\partial_x, \partial_y)$ is the gradient operator acting only in the horizontal. Fourier transforming the equation, effectively by substituting $\eta(x,y,t) = A \exp[\imath (k_x x + k_y y - \omega t)]$ in (2.159), gives the dispersion relation

$$\omega^2 = c^2 |\vec{k}|^2 + f_0^2, \qquad (2.160)$$

in which $\vec{k} = (k_x, k_y)$ and $c = \sqrt{g'\bar{H}}$ is the shallow water wave speed. This is plotted in Figure 2.21.

Alternately, (2.156)–(2.158) may be represented in the differential matrix form

$$\begin{pmatrix} \partial_t & -f_0 & g'\partial_x \\ f_0 & \partial_t & g'\partial_y \\ \bar{H}\partial_x & \bar{H}\partial_y & \partial_t \end{pmatrix} \begin{pmatrix} u \\ v \\ \eta \end{pmatrix} = \begin{pmatrix} 0 \\ 0 \\ 0 \end{pmatrix}. \qquad (2.161)$$

Fourier transforming gives the algebraic matrix equation

$$\begin{pmatrix} -\imath\omega & -f_0 & \imath k_x g \\ f_0 & -\imath\omega & \imath k_y g \\ \imath k_x H & \imath k_y H & -\imath\omega \end{pmatrix} \begin{pmatrix} u \\ v \\ \eta \end{pmatrix} = \begin{pmatrix} 0 \\ 0 \\ 0 \end{pmatrix}. \qquad (2.162)$$

Nontrivial solutions of this eigenvalue problem require the matrix determinant to be zero. Neglecting the trivial $\omega = 0$ root, we again find the dispersion relation (2.160) for inertial waves.

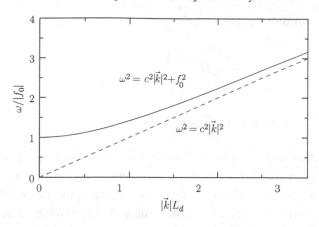

Fig. 2.21. Dispersion relation for inertial waves in shallow water (solid line). The dashed line shows the dispersion relation for shallow water waves in the absence of rotation. The wavenumber is given in nondimensional form through the deformation radius $L_d = c/|f_0|$ in which $c = \sqrt{g'\bar{H}}$ is the shallow water wave speed.

As expected, (2.160) becomes the dispersion relation for non-rotating shallow water waves in the limit $f_0 \to 0$. Rotational effects are significant only for very long-wavelength waves in which case their frequency is close to but not less than f_0.

The transition between the fast and slow frequency limits occurs for waves with horizontal extent of the order of

$$L_d \equiv \frac{c}{|f_0|} = \frac{\sqrt{g'\bar{H}}}{|f_0|}. \tag{2.163}$$

This is called the 'deformation radius' or sometimes the 'Rossby deformation radius'. In the description of interfacial waves, L_d is sometimes explicitly referred to as the 'internal deformation radius' in order to emphasize that the length scale corresponds to interfacial, not surface waves. The 'external deformation radius' is $\sqrt{gH}/|f_0|$. Crudely speaking, L_d is a characteristic measure of the distance travelled by an object moving at speed c before it is turned around by Coriolis forces in a time $|f_0|^{-1}$.

The eigenvectors of (2.162) can be used to express how the u and v field for plane waves relate to η. These are the polarization relations for inertial waves. Somewhat arbitrarily, we consider the special case with $k_y = 0$. That is, we orient the x- and y-axes so that x points in the direction of the wave propagation and y is oriented parallel to wave crests 90° counter-clockwise from the x-axis.

Supposing the actual vertical displacement is

$$\eta = A\cos(k_x x - \omega t), \tag{2.164}$$

the actual velocity components are

$$u = A \frac{\omega}{k_x \bar{H}} \cos(k_x x - \omega t) \qquad (2.165)$$

and

$$v = A \frac{f_0}{k_x \bar{H}} \sin(k_x x - \omega t). \qquad (2.166)$$

Thus the surface deflection of an inertial wave is in phase with the component of the velocity moving in the direction of the wave crests and is 90° out of phase with the smaller along-crest velocity component. The magnitudes of the along- and across-crest components are comparable if the waves are so long that the wave frequency is close to the inertial frequency. The associated horizontal velocity fields are sketched in Figure 2.22a.

Being shallow water waves, the horizontal velocity does not change with depth. From the velocity field we see that fluid parcels at all depths follow coherent elliptical paths with eccentricity that is larger for waves with faster frequency. The motion along the paths is 'anticyclonic', meaning that the parcels orbit in a direction opposite to the background rotation, as indicated by the sign of f_0.

The vertical component of vorticity is $\zeta \equiv \partial_x v - \partial_y u$. From (2.165) and (2.166), this is

$$\zeta = \frac{A f_0}{\bar{H}} \cos(k_x x - \omega t). \qquad (2.167)$$

This is in phase with the vertical displacement field, revealing that the vorticity is cyclonic below a wave crest and is anticyclonic below a trough, as shown in Figure 2.22b.

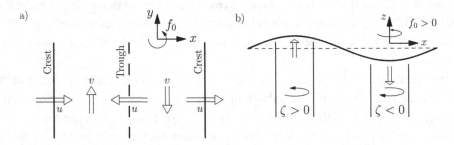

Fig. 2.22. a) Top view showing the flow field associated with inertial waves propagating rightwards in a counter-clockwise-rotating shallow one-layer fluid. b) Vertical cross-section showing the surface deflection and associated stretching (beneath the crest) and compression (beneath the trough) of relative vorticity.

This result also follows from the law for conservation of potential vorticity (1.88). Using (2.164) and setting $h = \eta$ as given by (2.167), we find that the potential vorticity of inertial waves is indeed constant everywhere and equals its value in the absence of any perturbation: $q = f_0/\bar{H}$.

2.7.3 Energetics of inertial waves

The energy equation for shallow water systems is derived through manipulation of the momentum and continuity equations. Taking u times (2.156), v times (2.157), adding the result together and multiplying by $\bar{H}$ gives the kinetic energy equation

$$\frac{\partial}{\partial t} E_K = -\nabla \cdot \vec{\mathcal{F}}_K + g'\bar{H}\,\eta\,\nabla \cdot \vec{u}, \tag{2.168}$$

in which $\vec{u} = (u, v)$, $\nabla = (\partial_x, \partial_y)$, $E_K \equiv \bar{H}|\vec{u}|^2/2$ is the vertically integrated kinetic energy per unit mass, and $\vec{\mathcal{F}}_K \equiv g'\bar{H}\vec{u}\eta$ is the corresponding kinetic energy flux.

The divergence of $\vec{u}$ on the right-hand side of (2.168) can be re-expressed in terms of η using the continuity equation (2.158). Recognizing $E_P = g'\eta^2/2$ as the available potential energy per unit mass, we have the total energy equation

$$\frac{\partial}{\partial t} E_T = -\nabla \cdot \vec{\mathcal{F}}_K, \tag{2.169}$$

in which $E_T = E_K + E_P$.

In Section 2.7.2 we found explicit expressions for the velocity fields of shallow water interfacial waves moving in the x-direction. Using (2.165) and (2.166) in the expression for the kinetic energy and averaging over a wavelength gives the mean vertically integrated kinetic energy per unit mass:

$$\langle E_K \rangle = \frac{1}{4}g'\frac{\omega^2 + f_0^2}{\omega^2 - f_0^2}A^2. \tag{2.170}$$

The corresponding mean available potential energy per unit mass is

$$\langle E_P \rangle = \frac{1}{4}g'A^2. \tag{2.171}$$

In the limit, $f_0 \to 0$, the energies are in equipartition, as expected for waves in a conservative system. But a rotating system is not conservative due to the external, albeit fictitious, force resulting from being in a non-inertial frame of reference. So the kinetic and potential energies are not equal. For waves with frequency ω near the inertial frequency, f_0, the kinetic energy is significantly larger than the available potential energy.

Using the dispersion relation, the mean total energy for a wave propagating in the x-direction is

$$\langle E_T \rangle = \frac{1}{2} \frac{\omega^2}{\bar{H} k_x} A^2. \tag{2.172}$$

As expected, because these are small-amplitude waves the mean vertically integrated energy flux per unit mass $\langle F_{Ex} \rangle \equiv g' \bar{H} \langle u\eta \rangle = c_{gx} \langle E_T \rangle$, in which c_{gx} is the horizontal group velocity.

Exercises

2.1 Derive the equations describing the evolution of one-layer deep water waves. Solve these and show the dispersion relation is the same as that for finite-depth waves in the limit of large kH.

2.2 (a) Derive the equations for the phase and group speed in (2.20) and (2.21), respectively.
   (b) By taking appropriate limits, derive their counterparts for deep and shallow water waves.

2.3 (a) From the linearized horizontal momentum equations, show that in general $p = \varrho_0 c_p u$.
   (b) Derive the formula for the dynamic pressure field (2.28) using the vertical momentum equation.

2.4 Compute the advection term of the u and w momentum equations, $\vec{u} \cdot \nabla \vec{u}$, for two-dimensional, small-amplitude waves in a one-layer fluid with finite-depth H. Use the dispersion relation to put your answer in terms of k, not ω, and use trigonometric double-angle formulae to simplify your result.

2.5 (a) Derive the formula (2.31), which shows that fluid parcels follow elliptical trajectories beneath waves in a one-layer fluid.
   (b) Determine the counterpart to this formula in the limit of deep and shallow water waves.

2.6 Through explicit integration, derive the formula for the mean kinetic and potential energy and for the energy flux due to surface waves.

2.7 (a) Following the outlined fluid parcel argument, derive the formula for the Stokes drift, u_S, as given by (2.57).
   (b) Show that the corresponding mean vertical velocity $w_S = 0$.

2.8 Derive the momentum conservation law (2.61) for quasi-monochromatic wavepackets and so derive the formula for the Stokes drift of deep surface waves given by (2.58).

Hint: To do this rigorously, explicitly assume the amplitude envelope of each field translates at the group velocity and varies slowly in space and time by writing $A_u = A_u(\epsilon(x - c_g t), \epsilon^2 t)$ for the horizontal velocity field and use the polarization relations to relate the amplitude envelopes of the vertical velocity and pressure fields to A_u. Rewrite the flux-form of the momentum equations in terms of the slow variables $X = \epsilon(x - c_g t)$ and $T = \epsilon^2 t$. Horizontally average locally about each x and, supposing $\epsilon \ll 1$, extract the leading terms which are of order ϵ.

2.9 Explicitly solving the wave equation (2.70), find the dispersion relation for shallow water waves. Show this result is identical to taking the limit $kH \ll 1$ in the finite-depth dispersion relation (2.17).

2.10 (a) Derive the dispersion relation for deep interfacial waves (2.84) from the linearized interface conditions (2.76) and (2.77).

(b) Likewise show the velocity potential amplitudes, $\mathcal{A}_1$ and $\mathcal{A}_2$, are related to the interface displacement amplitude A by (2.86).

(c) Find how the complex amplitudes of vertical and horizontal velocity relate to A and so determine the phase of the horizontal velocity in the lower and upper layers.

2.11 (a) Explicitly compute the mean kinetic energy associated with interfacial waves in an infinitely deep two-layer fluid.

(b) Derive an integral formula, analogous to (2.36), which gives the available potential energy of interfacial waves in an infinitely deep two-layer fluid. Evaluate this integral and, averaging, show it equals the corresponding mean kinetic energy.

2.12 Write the equations for the boundary value problem in which a two-layer fluid is bounded above by a rigid lid and the lower layer is infinitely deep. Solve the problem to find the dispersion relation and compare your solution with (2.92).

2.13 Find the mean energy of shallow water waves in a one-and-a-half-layer fluid. Compare your result with the corresponding energy for shallow water waves in a one-layer fluid.

2.14 Derive the dispersion relation, (2.94), for Boussinesq interfacial waves in a finite-depth two-layer fluid.

2.15 (a) Show that the mean kinetic and available potential energy of interfacial waves in a finite-depth, two-layer fluid are in equipartition.

(b) Compute the mean energy, $\langle E_T \rangle$, and mean energy flux, $\langle F_E \rangle$, and show that $\langle F_E \rangle = c_g \langle E_T \rangle$.

2.16 (a) Use separation of variables to find the spatial dependence of the velocity potentials ϕ_1 and ϕ_2 subject to the side-wall boundary conditions (2.123).

(b) A rectangular tank is filled 20 cm deep with salt water having density $\rho_2 = 1.02\,\text{g/cm}^3$ and 10 cm of fresh water is layered on top. A long wave ($k\bar{H} \ll 1$) propagates along the interface between the two fluids having displacement amplitude $A = 1$ cm. What is the horizontal velocity immediately above and below the crest.

2.17 (a) Write the partial differential equation in cylindrical co-ordinates with appropriate surface boundary conditions for one-layer small-amplitude waves oscillating in a fluid of mean depth H, contained in a cylinder of radius R.

(b) Using the method of separation of variables, show that the surface displacement of a particular mode of oscillation has the form

$$\eta(r,\theta,t) = A J_n(k_{rn}r)\cos(n\theta)\cos(\omega t),$$

in which J_n is the nth order Bessel function of the first kind and r and θ are the radial and azimuthal co-ordinates.

(c) From the surface boundary conditions, determine the dispersion relation in terms of the azimuthal and radial wavenumbers, n and k_{rn}, respectively.

(d) From the side-wall boundary conditions, write an implicit formula in terms of Bessel functions that limits k_{rn} to a discrete set of values.

(e) Suppose the cylinder is a coffee cup with radius $R = 4$ cm and it is filled to a depth $H = 8$ cm. What is the frequency of the lowest 'sloshing' mode? (These are the waves whose phase changes by 2π going 360° around the azimuth.) Compare this with a brisk walking frequency of $10\,\text{s}^{-1}$ and so speculate whether the coffee is likely to spill while walking. At what depth does resonance occur? Note: for this problem, tables or a symbolic algebra program should be used to evaluate the appropriate value of k_{rn} explicitly.

2.18 An experiment: fill a tall glass halfway with water and carefully pour a thin layer of oil on top. Practise sloshing the fluid so as to excite barotropic and baroclinic waves and compare their frequencies. How do their frequencies vary if the layer of oil is twice as deep? What relevant factors influence this experiment that are not represented by (2.117) and (2.119)?

2.19 (a) Derive the dispersion relation (2.112) for waves in a three-layer fluid.

(b) Show that it reduces to the dispersion relation for an infinitely deep two-layer fluid in the limit $H \to 0$.

(c) Show that (2.119) reduces to the dispersion relation for shallow water waves in a one-and-a-half-layer fluid in the limit $kH \to 0$.

2.20 Derive the formulae for the horizontal and vertical velocity fields associated with baroclinic waves in a three-layer fluid.

2.21 Find the dispersion relation for waves in a uniform-density fluid with velocity profile

$$\bar{U} = \begin{cases} U_0 + s_0(z - H) & z \geq H \\ U_0 & z < H \end{cases}.$$

Determine for which wavenumbers k, if any, the flow is unstable.

2.22 Find the dispersion relation for waves in a two-layer shear layer with velocity profile

$$\bar{U} = \begin{cases} -U_0 & z > 0 \\ U_0 & z \leq 0 \end{cases}$$

and density profile

$$\bar{\rho} = \begin{cases} \rho_1 & z > 0 \\ \rho_2 & z \leq 0 \end{cases},$$

in which $\rho_1 < \rho_2$. Determine for which wavenumbers k, if any, the flow is unstable.

2.23 Find the dispersion relation for waves in a piecewise-linear shear flow given by (2.138) and with a two-layer density profile given by (2.150). You may assume the density jumps are small so that the Boussinesq approximation is valid.

Hint: suppose the eigenfunctions in each layer assume the form given by (2.143). Substitute these into the interfacial conditions at $z = \pm H$ which amount to requiring $\hat{\psi}$ and the expression given by (2.137) to be continuous. Eliminate $\mathcal{A}$ and C from the resulting four equations to get two equations in the unknowns $\mathcal{B}_1$ and $\mathcal{B}_2$. The dispersion relation is found from the determinant of the matrix formed from the coefficients of $\mathcal{B}_1$ and $\mathcal{B}_2$ in these equations.

2.24 (a) Derive the partial differential equation (2.159) for rotating shallow water waves on an f-plane.
(b) Show that the resulting dispersion relation is the same as that determined from the matrix equation (2.160).

2.25 (a) Find the velocity field associated with inertial waves in a one-layer fluid with amplitude A and wavenumber vector $\vec{k} = (k_x, k_y)$.
(b) Use your result in (a) to compute the mean vertically integrated total energy and energy flux of the waves.

(c) Show that the mean energy flux is equal to the group velocity, $\vec{c}_g$, times the mean total energy.

2.26   In December 2004 a tsunami was launched near Indonesia and travelled westwards about 5°N across the Indian Ocean to southern India. Assuming the waves were barotropic with an amplitude of 2 cm in the open ocean and zonal wavelength $\lambda_x = 200$ km, estimate the frequency and mean vertically integrated energy flux associated with the waves. What power (in joules) was delivered by the waves along 1 km of their span.

2.27   A barotropic wave in the open ocean of depth 4 km has amplitude $A = 1$ cm and zonal wavenumber $k_x = 10^{-4}$ m^{-1}. It propagates eastwards at mid-latitudes where the Coriolis frequency is $f_0 = 10^{-4}$ s^{-1}. Find the amplitude of interfacial waves at the thermocline that have the equivalent mean vertically integrated energy. For this question, assume the thermocline is situated 100 m below the surface and the density jump across the interface is 1 kg/m^3.

3

Internal waves in uniformly stratified fluid

3.1 Introduction

If a fluid's potential density decreases continuously with height then it can support internal waves that, like interfacial waves, move up and down due to buoyancy forces but which are not confined to an interface: they can move vertically through the fluid. This chapter focuses upon the dynamics of small-amplitude internal waves in uniformly stratified, stationary fluid. It also examines some effects of shear in non-uniform stratification, with a more general treatment given in Chapter 6.

The key quantity that determines the temporal evolution and spatial structure of internal waves is the buoyancy frequency. Heuristic methods based on fluid-parcel arguments are used to define the buoyancy frequency. These help to anticipate the dispersion relation that describes internal waves that are not influenced by rotation. The dispersion relation for Boussinesq and non-Boussinesq waves is then derived for liquids and gases. The results are then extended by including the influence of boundaries, shear and rotation.

3.2 The buoyancy frequency

In our consideration of interfacial waves we assumed that the density changed discontinuously across the interface between, say, warm and cold or fresh and salty fluid. In reality, such interfaces are not infinitesimally thin. They are thick, and the density varies continuously from one side of the interface to the other. We can justifiably assume the interface thickness is negligibly small only if the horizontal scale of interfacial disturbances is relatively large. If this is not the case, then the effects of continuous vertical density variations must be considered.

The key physical parameter introduced in the discussion of interfacial waves was the reduced gravity g', given in the Boussinesq approximation by

$$g' \equiv g \frac{\rho_2 - \rho_1}{\varrho_0}. \tag{3.1}$$

141

Here ρ_1 and ρ_2 are the densities respectively above and below an interface and ϱ_0 is a characteristic density, which in the Boussinesq approximation we could take to be the lower layer density or the mean of ρ_1 and ρ_2.

If the background density decreases continuously with height, then oscillations of the fluid are not determined by g', but by a quantity known as the 'buoyancy frequency'. Conventionally this is denoted by the letter N and, unlike g', has units of inverse time. The buoyancy frequency is sometimes named after Brunt and Väisälä, who are often attributed with first deriving formulae for N (though Rayleigh preceded them). Thus it is also called the 'Brunt–Väisälä frequency', or a variation thereof.

Through a variety of approaches below, we will derive and interpret formulae for the buoyancy frequency. The simplest, heuristic approach is to extend the results found for waves at the interface between two liquids of comparable density, ϱ_0. Suppose the background density increases continuously from the upper to lower fluid by a total amount $\Delta\rho$ over a depth H. The characteristic frequency of motion for waves that are not too long or too short (that is, with $k \simeq H^{-1}$) is $\sqrt{g'/H}$. In the limit of small H we may approximate $\Delta\rho/H \simeq -d\bar\rho/dz$: the density difference is replaced by the background density gradient. By convention, z is taken to be upwards and so $-d\bar\rho/dz$ is positive for stably stratified liquids.

Using (3.1), we therefore estimate the characteristic frequency to be

$$N \sim \sqrt{\frac{g'}{H}} \simeq \sqrt{-\frac{g}{\varrho_0}\frac{d\bar\rho}{dz}}, \qquad (3.2)$$

in which we have used $g' = g\,\Delta\rho/\varrho_0$. This is the formula for the buoyancy frequency of a Boussinesq liquid.

Below we will see that the buoyancy frequency can take on various forms depending upon whether the fluid is a liquid or gas and upon whether the background density decreases only moderately with height, in which case it is said to be Boussinesq, or if the density decreases by a significant fraction of itself, in which case it is said to be non-Boussinesq or (specifically for a gas) anelastic. In all cases, the buoyancy frequency is taken to be a characteristic measure of the (angular) frequency of waves in continuously stratified fluid whose dynamics are primarily influenced by buoyancy.

In both the atmosphere and ocean, the buoyancy frequency has a characteristic value of $N_0 \simeq 10^{-2}\,\mathrm{s}^{-1}$. The corresponding characteristic buoyancy period is about 10 minutes.

3.2.1 Vertical oscillations of a liquid

We now show how the buoyancy frequency may be derived by considering the motion of a fluid parcel in an otherwise motionless ambient liquid, as illustrated

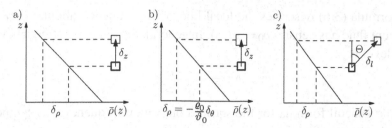

Fig. 3.1. Schematic illustrating forces acting upon a fluid parcel that is displaced vertically in a) a liquid and b) a gas. The forces acting on a diagonally displaced parcel of liquid are illustrated in c).

in Figure 3.1a. The background density decreases with height as prescribed by $\bar\rho(z)$ and we consider a fluid parcel of density $\rho_0 \equiv \bar\rho(z_0)$ situated initially at a vertical level z_0. If the parcel is displaced vertically by a small distance δ_z, our neglect of thermodynamics means that its density is unaltered.

Newton's laws predict the consequent motion:

$$\rho_0 \frac{d^2 \delta_z}{dt^2} = -\delta_\rho g. \tag{3.3}$$

Here g is the acceleration due to gravity and δ_ρ is the density difference between the fluid parcel and the surrounding fluid at its displaced position. The right-hand side represents the buoyancy force per volume and the left-hand side gives the corresponding mass per volume times acceleration.

The density difference can be written in terms of δ_z. Because $|\delta_z|$ is small:

$$\delta_\rho \simeq -\frac{d\bar\rho}{dz} \delta_z. \tag{3.4}$$

In a stably stratified fluid, $\bar\rho' < 0$. So (3.4) shows that an upward-displaced fluid parcel is heavier than its surroundings if the background density decreases with height.

Combining (3.3) and (3.4) gives the spring equation:

$$\frac{d^2 \delta_z}{dt^2} + N^2 \delta_z = 0, \tag{3.5}$$

in which

$$N^2 = -\frac{g}{\rho_0} \frac{d\bar\rho}{dz}. \tag{3.6}$$

Therefore vertically displaced fluid will oscillate at an angular frequency N, whose square is proportional to the background density gradient. The period of oscillation is $2\pi/N$.

The formula (3.6) describes the local buoyancy frequency about $z = z_0$ where the ambient fluid has density $\rho_0 = \bar{\rho}(z_0)$. In general, the buoyancy frequency at any vertical level is

$$N^2 = -\frac{g}{\bar{\rho}}\frac{d\bar{\rho}}{dz}. \tag{3.7}$$

This is the general formula for the squared buoyancy frequency of a temperature and/or salinity-stratified liquid.

More precisely, N^2 in (3.7) is given for a 'non-Boussinesq' liquid. It defines the buoyancy frequency even if the background density changes by a significant fraction of itself over the ambient fluid depth. In oceanographic applications the total density variation is small and it is sufficient to define

$$N^2 = -\frac{g}{\varrho_0}\frac{d\bar{\rho}}{dz}, \tag{3.8}$$

in which ϱ_0 is the characteristic density of the fluid. This corresponds to the result we found in deriving (3.2). Whereas (3.6) defines the local buoyancy frequency about z_0, (3.8) gives the buoyancy frequency at all depths in a Boussinesq liquid.

Equations (3.7) and (3.8) are derived assuming that density is independent of pressure. This is not the case in the oceanic abyss, below approximately 1 km from the surface. For dynamics at these depths, N^2 must involve the potential density rather than density. This uses the general equation of state for a liquid to relate density to pressure when the pressure is large. We do not derive the corresponding equation for N^2 here. However, the following derivation of the formula for N^2 in a gas is conceptually similar.

3.2.2 Vertical oscillations of an ideal gas

We now repeat the arguments above but account for the change in density of a parcel of gas as it moves through an ambient whose pressure varies significantly in the vertical. Because potential temperature, not density, is conserved during the parcel's motion, the formulae are recast in terms of $\bar{\theta}$ rather than $\bar{\rho}$.

First we need to consider how the background density depends upon the background potential temperature when these fields are in hydrostatic balance. The stability of the atmosphere is ultimately set by the rate of change with height of the background temperature, $\bar{T}(z)$. Assuming background hydrostatic balance (1.53) and using the ideal gas law (1.6), the corresponding background pressure profile is given by

$$\bar{p}(z) = P_0 \exp\left(-\int_0^z \frac{g}{R_a \bar{T}}\,dz\right). \tag{3.9}$$

Here we have defined P_0 to be the pressure at $z = 0$ and we now insist that $\bar{T}$ is given in Kelvin. The term in the integrand is often written in terms of

$$H_p(z) \equiv R_a \bar{T}(z)/g, \tag{3.10}$$

the pressure scale height.

Using (1.16), the background potential temperature is defined in terms of $\bar{T}$ and $\bar{p}$ by

$$\bar{\theta}(z) = \bar{T}\left(\frac{\bar{p}}{P_0}\right)^{-\kappa}, \tag{3.11}$$

in which $\kappa \simeq 2/7$ for the atmosphere. Taking z-derivatives on both sides of (3.11), the background potential temperature gradient is related to the background temperature gradient by

$$\frac{d\bar{\theta}}{dz} = \frac{\bar{\theta}}{\bar{T}}\left(\frac{d\bar{T}}{dz} + \Gamma\right), \tag{3.12}$$

in which

$$\Gamma \equiv g\kappa/R_a \tag{3.13}$$

is called the 'adiabatic lapse rate'. In the atmosphere $\Gamma \simeq 9.4\,\text{K/km}$. This is representative of the rate at which the temperature in the troposphere decreases with height. With $\bar{T}(z)$ prescribed, (3.12) can be solved so as to define $\bar{\theta}$ explicitly in terms of the background temperature.

The background density represented in terms of $\bar{T}$ and $\bar{p}$ is

$$\bar{\rho}(z) = \frac{\bar{p}}{R_a \bar{T}}. \tag{3.14}$$

Using hydrostatic balance, the background density gradient is given by

$$\frac{d\bar{\rho}}{dz} = -\frac{\bar{\rho}}{\bar{T}}\left(\frac{d\bar{T}}{dz} + \frac{1}{\kappa}\Gamma\right). \tag{3.15}$$

This expression can be used to determine $\bar{\rho}$ explicitly in terms of $\bar{T}$.

Equations (3.12) and (3.15) can be combined to relate the potential temperature scale height,

$$H_\theta \equiv \left(\frac{1}{\bar{\theta}}\frac{d\bar{\theta}}{dz}\right)^{-1}, \tag{3.16}$$

and the density scale height,

$$H_\varrho \equiv -\left(\frac{1}{\bar{\rho}}\frac{d\bar{\rho}}{dz}\right)^{-1}, \tag{3.17}$$

to the pressure scale height, H_p, given by (3.10). Explicitly, we find

$$\frac{1}{H_\theta} = \frac{1}{H_\varrho} - \frac{1}{\gamma H_p}, \tag{3.18}$$

in which $\gamma = 1 + R_a/C_v \simeq 7/5$. In particular, for an isothermal gas in hydrostatic balance, we found from (1.51) that $H_p = H_\varrho$. Hence the potential temperature scale height is directly related to the density scale height through $H_\theta = H_\varrho/\kappa \simeq (7/2)H_\varrho$. Typically H_θ is larger still compared with H_ϱ.

Equation (3.18) provides an explicit means through which to relate the density difference of a fluid parcel and the ambient after it has been vertically displaced a small distance δ_z. Suppose a fluid parcel situated initially at z_0 has potential temperature $\theta_0 \equiv \bar{\theta}(z_0)$ and density $\rho_0 \equiv \bar{\rho}(z_0)$. If displaced upwards by δ_z, its potential temperature does not change but its density decreases as it moves into lower pressure. Using (1.12), the density of the parcel displaced vertically by a small amount is given by

$$\rho = \rho_0 \left(\frac{\bar{p}(z_0 + \delta_z)}{\bar{p}(z_0)} \right)^{1/\gamma} \simeq \rho_0 \left(1 + \frac{1}{\gamma} \frac{\bar{p}'(z_0)}{\bar{p}(z_0)} \delta_z \right).$$

Using background hydrostatic balance, the relative density change of the lifted fluid parcel is

$$\frac{\rho - \rho_0}{\rho_0} = -\frac{1}{\gamma H_p} \delta_z.$$

The density of the parcel at $z_0 + \delta_z$ should then be compared with the ambient density at this level:

$$\bar{\rho}(z_0 + \delta_z) \simeq \rho_0 \left(1 + \bar{\rho}'(z_0)\delta_z \right) = \rho_0 \left(1 - \frac{\rho_0}{H_\varrho} \delta_z \right),$$

in which we have used the definition of the density scale height (3.17). Thus the relative density difference between the displaced parcel and the background at that height is

$$\frac{\delta_\rho}{\rho_0} = \left(\frac{1}{H_\varrho} - \frac{1}{\gamma H_p} \right) \delta_z.$$

Using (3.18), this is expressed in terms of the potential temperature scale height:

$$\frac{\delta_\rho}{\rho_0} = \frac{\delta_z}{H_\theta}. \tag{3.19}$$

This result should be compared with (1.78), which relates the fluctuation density to the fluctuation potential temperature.

Now, if a fluid parcel is vertically displaced, the difference between its potential temperature and that of its surroundings is

$$\delta_\theta \sim -\frac{d\bar{\theta}}{dz}\delta_z. \tag{3.20}$$

In a stably stratified gas, $\bar{\theta}' > 0$, so (3.20) shows that an upward-displaced fluid parcel has relatively lower potential temperature: $\delta_\theta < 0$. In other words, it is cooler and so more dense than its surroundings. Using (3.16) and (3.19), we arrive at the succinct relationship between density and potential temperature differences:

$$\frac{\delta_\rho}{\rho_0} = -\frac{\delta_\theta}{\theta_0}. \tag{3.21}$$

Putting this result into (3.3) and using (3.20), we have Newton's law written appropriately for a vertically displaced parcel of gas:

$$\rho_0 \frac{d^2\delta_z}{dt^2} = g\frac{\rho_0}{\theta_0}\delta_\theta \sim -g\frac{\rho_0}{\theta_0}\frac{d\bar{\theta}}{dz}\delta_z. \tag{3.22}$$

Again we find that a displaced fluid parcel obeys the spring equation (3.5) except that now the squared buoyancy frequency is given by

$$N^2 = \frac{g}{\theta_0}\frac{d\bar{\theta}}{dz}. \tag{3.23}$$

This is the local definition of N^2 based upon a fluid parcel with potential temperature $\theta_0 = \bar{\theta}(z_0)$. At any vertical location the squared buoyancy frequency is given by

$$N^2 = \frac{g}{\bar{\theta}}\frac{d\bar{\theta}}{dz}, \tag{3.24}$$

which defines N for an anelastic (non-Boussinesq) gas. If the background potential temperature changes by a small fraction of itself, then the squared buoyancy frequency is given in the Boussinesq approximation by

$$N^2 = \frac{g}{\vartheta_0}\frac{d\bar{\theta}}{dz}, \tag{3.25}$$

in which ϑ_0 is the characteristic value of the potential temperature.

The formulae (3.24) and (3.25) respectively differ from (3.7) and (3.8) in that N is defined in terms of the potential temperature rather than density, and a positive rather than negative sign precedes the derivatives. That is, N^2 is positive if the background potential temperature increases with height.

Using (3.24) and writing the potential temperature gradient in terms of the background temperature gradient using (3.12), we find

$$N^2 \simeq \frac{g}{\bar{T}} \left(\frac{d\bar{T}}{dz} + \Gamma \right),$$ (3.26)

in which Γ is the adiabatic lapse rate given by (3.13).

The background potential temperature is related to the background density by (1.69). Substituting this expression into (3.24), using the ideal gas law (1.6), and assuming background hydrostatic balance (1.53), we find

$$N^2 = \frac{g}{H_\varrho} - \left(\frac{g}{c_s} \right)^2,$$ (3.27)

in which the speed of sound, c_s, is given by (1.59) and the density scale height is given by (1.75). The same result can be derived by multiplying g on both sides of (3.18) and again using hydrostatic balance in the definition of the pressure scale height, H_p.

The first term on the right-hand side of (3.27) represents the buoyancy frequency as would be used for a liquid. Thus we see that the buoyancy frequency of a gas is smaller than that based upon determination by density variations alone. In particular, for an isothermal atmosphere the density scale height $H_\varrho = H_p = R_a T_0/g$ is constant. Then, the second term on the right-hand side of (3.27) evaluates to $g/(\gamma H_\varrho)$. That is,

$$N^2 = \frac{\kappa g}{H_\varrho} \simeq \frac{2}{7} \frac{g}{H_\varrho}.$$ (3.28)

3.2.3 Stable, neutral and uniform stratification

We are now in a position to give a rigorous mathematical definition of stable stratification. A fluid, whether a liquid or gas, is stably stratified if $N^2(z) > 0$ for all z. This being the case, (3.5) predicts that a displaced fluid parcel always feels a restoring force directed back to its equilibrium position and its motion is oscillatory. Conversely, where $N^2 < 0$ the stratification is unstable: displaced fluid feels a force that drives the parcel away from its initial position at an exponential rate. A fluid is neutrally stratified if $N^2 = 0$, in which case a displaced fluid parcel feels no buoyancy forces when it is vertically displaced: it is driven neither towards nor away from its initial position.

From (3.7) or (3.8), we see that in bodies of water where density is not affected by pressure (which precludes the oceanic abyss), the fluid is stably stratified if its density decreases with height. It is neutrally stratified if the density is constant.

From (3.24) or (3.25), we see that the atmosphere is stably stratified if the potential temperature increases with height or, equivalently using (3.26), if the temperature decreases with height no faster than the adiabatic lapse rate, Γ. The troposphere, being relatively weakly stratified, characteristically exhibits this decrease in temperature with height.

Unlike a uniform liquid at constant temperature, an isothermal gas is stably stratified and supports internal waves. However, from (3.28), the buoyancy frequency of an isothermal room at 20°C is approximately $0.02\,\mathrm{s}^{-1}$ (a buoyancy period of about six minutes). Therefore their motion is likely obfuscated by the more rapid motion of ventilated air. Their presence is significant on larger spatial scales where the waves can be thought of as a perturbation to an otherwise stationary or slowly evolving ambient.

A stratified fluid is said to be uniformly stratified if N^2 is constant with height. In particular, a Boussinesq liquid is uniformly stratified if its density decreases linearly with height as

$$\bar{\rho} = \varrho_0 \left(1 - \frac{z}{H}\right), \tag{3.29}$$

in which H is a constant and $|z| \ll H$ over the domain of interest. From (3.8), the corresponding squared buoyancy frequency is

$$N^2 = \frac{g}{H}. \tag{3.30}$$

A uniformly stratified non-Boussinesq liquid has exponentially decreasing density with height. Thus

$$\bar{\rho} = \varrho_0 e^{-z/H}, \tag{3.31}$$

and N^2 is again given by (3.30). Here H is the e-folding depth of the background density profile and the magnitude of z is unrestricted. However, if we restrict $z \ll H$, then a Taylor series expansion of the right-hand side of (3.31) gives $\bar{\rho} \simeq \varrho_0(1 - z/H)$, which is the corresponding Boussinesq density profile (3.29).

Similarly, in the Boussinesq limit for a uniformly stratified gas the potential temperature increases linearly with height and, for an anelastic gas, the potential temperature increases exponentially with height.

3.2.4 Diagonal oscillations of a liquid

Extending the results in Section 3.2.1, we consider the case in which a fluid parcel is displaced a small distance δ_ℓ along a line at an angle Θ to the vertical, as shown in Figure 3.1c. We further assume that the consequent parcel motion is constrained to move along this line. Although this seems an unphysical assumption, we will show in Section 3.3.1 that a fluid parcel displaced by an internal wave undergoes exactly this diagonal linear motion.

Newton's laws for the force along the line give

$$\varrho_0 \frac{d^2 \delta_\ell}{dt^2} = -\delta_\rho g \cos \Theta. \tag{3.32}$$

The $\cos \Theta$ term enters by resolving the downward gravitational force into the along-line direction. For a liquid, δ_ρ can be written in terms of the background density gradient and δ_ℓ by the relation $\delta_\rho = -\bar{\rho}' \delta_\ell \cos \Theta$. Thus (3.32) becomes

$$\frac{d^2 \delta_\ell}{dt^2} + N^2 \cos^2 \Theta \, \delta_\ell = 0. \tag{3.33}$$

Therefore fluid that moves along a line at an angle Θ to the vertical oscillates with frequency

$$\omega = N \cos \Theta. \tag{3.34}$$

As expected, if the oscillations are vertical so that $\Theta = 0$, then $\omega = N$.

It turns out that (3.34) gives the dispersion relation for internal waves in uniformly stratified non-rotating fluids. We will see that the diagonal oscillations are parallel to lines of constant phase. As illustrated in Figure 3.2, whereas the phase lines form an angle Θ to the vertical, the wavenumber vector, $\vec{k}$, is perpendicular to these lines and so forms an angle Θ to the horizontal in $\vec{k}$-space.

Hence we can write

$$\cos \Theta = |\vec{k}_h| / |\vec{k}|, \tag{3.35}$$

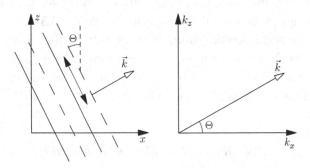

Fig. 3.2. a) Schematic illustrating motion (solid double arrows) parallel to lines of constant phase (solid and dashed lines) and the corresponding orientation of the wavenumber vector, $\vec{k}$ (open-headed arrow). Lines of constant phase form an angle Θ with the vertical and $\vec{k}$ forms an angle Θ with the horizontal. b) The wavenumber vector, $\vec{k} = (k_x, k_z)$, shown in wavenumber space.

in which $\vec{k}_h$ is the horizontal component of $\vec{k}$. In particular, the dispersion relation for internal waves in the x–z plane is

$$\omega^2 = N^2 \frac{k_x^2}{k_x^2 + k_z^2}. \tag{3.36}$$

This relation is derived rigorously in the next section.

3.3 Boussinesq internal waves

Here we consider the dynamics of small-amplitude periodic internal waves in a uniformly stratified, non-rotating fluid with zero mean flow. For mathematical convenience, we assume the motion is two-dimensional occurring in the x–z plane. For now, we invoke the Boussinesq approximation, meaning that the background density varies by a small fraction of its characteristic value. Non-Boussinesq effects are considered in Section 3.7.

3.3.1 Equations of motion in a liquid

For internal waves in oceans and lakes, the fully nonlinear equations of motion are given in terms of the velocity $\vec{u} \equiv (u, w)$, dynamic pressure p, and fluctuation density ρ by the momentum conservation equations (1.64), the internal energy equation for a liquid (1.67), and the continuity equation (1.66) for a Boussinesq, and hence incompressible, fluid. These are reproduced below for non-rotating, two-dimensional motion:

$$\varrho_0 \frac{Du}{Dt} = -\frac{\partial p}{\partial x},$$
$$\varrho_0 \frac{Dw}{Dt} = -\frac{\partial p}{\partial z} - g\rho.$$
$$\frac{D\rho}{Dt} = -w \frac{d\bar{\rho}}{dz}. \tag{3.37}$$
$$\nabla \cdot \vec{u} = 0.$$

Here ϱ_0 is a characteristic density of the fluid and the background pressure $\bar{p}$ is assumed to be in hydrostatic balance with the background density $\bar{\rho}$ through (1.53). If pressure non-negligibly affects the density, as in the oceanic abyss, then the density should be replaced by the potential density.

Examining the behaviour of small-amplitude internal waves in a fluid with no mean background flow, we can discard the advection terms in the material derivative because these nonlinear terms, which are proportional to the squared amplitude of the waves, are negligibly small compared with the linear terms. Thus we replace D/Dt with $\partial/\partial t$ in the first three equations of (3.37).

We wish to reduce the resulting set of four coupled partial differential equations in the unknowns u, w, p and ρ to a single equation in one unknown. This can be done in various ways.

In a direct approach, one can take the curl of the momentum equations and so derive the vorticity equation, which is independent of pressure:

$$\varrho_0 \frac{\partial \zeta}{\partial t} = g \frac{\partial \rho}{\partial x}. \tag{3.38}$$

Here $\zeta = \partial_z u - \partial_x w$ is the y-component of the vorticity vector. The equation shows that vorticity is created within the fluid as a result of a baroclinic torque.

The continuity equation for an incompressible fluid allows us to write $\vec{u}$ in terms of gradients of the streamfunction ψ. Explicitly, $u = -\partial_z \psi$ and $w = \partial_x \psi$. Therefore $\zeta = -\nabla^2 \psi$, in which $\nabla^2 = \partial_{xx} + \partial_{zz}$ is the Laplacian operator in the x–z plane. Thus the vorticity equation involves ψ and ρ alone.

Likewise the internal energy equation can be written in terms of ψ and ρ according to

$$\frac{\partial^2 \rho}{\partial t^2} = -\frac{\partial \psi}{\partial x} \frac{d \bar{\rho}}{dz}. \tag{3.39}$$

Eliminating ρ from (3.38) and (3.39) gives a single equation for ψ:

$$\frac{\partial^2}{\partial t^2} \nabla^2 \psi + N_0^2 \frac{\partial^2}{\partial x^2} \psi = 0. \tag{3.40}$$

Here the background density gradient has been written in terms of the squared buoyancy frequency, N_0^2, using the definition (3.8). The subscript in N_0 is included to emphasize that the buoyancy frequency is constant in a uniformly stratified fluid.

An alternate and more general approach to arriving at this differential equation is through matrix algebra. The formulae in (3.37) are represented in the matrix form by

$$\begin{pmatrix} \partial_t & 0 & 0 & \frac{1}{\varrho_0}\partial_x \\ 0 & \partial_t & \frac{g}{\varrho_0} & \frac{1}{\varrho_0}\partial_z \\ 0 & -\frac{\varrho_0}{g}N_0^2 & \partial_t & 0 \\ \partial_x & \partial_z & 0 & 0 \end{pmatrix} \begin{pmatrix} u \\ w \\ \rho \\ p \end{pmatrix} = \vec{0}, \tag{3.41}$$

in which we have again used (3.8) to define $d\bar{\rho}/dz$ in terms of N_0^2.

Because the coefficients of the derivatives of the matrix are constant (and so differentiation is commutative) we can take the determinant of the differential matrix operator to get a single partial differential equation for the motion of small-amplitude Boussinesq internal waves:

$$\left[\partial_{tt} \nabla^2 + N_0^2 \partial_{xx} \right] b = 0. \tag{3.42}$$

Here $b(\vec{x}, t)$ can be any one of the basic state fields u, w, ρ or p. More generally b can be any quantity linearly related to these basic state fields, and so can represent the streamfunction, as in (3.40), the vorticity or the displacement fields. So the structures of the fields are the same; only their amplitude and phase may differ.

The fully three-dimensional analogue of (3.42) is

$$\left[\partial_{tt}\nabla^2 + N_0{}^2\nabla_h{}^2\right]b = 0, \tag{3.43}$$

in which $\nabla_h \equiv (\partial_x, \partial_y)$ is the horizontal component of the three-dimensional gradient operator $\nabla \equiv (\partial_x, \partial_y, \partial_z)$.

3.3.2 Equations of motion in a Boussinesq gas

Although density can change during the motion of a gas, due to energy conservation we find that the potential temperature of a parcel of air does not change if its evolution is adiabatic. Thus we will develop the equations for Boussinesq internal waves in the atmosphere using the fluctuation potential temperature θ rather than ρ as a basic state variable. The Boussinesq approximation is reasonable if we are concerned with the dynamics of waves propagating over vertical distances much smaller than the density scale height, H_ϱ.

The motion of internal waves is given by the momentum and internal energy equations for an incompressible fluid, but cast in terms of the potential temperature. In particular, from (1.77), the vertical momentum equation is

$$\varrho_0\frac{Dw}{Dt} = -\frac{\partial p}{\partial z} + g\frac{\varrho_0}{\vartheta_0}\theta, \tag{3.44}$$

in which θ is the fluctuation potential temperature. This result amounts to replacing the fluctuation density with the fluctuation potential temperature using (3.21)

$$\frac{\theta}{\vartheta_0} = -\frac{\rho}{\varrho_0}. \tag{3.45}$$

The negative sign on the right-hand side reflects the fact that relatively warmer air is less dense. The characteristic potential temperature, ϑ_0 could be taken as T_0, the background temperature evaluated at the reference pressure P_0. Likewise, for the characteristic density we have $\varrho_0 = P_0/(R_a T_0)$.

The internal energy equation, $D\vartheta/Dt = 0$, describes the change of the total potential temperature, $\vartheta = \bar{\theta} + \theta$. Explicitly, the fluctuation potential temperature changes according to

$$\frac{D\theta}{Dt} = -w\frac{d\bar{\theta}}{dz}. \tag{3.46}$$

For small-amplitude waves, (3.44) and (3.46) become

$$\frac{\partial w}{\partial t} = -\frac{1}{\varrho_0}\frac{\partial p}{\partial z} + \frac{g}{\vartheta_0}\theta, \tag{3.47}$$

$$\frac{\partial \theta}{\partial t} = -w\frac{d\bar{\theta}}{dz}. \tag{3.48}$$

Combining these results with the horizontal momentum equation and the continuity equation for incompressible fluid, gives the system of coupled differential equations, written in matrix form as

$$\begin{pmatrix} \partial_t & 0 & 0 & \frac{1}{\varrho_0}\partial_x \\ 0 & \partial_t & -\frac{g}{\vartheta_0} & \frac{1}{\varrho_0}\partial_z \\ 0 & \frac{\vartheta_0}{g}N_0{}^2 & \partial_t & 0 \\ \partial_x & \partial_z & 0 & 0 \end{pmatrix}\begin{pmatrix} u \\ w \\ \theta \\ p \end{pmatrix} = \vec{0}. \tag{3.49}$$

Here $N_0{}^2$ is the squared buoyancy frequency defined in terms of the gradient of the background potential temperature for a gas (3.25).

The structure of (3.49) is similar to the corresponding equations for a liquid. Indeed, taking the determinant of the matrix operator, the resulting partial differential equation is identical to (3.42).

A clear way to show the equivalence of the equations for a Boussinesq liquid and gas is to recast them in terms of the vertical displacement field, ξ. In a uniformly stratified liquid $\rho = -\bar{\rho}'\xi = (\varrho_0 N_0{}^2/g)\xi$. Substituting this into the vertical momentum equation of (3.37) and simplifying gives

$$\frac{Dw}{Dt} = -\frac{1}{\varrho_0}\frac{\partial p}{\partial z} - N_0{}^2\xi. \tag{3.50}$$

The internal energy equation becomes

$$\frac{D\xi}{Dt} = w, \tag{3.51}$$

which is just the statement that the vertical displacement of a fluid parcel changes at the rate given by the vertical velocity.

For a gas we have $\theta = -\bar{\theta}'\xi = -(\vartheta_0 N_0{}^2/g)\xi$. Substituting this into (3.44) and (3.46) again gives (3.50) and (3.51), respectively.

So the essential motion and structure of Boussinesq internal waves are the same whether they exist in a liquid or a gas. The governing equations differ significantly only in the definition of $N_0{}^2$, which is proportional to the negative background density gradient for a liquid and the positive background potential temperature gradient for a gas.

3.3.3 Dispersion relation

For a uniformly stratified fluid, the buoyancy frequency N_0 is constant. Therefore, because all the coefficients in the differential matrix operator of (3.41) are constants and assuming the domain is unbounded, we can represent solutions by

$$b(x,z,t) = \mathcal{A}_b \exp[\iota(k_x x + k_z z - \omega t)], \tag{3.52}$$

in which b represents any of the basic state fields, u, w, p and ρ, or fields linearly related to them. For real-valued k_x, k_z and ω, substituting fields of the form (3.52) into (3.41) is equivalent to performing a Fourier transform.

The coupled differential equations in differential matrix form (3.41) are thus converted into an algebraic matrix problem:

$$\begin{pmatrix} -\iota\omega & 0 & 0 & \frac{1}{\varrho_0}\iota k_x \\ 0 & -\iota\omega & \frac{g}{\varrho_0} & \frac{1}{\varrho_0}\iota k_z \\ 0 & -\frac{\varrho_0}{g}N_0^2 & -\iota\omega & 0 \\ \iota k_x & \iota k_z & 0 & 0 \end{pmatrix} \begin{pmatrix} \mathcal{A}_u \\ \mathcal{A}_w \\ \mathcal{A}_\rho \\ \mathcal{A}_p \end{pmatrix} = \vec{0}. \tag{3.53}$$

For nontrivial solutions, the determinant of the matrix must be zero. This gives the dispersion relation for Boussinesq internal waves in the x–z plane:

$$\omega^2 = N_0^2 \frac{k_x^2}{k_x^2 + k_z^2}. \tag{3.54}$$

This is the result (3.36) found by a fluid parcel argument. It can likewise be derived by Fourier transforming (3.42) through substitution of (3.52) into that equation.

Taking the square root on both sides of (3.54) gives the explicit form of the dispersion relation for two-dimensional Boussinesq internal waves in uniformly stratified fluid. It is conceptually convenient to require ω to be positive so that the sign of the components of $\vec{k}$ determines the direction of wave propagation. Thus for rightward- and leftward-propagating waves, respectively, we have

$$\omega = N_0 k_x/|\vec{k}| \qquad \text{for } k_x > 0,$$
$$\omega = -N_0 k_x/|\vec{k}| \qquad \text{for } k_x < 0. \tag{3.55}$$

In a more convenient representation of the dispersion relation, we define

$$\Theta \equiv \tan^{-1}(k_z/k_x) \quad \text{for } -\pi/2 \leq \Theta \leq \pi/2, \tag{3.56}$$

in which case (3.55) is simply given by

$$\omega = N_0 \cos\Theta. \tag{3.57}$$

This is the result (3.34) determined using a fluid parcel argument.

In (3.56) we restrict Θ to lie between $-\pi/2$ and $\pi/2$ so that the frequency given by (3.57) is non-negative. The sign of Θ is determined not only by k_x but more generally by the sign of k_z/k_x. For rightward-propagating waves ($k_x > 0$), Θ is positive if crests move upwards ($k_z > 0$) and Θ is negative if crests move downwards ($k_z < 0$).

The angle Θ has three interrelated interpretations: it is the angle formed between the wavenumber vector and the k_x-axis, as shown in Figure 3.2b; in real space it is the angle formed between lines of constant phase and the z-axis, as shown in Figure 3.2a; and the magnitude of Θ is a measure of the wave frequency relative to N_0.

If $\vec{k}$ is known, Θ is given by (3.56) and the frequency is given by (3.57). Alternately, if ω is known, then from (3.57)

$$\Theta = \pm\cos^{-1}(\omega/N_0), \tag{3.58}$$

in which the sign is positive if k_z and k_x have the same sign and it is negative if k_z and k_x have opposite signs. Additionally knowing k_x, the vertical wavenumber is given by (3.56). The result could alternately be found by rearranging the dispersion relation (3.54) to isolate k_z:

$$k_z{}^2 = k_x{}^2 \left(\frac{N_0{}^2}{\omega^2} - 1 \right) = k_x{}^2 \tan^2 \Theta, \tag{3.59}$$

in which the last expression makes use of (3.57).

An immediate consequence of the dispersion relation (3.57) is that propagating internal waves cannot have a frequency faster than N_0. The fastest frequency waves correspond to those with $\Theta = 0$, which are waves with infinitely large vertical wavelength, their crests lying parallel to the z-axis. As the wave frequency decreases to zero, $|\Theta|$ increases to $\pi/2$ (90°). Waves with nearly zero frequency have crests that lie nearly parallel to the horizontal axis.

A peculiar feature of internal waves is that the orientation of their lines of constant phase depends upon the frequency of the waves and not their spatial scale. Internal waves moving with a fixed frequency have phase lines that lie at a fixed angle Θ to the vertical. This is illustrated dramatically in laboratory experiments that oscillate a horizontally aligned cylinder in a tank filled with uniformly salt-stratified water. The oscillation and buoyancy frequency being fixed means that the phase lines must be oriented at a fixed angle to the vertical, independent of the cylinder size or cross-sectional shape. The resulting cross-shaped wave patterns are illustrated schematically in Figures 3.3a and b. Because of its resemblance to the Scottish flag, the internal wave pattern is sometimes referred to as a 'St Andrews

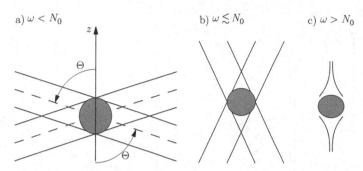

Fig. 3.3. Schematic of internal wave beams emanating along the arms of a cross with a cylinder at the centre oscillating vertically with frequency, ω, such that a) $\omega < N_0$, b) $\omega \lesssim N_0$, c) $\omega > N_0$.

Cross'. The structure of the wave beams is a superposition of plane waves having different spatial structure but identical frequencies. This is discussed in more detail in Section 5.2.

Figure 3.3a shows that Θ is measured counter-clockwise from the positive z-axis in the upper half-plane. In the lower half-plane Θ is measured counter-clockwise from the negative z-axis. Thus Θ is positive for the upper left-hand and lower right-hand beams, and Θ is negative for the other two beams.

At relatively slow oscillation frequencies ($\omega \ll N_0$), the arms of the cross lie close to the horizontal corresponding to $|\Theta| \simeq \pi/2$. The arms are more vertical if the frequency is faster, as shown in Figure 3.3b. If $\omega = N_0$ the beams emanate vertically above and below the cylinder. If the frequency of the cylinder exceeds the buoyancy frequency, then the disturbances are evanescent, as discussed in Section 3.3.7. This means that the amplitude of the waves decreases exponentially with distance from the source, as illustrated in Figure 3.3c.

3.3.4 Phase and group velocity

Applying the definition of the phase velocity (1.115) to the dispersion relation (3.54) gives

$$\vec{c}_p = \text{sign}(k_x)\, N_0 \frac{k_x}{|\vec{k}|^3}(k_x, k_z) = \frac{N_0}{k_x} \cos^2 \Theta \, (\cos \Theta, \sin \Theta). \qquad (3.60)$$

The last expression has cast the result in terms of the angle Θ which depends upon frequency by (3.58) or upon the relative vertical wavenumber by (3.56). The form of $\vec{c}_p$ in (3.60) is particularly convenient because the sign of each component of the phase velocity is given by the sign of k_x and Θ. Requiring $\omega > 0$, crests move upwards if $k_x \Theta > 0$ ($k_x k_z > 0$) and downwards if $k_x \Theta < 0$ ($k_x k_z < 0$). For any Θ they move rightwards if $k_x > 0$ and leftwards if $k_x < 0$.

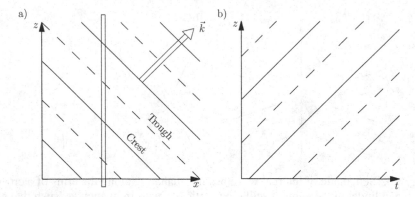

Fig. 3.4. a) Snapshot of internal waves with crests that move upwards and to the right. b) Vertical time series of the waves formed by examining the disturbances along the line indicated in a) as they evolve over time. The slope of the lines in b) is $c_{Pz} \equiv \omega/k_z$, which is different from the vertical component of $\vec{c}_p$.

The components of $\vec{c}_p$ are different from the phase speeds measured by an observer at a fixed location. For example, if a thermistor chain at a fixed horizontal position measures the passage of waves over the vertical extent of the chain, the wave crests would move vertically along the chain at speed ω/k_z. Explicitly, using (1.116) and (3.57), this speed is

$$c_{Pz} = \frac{N_0}{k_z} \cos \Theta. \tag{3.61}$$

In this formula, c_{Pz} is determined by quantities that can be measured from a vertical time series.

Figure 3.4 compares a spatial snapshot of a plane wave with a vertical time series constructed by recording the evolution of disturbances along a vertical line. Note that, while the phase lines in the spatial snapshot are oriented downwards and to the right, the corresponding phase lines slant upwards and to the right in the time series if the waves move rightwards. The slope of the lines in Figure 3.4b gives c_{Pz}. For a disturbance measured in z and t, Fourier transforming immediately gives k_z and ω. From these, $\cos \Theta$ is determined from (3.57).

Whereas crests move at the phase velocity, a wavepacket moves at the group velocity, $\vec{c}_g$. Using the general definition of the group velocity (1.120) and the dispersion relation (3.54), the group velocity is

$$\vec{c}_g = \text{sign}(k_x) N_0 \frac{k_z}{|\vec{k}|^3} (k_z, -k_x) = \frac{N_0}{k_x} \sin \Theta \cos \Theta \, (\sin \Theta, -\cos \Theta). \tag{3.62}$$

A surprising consequence of (3.60) and (3.62) is that the phase and group velocities are orthogonal to each other: $\vec{c}_p \cdot \vec{c}_g = 0$. Whereas the crests move in the direction

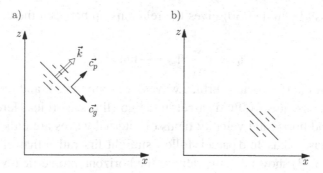

Fig. 3.5. Evolution of a wavepacket at a) early and b) late times. The phase velocity points in the same direction as the wavenumber vector. The group velocity points parallel to wave crests so that, if the crests move upwards, the wavepacket moves downwards.

of the wavenumber vector, the wavepacket as a whole moves in a direction parallel to the crests. Also surprising is that the vertical components c_{pz} and c_{gz} are opposite-signed: if the wavepacket moves upwards, crests move downwards and vice versa.

In this context, consider the internal waves generated by an oscillating cylinder as illustrated in Figures 3.3a and b. The upper right-hand arm of the cross has phase lines that move downwards while the waves themselves emanate radially outwards and upwards along the beam. The motion of a spatially localized wavepacket is illustrated schematically in Figure 3.5. Here the crests move upwards and to the right while the wavepacket as a whole moves downwards and to the right along a path aligned parallel to the crests. The rate of vertical advance of the wavepacket equals the vertical group velocity.

The vertical component of the group velocity is zero for waves with $\Theta = 0°$ and $|\Theta| = 90°$. The vertical group velocity for waves with a fixed horizontal wavenumber is largest if $|\Theta| = \tan^{-1}(1/\sqrt{2}) \simeq 35°$. The horizontal group velocity is largest if $|\Theta| = \tan^{-1}(\sqrt{2}) \simeq 55°$. For waves with a fixed total wavenumber $|\vec{k}|$, the vertical group velocity is largest if $\Theta = 45°$, and the horizontal group velocity is largest in the limit $\Theta \to 90°$.

3.3.5 Polarization relations

The eigenvector of the algebraic matrix problem (3.53) relates the amplitudes of the basic state fields to each other and so gives the polarization relations for plane internal waves in uniformly stratified Boussinesq fluid. Alternately, the polarization relations can be determined directly from the differential relationships given in (3.41). In particular, substituting (3.52) into the continuity equation for incompressible fluid (1.30) gives the relationship between the components of velocity:

$$\mathcal{A}_u = -\frac{k_z}{k_x}\mathcal{A}_w = -\tan\Theta\,\mathcal{A}_w. \tag{3.63}$$

In our study of interfacial and surface waves, we found that u and w were 90° out of phase and, consequently, fluid parcels undergo elliptical orbits. Here we find that the vertical and horizontal velocity fields of internal waves are either in phase or 180° out of phase. Thus fluid parcels follow straight-line rather than elliptical paths. Specifically, (3.63) shows that the ratio of the horizontal to vertical velocity fields (and hence displacement fields) is $-\tan\Theta$. That is, fluid parcels oscillate along a line forming an angle Θ with the vertical. This justifies the fluid parcel argument presented in Section 3.2.4, which was used to derive the dispersion relation for internal waves.

By analogy with interfacial waves, the vertical displacement field ξ can be a useful and intuitive measure of internal wave amplitudes. For small-amplitude waves, it is defined implicitly through the relation

$$w = \partial\xi/\partial t. \tag{3.64}$$

Thus

$$\mathcal{A}_\xi = \frac{\imath}{\omega}\mathcal{A}_w = -\frac{\imath}{N_0\sin\Theta}\mathcal{A}_u, \tag{3.65}$$

in which we have used (3.57) and (3.63). So, assuming $A_\xi \equiv \mathcal{A}_\xi$ is real and using (3.52), (3.56) and (3.58), the actual values of the fields are

$$\xi(x,z,t) = \Re\{A_\xi\exp[\imath(k_x x + k_z z - \omega t)]\} = A_\xi\cos(k_x x + k_z z - \omega t)$$

$$u(x,z,t) = -A_\xi\,N_0\sin\Theta\,\cos(k_x x + k_z z - \omega t)$$

$$w(x,z,t) = = A_\xi\,N_0\cos\Theta\,\sin(k_x x + k_z z - \omega t).$$

Likewise, we can relate ξ to the fluctuation density field, ρ, by analogy with (3.4):

$$\rho = -\frac{d\bar{\rho}}{dz}\xi = \frac{\varrho_0}{g}N_0{}^2 A_\xi\cos(k_x x + k_z z - \omega t). \tag{3.66}$$

Two other dynamically relevant fields are the dynamic pressure and vorticity. The former is determined from the horizontal momentum equation to give

$$\mathcal{A}_p = \frac{\omega}{k_x}\mathcal{A}_u \;\Rightarrow\; p(x,z,t) = -A_\xi\frac{N_0{}^2}{k_x}\sin\Theta\,\cos(k_x x + k_z z - \omega t). \tag{3.67}$$

The latter is determined from the definition $\zeta \equiv \partial_z u - \partial_x w$ to give

$$\mathcal{A}_\zeta = -\frac{\omega}{k_x}|\vec{k}|^2\mathcal{A}_\xi. \;\Rightarrow\; = -AN_0 k_x\sec\Theta\cos(k_x x + k_z z - \omega t). \tag{3.68}$$

In particular, the vorticity at the crest of rightward-propagating waves is negative. As shown in Figure 3.6a, this corresponds to counter-clockwise rotation in the x–z plane.

These results are summarized in Table 3.1. Figure 3.6 schematically represents the relationship between dynamically relevant fields for upward and downward-propagating waves.

3.3.6 Hydrostatic and nonhydrostatic waves

We have already insisted that the background pressure and density are in hydrostatic balance. In addition, if the dynamic pressure and fluctuation density are in hydrostatic balance so that

$$\frac{\partial p}{\partial z} = -\rho g, \tag{3.69}$$

then the fluid is in total hydrostatic balance and the waves themselves are said to be 'hydrostatic'. From the vertical momentum equation in (3.37), this means that vertical accelerations are negligibly small.

To derive the corresponding dispersion relation, we replace the vertical momentum equation in the second row of (3.41) with (3.69) and then take the determinant of the resulting matrix equations. This gives the differential equation

$$\left[\partial_{tt}\partial_{zz} + N_0^2 \partial_{xx}\right] b = 0, \tag{3.70}$$

for some basic state field, b. Fourier transforming gives

$$\omega^2 = N_0^2 \frac{k_x^2}{k_z^2} = N_0^2 \cot^2 \Theta, \tag{3.71}$$

in which Θ is given by (3.56).

Equation (3.71) is a special case of the exact dispersion relation (3.57). Taking the limit $|\Theta| \to \pi/2$, we have $\cot \Theta \simeq \cos \Theta$. From (3.57) we see that hydrostatic internal waves have low frequency compared to N_0. Also, from (3.56), we see that $|k_x| \ll |k_z|$, so the horizontal wavelength is long compared with the vertical wavelength and lines of constant phase are approximately horizontal.

Conversely, if the wave frequency is close to the buoyancy frequency and, equivalently, if the slope of phase lines is significantly different from zero, then the vertical pressure gradient and buoyancy forces are not in balance and the waves are said to be 'nonhydrostatic'.

Table 3.1. *Dispersion relation, phase and group velocities, polarization relations and correlations between fields for small-amplitude internal waves in the x–z plane in stationary, uniformly stratified Boussinesq fluids with no background rotation. As well as the basic state fields, the table lists values of the streamfunction (ψ), vorticity (ζ), the change in the squared buoyancy frequency due to stretching of isopycnals by waves (ΔN^2) and its time derivative ($N^2{}_t$). Each field b is characterized by the magnitude and phase of its complex amplitude $\mathcal{A}_b$. Results are given in terms of N_0, k_x, the vertical displacement amplitude A_ξ, which is assumed to be real and positive, and the angle $\Theta \equiv \tan^{-1}(k_z/k_x)$, which is defined so that $-90° \le \Theta \le 90°$. In the rightmost column, the real parts of the fields are given in terms of the phase $\varphi = k_x x + k_z z - \omega t$.*

Dispersion relation:
$$\omega = N_0 k_x / |\vec{k}| = N_0 \cos\Theta$$

Phase velocity:
$$c_{px} = \frac{\omega}{|\vec{k}|}\frac{k_x}{|\vec{k}|} = \frac{N_0}{k_x}\cos^3\Theta, \qquad c_{pz} = \frac{\omega}{|\vec{k}|}\frac{k_z}{|\vec{k}|} = \frac{N_0}{k_x}\cos^2\Theta\sin\Theta$$

$$c_{Px} = \omega/k_x = \frac{N_0}{k_x}\cos\Theta, \qquad c_{Pz} = \omega/k_z = \frac{N_0}{k_x}\cos\Theta\cot\Theta$$

Group velocity:
$$c_{gx} = \frac{\partial\omega}{\partial k_x} = \frac{N_0}{k_x}\cos\Theta\sin^2\Theta, \qquad c_{gz} = \frac{\partial\omega}{\partial k_z} = -\frac{N_0}{k_x}\cos^2\Theta\sin\Theta$$

Defining formula	Relationship to vertical displacement	
ξ	A_ξ	$\xi = A_\xi\cos\varphi$
$\rho = -\dfrac{d\bar\rho}{dz}\xi$	$\mathcal{A}_\rho = -\dfrac{d\bar\rho}{dz}A_\xi$	$\rho = \rho_0\dfrac{1}{H_\rho}A_\xi\cos\varphi$
$\vec{u} = \nabla\times(\psi\hat{y})$	$\mathcal{A}_\psi = -\dfrac{\omega}{k_x}A_\xi$	$\psi = -\dfrac{N_0}{k_x}\cos\Theta\,A_\xi\cos\varphi$
$w = \dfrac{\partial\xi}{\partial t}$	$\mathcal{A}_w = -\imath\omega A_\xi$	$w = N_0\cos\Theta\,A_\xi\sin\varphi$
$u = -\dfrac{\partial\psi}{\partial z}$	$\mathcal{A}_u = \imath\dfrac{k_z\omega}{k_x}A_\xi$	$u = -N_0\sin\Theta\,A_\xi\sin\varphi$
$\zeta = -\nabla^2\psi$	$\mathcal{A}_\zeta = -\dfrac{\omega}{k_x}|\vec{k}|^2 A_\xi$	$\zeta = -N_0 k_x\sec\Theta\,A_\xi\cos\varphi$
$\rho_0\dfrac{\partial u}{\partial t} = -\dfrac{\partial p}{\partial x}$	$\mathcal{A}_p = \imath\rho_0\dfrac{k_z\omega^2}{k_x{}^2}A_\xi$	$p = -\rho_0\dfrac{N_0{}^2}{k_x}\sin\Theta\cos\Theta\,A_\xi\sin\varphi$
$\Delta N^2 = -\dfrac{g}{\rho_0}\dfrac{\partial\rho}{\partial z}$	$\mathcal{A}_{\Delta N^2} = -\imath k_z N_0{}^2 A_\xi$	$\Delta N^2 = N_0{}^2 k_x\tan\Theta\,A_\xi\sin\varphi$
$N^2{}_t = \dfrac{\partial\Delta N^2}{\partial t}$	$\mathcal{A}_{N^2_t} = -k_z\omega N_0{}^2 A_\xi$	$N^2{}_t = N_0{}^3 k_x\sin\Theta\,A_\xi\cos\varphi$

Correlations

$$\langle u\rangle_L = -\langle\xi\zeta\rangle = \frac{1}{2}N_0 k_x\sec\Theta\,A_\xi{}^2$$

$$\langle\mathcal{F}_{Mz}\rangle = \varrho_0\langle uw\rangle = -\frac{1}{4}\varrho_0 N_0{}^2\sin 2\Theta\,A_\xi{}^2 = \varrho_0 c_{gz}\langle u\rangle_L$$

$$\langle E\rangle = \langle E_K + E_P\rangle = \frac{1}{2}\varrho_0 N_0{}^2 A_\xi{}^2$$

$$\langle\mathcal{F}_{Ez}\rangle = \langle pw\rangle = -\frac{1}{2}\varrho_0\frac{N_0{}^3}{k_x}\sin\Theta\cos^2\Theta\,A_\xi{}^2 = c_{gz}\langle E\rangle$$

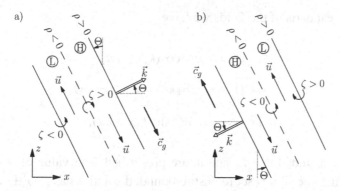

Fig. 3.6. Schematic showing polarization relations for internal waves with a) up and rightward-propagating and b) down and leftward-propagating crests. Solid lines correspond to crests, where the vertical displacement is largest, and dashed lines correspond to troughs, where the vertical displacement is smallest. Extremes of high and low pressure are shown by the circled H and L symbols, respectively.

3.3.7 Evanescent disturbances

In the derivation of (3.41) and in the solution of these equations, we made no assumptions about the value of ω relative to N_0. And so the structure of fast-frequency disturbances in theory can be represented by (3.52), in which the frequency is related to the wavenumber vector by (3.54).

In practice, this is not realistic because, in order to satisfy the dispersion relation, the wavenumber components cannot both be real-valued. This is evident from the expression for k_z in (3.59). If $\omega > N_0$, the term in parentheses of the middle expression is negative and so k_z is imaginary for real k_x.

From (3.52) this means that the waves exhibit oscillatory structure in x but exponential structure in z. Thus, it is more realistic to represent such disturbances not by (3.52) but by

$$b(x,z,t) = A_b e^{\imath(k_x x - \omega t)} e^{\gamma z}, \qquad (3.72)$$

in which

$$\gamma = \pm k_x \sqrt{1 - \frac{N_0{}^2}{\omega^2}}, \text{ for } \omega > N_0. \qquad (3.73)$$

The sign of γ in (3.73) depends upon the boundary conditions so that, for example, the waves decrease exponentially above and below a rapidly oscillating cylinder as illustrated in Figure 3.3c.

In (3.73) we have effectively replaced k_z, the vertical wavenumber for propagating waves, with $\omega < N_0$, by $\gamma = \imath k_z$, the inverse e-folding distance describing the exponential decay in the vertical of waves with $\omega > N_0$. Likewise, we can perform this substitution in the corresponding polarization relations. In particular, from (3.63) and (3.65) we have $\mathcal{A}_u = -\imath \frac{\gamma}{k_x} \mathcal{A}_w = -\frac{\gamma}{k_x} A_\xi$, in which A_ξ is assumed real. Taking real parts of the fields we have

$$\xi(\vec{x},t) = A_\xi e^{\gamma z} \cos(k_x x - \omega t)$$

$$u(\vec{x},t) = -\frac{\gamma}{k_x} A_\xi e^{\gamma z} \cos(k_x x - \omega t) \qquad (3.74)$$

$$w(\vec{x},t) = \omega A_\xi e^{\gamma z} \sin(k_x x - \omega t),$$

in which it is assumed that k_x and ω are prescribed. The value of γ is given by (3.73), in which the sign is set to ensure bounded solutions as $z \to \pm\infty$.

So, as with the case of interfacial waves, the components of the velocity field are now 90° out of phase and the fluid follows elliptical rather than straight-line paths. In the limit $N_0 \to 0$, $\gamma \to \pm k_x$, so (3.72) is consistent with the structure for interfacial waves given by (2.81) and (2.82), in which the velocity potential is a product of an exponential function of z with periodic functions of x and t. Indeed, if the fluid surrounding an interface is weakly stratified, we expect the motion nonetheless to behave approximately as if it were irrotational.

Disturbances with exponentially decreasing amplitude in stratified fluids are said to be 'evanescent', meaning vanishing. This is the same terminology used in electromagnetic theory to describe light waves. When light is incident upon a lens at a glancing angle, its amplitude decreases exponentially towards the lens interior. Sometimes evanescent disturbances are called 'evanescent waves'. But this ambiguous terminology will be avoided here.

3.4 Transport by Boussinesq internal waves

As internal waves propagate vertically through a continuously stratified fluid they carry momentum with them. Momentum is conserved and so where the waves are generated they exert drag and accelerate the fluid where they break. Likewise, the waves transport energy, extracting it where they are created and depositing it where they dissipate. As such, the waves act as a conduit through which the middle and upper atmosphere can be influenced by processes near the ground and through which the deep ocean can respond to the surface mixed region.

In the atmosphere, the transport of momentum by internal waves is most significant dynamically. This is because the large-scale flows of the atmosphere are primarily zonal, travelling predominantly eastwards at mid-latitudes and westwards near the equator in the troposphere. Thus waves generated by unidirectional flow over topography, for example, carry momentum upwards and accelerate the flow in a direction opposing the ambient wind where they break (see Section 5.4). Energy transport is less significant because, although energy deposition results in turbulence, the absorption of solar radiation in the atmosphere acts efficiently to restratify the mixed region.

Conversely, the transport of energy by internal waves in the ocean is usually the most significant dynamically. Internal waves generated by oscillatory tidal flows over topography (see Section 5.5) excite waves travelling in opposing directions so that the net momentum transport is zero. Even beneath strong unidirectional currents, the flow has embedded within it eddies that act to redirect momentum. On the other hand, where the waves break the energy they transport acts to mix the fluid. Because sunlight does not penetrate significantly below about 100 m of the surface, solar radiation does not act to restratify this mixed fluid. Therefore energy transport by waves and consequent mixing is crucial for understanding the salinity and temperature structure of the oceans.

Here we consider the transport of momentum and energy by small-amplitude, two-dimensional plane waves in uniformly stratified, stationary fluid with no background rotation. For now we will suppose the waves are Boussinesq, though this will be relaxed in Section 3.7.

3.4.1 Zero mass/internal energy transport

For an incompressible fluid, the fully nonlinear flux-form of the internal energy equation is given by (1.67). In the x–z plane, this becomes

$$\frac{\partial \rho}{\partial t} = -\frac{\partial u\rho}{\partial x} - \frac{\partial w\rho}{\partial z} - w\frac{d\bar{\rho}}{dz}, \qquad (3.75)$$

in which the total density ϱ has explicitly been written in terms of the background and fluctuation components: $\bar{\rho}(z) + \rho(\vec{x}, t)$.

Averaging (3.75) over one horizontal wavelength gives the mass conservation equation

$$\frac{\partial \langle \rho \rangle}{\partial t} = -\frac{\partial \langle w\rho \rangle}{\partial z}. \qquad (3.76)$$

The averaging in x has eliminated the first and last terms on the right-hand side of (3.75). The left-hand side of (3.76) symbolically represents the mean change in density at a point in space. The equation thereby states that, on average, the density changes in time due to the divergence of the vertical mass flux $\langle w\rho \rangle$.

The polarization relations for small-amplitude, Boussinesq internal waves show that the vertical velocity and perturbation density are 90° out of phase, as illustrated in Figure 3.6. In Section 1.15.8 we found that the correlation is zero between two periodic disturbances that are 90° out of phase. And so we have

$$\langle w\rho \rangle = 0. \tag{3.77}$$

Therefore the average mass flux due to internal waves is identically zero. The waves may periodically change the density at a point, but on average the density remains the same. Even if the waves break, there is no net density change, only a redistribution of the background density due to mixing. Here the statement of zero mass transport is equivalent to the statement that internal waves do not transport heat or salinity.

3.4.2 Mechanical energy transport

The flux-form of the kinetic energy equation for an incompressible fluid is derived by taking the dot-product of $\vec{u}$ with the vector form of the momentum equations (1.64) and then bringing each component of $\vec{u}$ inside the derivatives using the incompressibility relation (1.30).

If there is no background horizontal flow, for a Boussinesq fluid we find

$$\frac{\partial E_K}{\partial t} = -\nabla \cdot [(E_K + p)\vec{u}] - gw\rho, \tag{3.78}$$

in which

$$E_K = \frac{1}{2}\varrho_0 |\vec{u}|^2 \tag{3.79}$$

is the kinetic energy density, with units of energy per unit volume.

From (1.94), the total potential energy is classically defined as the volume integral of ϱgz. But this is not a particularly useful quantity for studying internal waves. Under this definition, potential energy is converted entirely into kinetic energy only if the fluid as a whole is brought down to a reference level, $z = 0$ say. Clearly incompressibility does not allow this to occur. It is the potential energy difference between the perturbed and unperturbed states that is dynamically relevant for internal waves. This is the available potential energy.

Although the formula for the available potential energy density can be derived by computing the difference between the potential energy of the disturbed and undisturbed states, here we will derive it heuristically by a fluid parcel argument. This is done for a liquid, but a similar argument can be used for a gas. Suppose the fluid has approximately constant buoyancy frequency about some arbitrary level, which we denote by $z = 0$. The background density near $z = 0$ can be represented approximately by $\bar{\rho} \simeq \varrho_0(1 - z/H)$, in which H is the local value of the density scale height. We wish to exchange two parcels located at $z = \pm\delta$. This costs potential energy given by the difference between the potential energy after and before the exchange. That is, the energy required is

$$\Delta E = g\varrho_0[(1 - \delta/H)(-\delta) + (1 + \delta/H)(\delta)]$$
$$- g\varrho_0[(1 + \delta/H)(-\delta) + (1 - \delta/H)(\delta)] \quad (3.80)$$
$$= 4\varrho_0\delta^2 g/H.$$

The available potential energy density associated with one of the two fluid parcels is half this amount, $\Delta E/2$.

An explicit definition of the available potential energy density is derived from (3.80) by replacing g/H with N_0^2 and setting the total displacement 2δ to be ξ. Thus $\Delta E/2$ becomes

$$E_P = \frac{1}{2}\varrho_0 N_0^2 \xi^2. \quad (3.81)$$

This form is analogous to the depth-integrated available potential energy (2.36) associated with surface waves. Equation (3.81) shows that the available potential energy is larger if the waves vertically displace fluid to a greater extent against the background stratification, or if the background stratification itself is stronger.

Sometimes it is more useful to represent E_P in terms of the fluctuation density, ρ, rather than ξ. Using (3.66) to relate ξ to ρ for small-amplitude disturbances we have

$$E_P = \frac{1}{2}\frac{1}{\varrho_0}\frac{g^2}{N_0^2}\rho^2. \quad (3.82)$$

Despite possible computational advantages, this form of the available potential energy density is not all that intuitive. The presence of N_0^2 in the denominator of (3.82) seems to imply that E_P becomes large as N_0^2 becomes small. However, for a wave with a given vertical displacement amplitude, A_ξ, the fluctuation density is proportional to N_0^2 and so E_P in fact decreases as N_0^2 decreases, a fact made obvious by (3.81).

The evolution of available potential energy is given by multiplying $g\rho/(-d\bar\rho/dz)$ on both sides of the internal energy equation (1.67) and converting the result to flux-form. In uniformly stratified fluid, the background density gradient is constant and may be brought inside the partial z-derivative terms. Using (3.82), we thereby find

$$\frac{\partial E_P}{\partial t} = -\nabla \cdot (\vec{u}E_P) + gw\rho. \quad (3.83)$$

Comparing (3.78) and (3.83), we realize that $gw\rho$ represents the conversion from kinetic to available potential energy when vertical motions carry dense fluid upwards or light fluid downwards against buoyancy forces. Conversely, kinetic energy increases at the expense of available potential energy when relatively heavy fluid descends or light fluid rises. As such, the term $gw\rho$ is sometimes called the 'buoyancy flux'.

Adding together the kinetic and available potential energy equations gives the total energy conservation law

$$\frac{\partial E}{\partial t} = -\nabla \cdot (\vec{u}(E+p)), \tag{3.84}$$

in which

$$E \equiv E_K + E_P. \tag{3.85}$$

Equation (3.84) shows that energy in an incompressible fluid changes solely due to the divergence of the energy flux, $\vec{u}(E+p)$.

Now suppose that perturbations are due to small-amplitude periodic internal waves. Horizontally averaging over one wavelength and using the polarization relations in Table 3.1, the mean kinetic energy represented in terms of the vertical displacement amplitude is

$$\langle E_K \rangle = \frac{1}{4}\varrho_0 N_0{}^2 A_\xi{}^2. \tag{3.86}$$

Likewise, from the formula for the available potential energy density given by (3.81), the mean available potential energy is identical to the mean kinetic energy. So the total mean energy is

$$\langle E \rangle = 2\langle E_K \rangle = 2\langle E_P \rangle = \frac{1}{2}\varrho_0 N_0{}^2 A_\xi{}^2. \tag{3.87}$$

This equipartitioning of kinetic and available potential energy, which was also seen for interfacial waves, is a general result for waves in a non-inertial (non-rotating) reference frame.

Horizontally averaging (3.78) and (3.83) and keeping only the leading-order terms, which are quadratic in amplitude, gives the conservation laws

$$\frac{\partial}{\partial t}\langle E_K \rangle = -\frac{\partial}{\partial z}\langle wp \rangle$$
$$\frac{\partial}{\partial t}\langle E_P \rangle = 0. \tag{3.88}$$

In formulating these equations the buoyancy flux term has vanished as a result of (3.77).

Horizontally averaging the total energy equation (3.84), or equivalently summing (3.88), gives the energy conservation law

$$\frac{\partial \langle E \rangle}{\partial t} = -\frac{\partial \langle \mathcal{F}_E \rangle}{\partial z}, \tag{3.89}$$

in which the vertical energy flux is given by

$$\langle \mathcal{F}_E \rangle \equiv \langle wp \rangle = \frac{1}{2} \varrho_0 \frac{N_0^3}{k_x} \sin \Theta \cos^2 \Theta \, A_\xi^2. \qquad (3.90)$$

Comparing this with the vertical component of the group velocity in (3.62) and with (3.87), we arrive at the expected result

$$\langle \mathcal{F}_E \rangle = c_{gz} \langle E \rangle. \qquad (3.91)$$

Horizontally averaged energy is vertically transported at the vertical group speed of the waves. In particular, for waves with fixed amplitude and horizontal extent, the vertical energy transport is greatest for waves with frequency $\omega = \sqrt{2/3} N_0 \simeq 0.82 N_0$ ($\Theta \simeq 35^o$), corresponding to waves with the maximum vertical group velocity.

Some care must be taken in the interpretation of (3.89). For vertically and horizontally periodic internal waves, (3.90) shows that the energy flux is independent of z. Hence (3.89) predicts that the energy is unchanging with time. This does not imply that the mean energy is zero, only that as much energy is transported into a vertical slab from below as is transported out of it from above.

A clearer illustration of what determines $\langle E \rangle$ is shown in Figure 3.7a. Here the internal waves are manifest as a quasi-monochromatic wavepacket whose amplitude envelope moves upwards at speed c_{gz}. The energy flux is given approximately

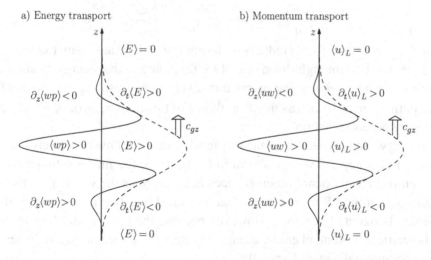

a) Energy transport   b) Momentum transport

Fig. 3.7. a) Energy transport associated with an upward-propagating, vertically localized wavepacket and b) the corresponding transport of momentum per unit mass. The latter results in a wave-induced mean flow $\langle u \rangle_L$ that moves upwards with the wavepacket.

by (3.90), where now it is understood that $A_\xi \equiv A_\xi(z)$ is the amplitude envelope of the vertical displacement field. At the leading edge of the wavepacket, the amplitude envelope decreases with height meaning that the divergence of the flux is negative. Hence, according to (3.89), the mean energy at a fixed vertical position increases as the leading edge moves past it. Similarly the mean energy decreases at the trailing edge where the flux divergence is positive. The energy change due to a passing wavepacket that does not break is reversible in the sense that the change in energy after the wavepacket has passed is zero. When wave breaking occurs, the energy due to the wave's flux goes to zero and (3.89) predicts the resulting irreversible deposition of energy to the ambient.

3.4.3 Wave action and pseudoenergy

In stationary fluid, energy as defined by (3.87) is conserved in the sense that its time rate of change at a point is given in terms of the divergence of the energy flux. But energy, so defined, is not conserved for waves that propagate in a background horizontal flow whose speed changes with height. To see this, we represent the total horizontal velocity field by the sum of the background flow $\bar{U}(z)$ and the fluctuation horizontal velocity due to waves $u(\vec{x}, t)$. The energy equation is found by substituting $\bar{U} + u$ into the momentum equations (as done explicitly in Section 3.6.1), taking the dot-product of these equations with $\vec{u} = (u, w)$, horizontally averaging, and keeping only the leading-order quadratic terms. Further assuming the average buoyancy flux is zero gives

$$\frac{\partial \langle E \rangle}{\partial t} = -\varrho_0 \langle uw \rangle \frac{d\bar{U}}{dz} - \frac{\partial \langle wp \rangle}{\partial z}. \tag{3.92}$$

If $\bar{U} = U_0$ is constant, (3.92) reduces to the energy conservation law (3.89).

The first term on the right-hand side of (3.92) indicates that energy changes in a shear flow not only because the energy flux diverges, but also because momentum is transported vertically across the shear flow. For this reason, $\varrho_0 \langle uw \rangle \frac{d\bar{U}}{dz}$ is called the 'energy production term'.

The energy changes because of the way in which a shear flow effectively changes the wave frequency. As will be shown in Section 6.3, the intrinsic frequency, ω, (that seen from a stationary observer) does not change as the waves pass through a steady background flow. Likewise, that section shows that k_x is constant if the ambient is horizontally uniform. However, because the background flow changes in z the vertical wavenumber can change and does so according to the dispersion relation for internal waves. Explicitly,

$$k_z = \pm k_x \sqrt{\frac{N_0^2}{(\omega - \bar{U} k_x)^2} - 1} = \pm k_x \sqrt{\frac{N_0^2}{\Omega^2} - 1}, \tag{3.93}$$

in which the sign of the square root depends upon the direction of wave propagation. Here we have introduced the extrinsic frequency, $\Omega = \omega - \bar{U}k_x$, which does change with height. By analogy to the apparent change in the frequency of sound that originates from a moving object, Ω is sometimes called the Doppler-shifted frequency.

We have already seen that the group velocity depends upon the wavenumber and frequency of the waves. So as the background flow Doppler-shifts the waves, the speed at which they vertically transport energy changes. If the resulting vertical group velocity of upward-propagating waves decreases with height, the energy piles up, as predicted by the energy production term.

We would like to define a conserved quantity analogous to energy for internal waves in a shear flow. It should be conserved in the sense that it remains constant for waves propagating through shear and it changes only when the waves are generated or dissipated. This quantity is called the 'wave action'.

For small-amplitude waves, it is given by

$$\langle \mathcal{A} \rangle = \frac{\langle E \rangle}{\Omega} = \frac{\langle E \rangle}{\omega - k_x \bar{U}},$$ (3.94)

the ratio of the mean energy and extrinsic frequency.

Wave action satisfies the conservation law

$$\frac{\partial \langle \mathcal{A} \rangle}{\partial t} = -\frac{\partial \langle \mathcal{F}_A \rangle}{\partial z},$$ (3.95)

in which the flux of wave action is given by the product of the vertical group velocity and the wave action:

$$\langle \mathcal{F}_A \rangle = c_{gz} \langle \mathcal{A} \rangle = \frac{\langle \mathcal{F}_E \rangle}{\omega - k_x \bar{U}}.$$ (3.96)

Traditionally, wave action has been used as a proxy for energy conservation for waves in the presence of shear. A closely related conserved quantity has also been proposed in its stead. This is called the 'pseudoenergy', which is the appropriate conserved quantity for waves in a time-invariant system.

For small-amplitude waves, the pseudoenergy is simply the product of the wave action and the intrinsic frequency:

$$\langle \mathcal{E} \rangle = \omega \langle \mathcal{A} \rangle = \langle E \rangle \frac{\omega}{\Omega}.$$ (3.97)

Because wave action is conserved and the intrinsic frequency ω does not change with time, the product must also be conserved. Unlike the definition for wave action, pseudoenergy has the same units as energy and indeed equals the wave energy in the absence of any background flow. As such, pseudoenergy is more intuitive as a conserved quantity than wave action.

Multiplying both sides of (3.95) by ω gives the small-amplitude pseudoenergy conservation law

$$\frac{\partial \langle \mathcal{E} \rangle}{\partial t} = -\frac{\partial \langle \mathcal{F_E} \rangle}{\partial z}, \tag{3.98}$$

in which

$$\langle \mathcal{F_E} \rangle = c_{gz} \langle \mathcal{E} \rangle. \tag{3.99}$$

Pseudoenergy and its flux can be written generally for large-amplitude waves. However, the addition of terms of the order of the amplitude-cubed and higher do not significantly change $\langle \mathcal{E} \rangle$, even for waves close to breaking amplitude. We therefore expect (3.97) and (3.99) sufficiently characterize energy transport in shear for most practical applications.

3.4.4 Momentum transport

The flux-form of the horizontal momentum equation restricted to the x–z plane is given for an incompressible fluid by

$$\varrho \left(\frac{\partial u}{\partial t} + \frac{\partial uu}{\partial x} + \frac{\partial uw}{\partial z} \right) = -\frac{\partial p}{\partial x}, \tag{3.100}$$

in which $\varrho = \bar{\rho} + \rho$ is the total density, and we have assumed the fluid is incompressible.

In the Boussinesq approximation, we assume that ϱ can be treated as a constant, ϱ_0, on the left-hand side of (3.100). Horizontally averaging eliminates the x-derivatives and so gives the momentum conservation law

$$\frac{\partial \langle M \rangle}{\partial t} = -\frac{\partial \langle \mathcal{F}_M \rangle}{\partial z}, \tag{3.101}$$

in which $M \equiv \varrho_0 u$ and

$$\mathcal{F}_M \equiv \varrho_0 \langle uw \rangle. \tag{3.102}$$

The momentum flux $\mathcal{F}_M$, sometimes denoted by the symbol τ, is one component of the Reynolds stress tensor.

Dividing by ϱ_0 on both sides of (3.101) gives

$$\frac{\partial \langle u \rangle}{\partial t} = -\frac{\partial \langle uw \rangle}{\partial z}. \tag{3.103}$$

This states that the horizontally averaged flow accelerates if the mean vertical transport (w) of horizontal momentum per unit mass (u) diverges.

Because the horizontal and vertical velocity fields of propagating waves are directly correlated, they do indeed transport momentum. Explicitly, in terms of the amplitude of the vertical displacement field for Boussinesq waves, the momentum flux per unit mass is

$$\langle uw \rangle = -\frac{1}{2}\text{sign}(k_x k_z)\, \omega^2 \sqrt{\frac{N_0^2}{\omega^2} - 1}\, |A_\xi|^2 = -\frac{1}{4}N_0^2 \sin 2\Theta\, |A_\xi|^2. \qquad (3.104)$$

The last expression has used the definition (3.56) for Θ, the sign of which equals the sign of $k_z k_x$. The negative sign in both expressions in (3.104) indicates that the momentum transport is upwards, for example, if k_x is positive (the wave crests move from left to right) and k_z is negative (the crests move downwards and the vertical component of the group velocity is upwards). A surprising result of (3.104) is that for waves of fixed amplitude, their momentum transport is largest if $|\Theta| = 45°$ ($\omega = N_0/\sqrt{2}$), independent of their wavenumber.

Our choice of writing the momentum flux in terms of the vertical displacement amplitude is somewhat arbitrary. Expressions similar to those in (3.104) can be written in terms of the amplitude of the vertical velocity or any other field for that matter. One reason the expressions are written here in terms of the vertical displacement is because A_ξ is related directly to hill height if the waves are generated by flow over topography.

Equation (3.104) is appropriate only for propagating internal waves. If the disturbance fields are evanescent as a result of being forced at a frequency larger than the buoyancy frequency, then their associated momentum flux is zero. Mathematically this follows from the fact that k_z is imaginary if $\omega > N_0$. Hence, as shown in (3.74), the horizontal velocity field is 90° out of phase with the vertical velocity field.

For vertically propagating internal waves, the usual interpretation of (3.103) is that the flux $\langle uw \rangle$ diverges when internal waves break and this acts irreversibly to accelerate the background flow. However, as shown in Figure 3.7b, an alternative interpretation is that a wavepacket accelerates the background at its leading edge and decelerates it at its trailing edge, so that there is no change in the flow speed before and after the passage of the waves. The horizontally averaged, reversible flow associated with a wavepacket is referred to as the wave-induced mean flow.

3.4.5 Wave-induced mean flow

In our study of deep water surface waves, we found that the mean horizontal displacement of a fluid parcel due to the passage of waves was zero at leading order. But accounting for the variation in velocity with depth at next order, the mean horizontal parcel displacement was found to move in the same direction as the phase velocity of the waves. This time-averaged mean flow, known as the Stokes drift, is the appropriate measure of momentum per unit mass associated with a deep surface wave.

Likewise, there is a wave-induced mean flow associated with internal waves. An explicit formula for this in terms of the properties of the waves can be derived using Stokes' theorem, which connects the vorticity in a domain to the circulation surrounding it. Generally, Stokes' theorem is written $\int\int_S \nabla \times \vec{F} \cdot \hat{n}\, dS = \oint_C \vec{F} \cdot d\vec{r}$, in which S is an open surface and C is a closed curve circumscribing the opening in the surface. Taking the vector function $\vec{F}$ to be the velocity and collapsing the surface to the x–z plane we have

$$\int\int_{\mathcal{D}} (\nabla \times \vec{u}) \cdot \hat{y}\, dA = \oint_C \vec{u} \cdot d\vec{r}, \qquad (3.105)$$

in which we take $\mathcal{D}$ to be the area between upward-displaced fluid and its equilibrium position in the first half-period of the wave's oscillation. C is the counter-clockwise closed curve oriented along the equilibrium position for the first half-period and then tracing back along the displaced isopycnal, as illustrated in Figure 3.8.

Assuming that the vertical displacement is small so that the vorticity is independent of z, the area integral on the left-hand side of (3.105) becomes $-\int_0^{\lambda_x/2} \xi \zeta\, dx$. In a right-handed co-ordinate system, the x-axis points leftwards when looking at the x–z plane down the y-axis. So we have introduced a negative sign in front of the integral as a result of reversing the direction of x-integration.

On the right-hand side of (3.105), the circulation over one quarter-period along the isopycnal cancels that over the other quarter-period. The only part of the line integral that remains is the flow in the x-direction along the equilibrium position. Defining the average flow to be $\langle u \rangle_L$, the circulation integral is simply $\langle u \rangle_L \lambda_x/2$.

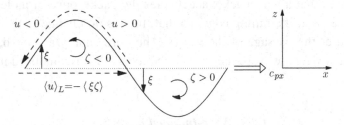

Fig. 3.8. Schematic illustrating the derivation of the wave-induced mean flow through the correlation of the vertical displacement and vorticity fields: $\langle u \rangle_L = -\langle \xi \zeta \rangle$. The dashed curves indicate the closed path C enclosing an area $\mathcal{D}$ of the fluid between the upward-displaced fluid parcels and the equilibrium position (dotted line). Stokes' theorem relates the circulation around the path to the area integral of the vorticity ζ, which is negative below the crest for rightward-moving waves.

Comparing the right- and left-hand integrals we therefore have

$$\langle u \rangle_L = -\frac{1}{\lambda_x/2} \int_0^{\lambda_x/2} \xi\zeta \, dx = -\langle \xi\zeta \rangle. \qquad (3.106)$$

Repeating the procedure over the second half-period gives the same result. The sign of $\langle u \rangle_L$ does not change because the vorticity changes sign as the displacement changes sign.

Using the polarization relations for Boussinesq internal waves of vertical displacement amplitude A_ξ, (3.106) becomes

$$\langle u \rangle_L = \frac{1}{2}\frac{N_0{}^2}{\omega} k_x |A_\xi|^2 = \frac{1}{2}N_0 k_x \sec\Theta |A_\xi|^2, \qquad (3.107)$$

in which Θ is given by (3.56). This is the Stokes drift for internal waves.

In (3.106) the subscript L has been introduced as a reminder that this is the horizontally averaged Lagrangian motion of fluid parcels due to the passage of waves. From a Lagrangian perspective, fluid parcels do not simply zig-zag back and forth at an angle Θ to the vertical, but over each cycle they drift in the same direction as the horizontal phase speed of the waves. Thus we should interpret $\langle u \rangle$ in (3.109), not as the average of a sinusoidal motion in an Eulerian frame, but as the Lagrangian mean motion, $\langle u \rangle_L$.

In the language of Hamiltonian fluid dynamics, $\mathcal{M} \equiv \varrho_0 \langle u \rangle_L$ is called the 'pseudo-momentum', which is interpreted as the appropriate conserved quantity for waves in a horizontally invariant system. Hamiltonian fluid dynamics predicts that the vertical flux of horizontal pseudomomentum is

$$\langle \mathcal{F}_M \rangle = \varrho_0 \langle uw \rangle. \qquad (3.108)$$

This is identical to the Reynolds stress (3.102) and it is consistent with the momentum conservation formula (3.101).

The wave-induced mean flow for internal waves is alternately and, perhaps, more intuitively derived if one considers the motion of a wavepacket. The approach is similar to that used in the derivation of (2.63) to get the formula (2.58) for the Stokes drift of deep surface waves. We consider horizontally periodic internal waves whose amplitude envelope varies slowly in the vertical and in time and which translates upwards at the vertical group velocity, c_{gz}. Explicitly, we write

$$u = \Re\{A_u(Z,T) \exp[i(\vec{k}\cdot\vec{x} - \omega t)]\} \quad \text{and} \quad w = \Re\{A_w(Z,T) \exp[i(\vec{k}\cdot\vec{x} - \omega t)]\},$$

in which $Z = \epsilon(z - c_{gz}t)$ describes the translation of the wavepacket and $T = \epsilon^2 t$ describes the relatively slow change of the amplitude envelope due to dispersion, as represented by Schrödinger's equation (1.136). The quantity $\epsilon \equiv (\sigma_z k_z)^{-1} \ll 1$ measures the vertical wavelength relative to the vertical extent of the wavepacket.

Putting these expressions into the flux-form of the horizontal momentum equations (3.100) and averaging over one horizontal wavelength gives the momentum conservation law (3.103), but now written in terms of the vertically varying amplitude envelopes of the u and w fields:

$$\frac{\partial}{\partial t} \langle u \rangle = -\frac{\partial}{\partial z} \langle uw \rangle = -\frac{\partial}{\partial z} \left(\frac{1}{2} \Re\{A_u(Z,T)\, A_w^\star(Z,T)\} \right), \qquad (3.109)$$

in which the star denotes the complex conjugate.

Because the wave-induced mean flow $\langle u \rangle$ moves upwards with the wavepacket, it should depend upon the translating co-ordinate Z. Thus on the left-hand side of (3.109) we have $\partial_t = -\epsilon c_{gz} \partial_Z + \epsilon^2 \partial_T$ and on the right-hand side we have $\partial_z = \epsilon \partial_Z$. Neglecting the ϵ^2 term and vertically integrating the resulting equation in Z gives an implicit expression for the wave-induced mean flow due to a horizontally periodic internal wave of finite vertical extent:

$$\langle uw \rangle = \frac{1}{2} \Re\{A_u A_w^\star\} = c_{gz} \langle u \rangle_L. \qquad (3.110)$$

Rearranging (3.110) to isolate $\langle u \rangle_L$, using the polarization relations and the explicit expression for c_{gz} in (3.62), we once again find the explicit formula (3.107) for the wave-induced mean flow of Boussinesq internal waves having vertical displacement amplitude A_ξ.

Thus we have found that the mean momentum flux per unit mass $\langle uw \rangle$ equals the wave-induced mean flow times the vertical group velocity. The formula is analogous to (3.91), which states that the mean energy flux equals the mean energy times the vertical group velocity. In this sense, the wave-induced mean flow can be identified with the momentum per unit mass of horizontally periodic internal waves.

This consideration of the wave-induced mean flow provides a different interpretation of the momentum conservation law (3.103), as illustrated schematically in Figure 3.9. The usual interpretation is that internal waves transport momentum, having no effect upon the mean flow until they break, at which point the background flow accelerates where momentum is deposited. This makes sense only if the incident waves are plane waves. In more realistic circumstances, internal waves propagate as wavepackets that change the mean flow as they propagate without breaking.

Like the Stokes drift, the wave-induced mean flow should not be confused with the background mean flow, $\bar{U}(z)$, which exists both in the presence and in the absence of waves. $\bar{U}(z)$ can be prescribed arbitrarily, but $\langle u \rangle_L$ is a function of the

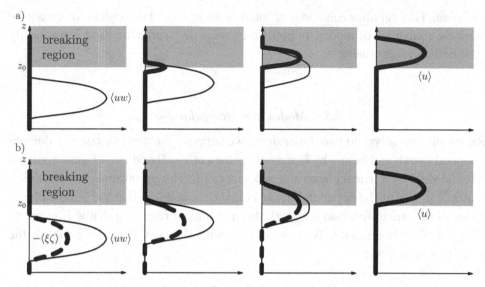

Fig. 3.9. a) Schematic illustrating the position of an upward-propagating wavepacket encountering a region where it breaks (shaded grey). The thin solid line represents the momentum flux per unit mass, $\langle uw \rangle$. The thick line represents the horizontally averaged flow. b) Before the waves break (dashed line) this is the wave-induced mean flow, $\langle u \rangle_L = -\langle \xi \zeta \rangle = \langle uw \rangle / c_{gz}$. After the waves break (solid line) momentum is permanently deposited to the background horizontal flow.

structure and amplitude of the internal waves. Thus the total horizontally averaged flow at any level z is the superposition $\langle u \rangle = \bar{U}(z) + \langle u \rangle_L$.

When internal waves break, their wave-induced mean flow is irreversibly transformed into a steady background mean flow. Momentum conservation requires that breaking internal waves accelerate the background flow so that the vertically integrated wave-induced mean flow, $\int \langle u \rangle_L \, dz$, equals the vertically integrated change to the background flow, $\int \Delta \bar{U} \, dz$, after they break.

3.5 Bounded internal waves

As for interfacial waves in domains with vertical side-walls, the effect of the side boundaries is to limit the allowable horizontal wavenumbers of internal waves to a discrete spectrum. Similarly, internal waves bounded above and below by horizontal boundaries are confined to a discrete vertical wavenumber spectrum. They form a discrete set of modes.

If the basin is not rectangular, the waves behave in a remarkably different fashion. They develop into beams rather than modes through a process called geometric focusing. This peculiar behaviour of internal waves will be considered separately from the dynamics of modes. In both cases we will assume the fluid is uniformly stratified and Boussinesq.

3.5.1 Modes in rectangular box

Restricting ourselves to two dimensions, we suppose the waves exist in a domain bounded above and below by horizontal planes at $z = 0$ and $z = H$, and bounded left and right by vertical planes at $x = 0$ and $x = L$. The no-normal-flow condition requires $w = 0$ at the upper and lower boundaries and $u = 0$ at the side-walls. In terms of the streamfunction $\psi(x,z,t)$, this amounts to requiring that ψ is constant along all four boundaries. Without loss of generality we therefore establish the boundary conditions

$$\psi(x,0,t) = \psi(x,H,t) = 0 \quad \text{for} \quad 0 \le x \le L,\, t > 0$$
$$\psi(0,z,t) = \psi(L,z,t) = 0 \quad \text{for} \quad 0 \le z \le H,\, t > 0. \tag{3.111}$$

Within the interior of the domain small-amplitude motions satisfy (3.42), in which we assign the arbitrary eigenfunction b to be the streamfunction ψ so that we can easily apply these boundary conditions. Thus we have

$$\left[\partial_{tt} \nabla^2 + N_0^2 \partial_{xx} \right] \psi = 0. \tag{3.112}$$

Its solution is generally given by the real part of $\mathcal{A}_\psi \exp[\imath (k_x x + k_z z - \omega t)]$. However, the boundary conditions (3.111) restrict the allowable solutions to those which effectively act as standing waves: superpositions of left–right and up–down propagating waves with fixed nodes at the boundaries. Thus the spatial dependence of ψ is given by pure sinusoids in x and z so that

$$\psi(x,z,t) = \sum_{i=1}^{\infty} \sum_{j=1}^{\infty} \mathcal{A}_{ij} \sin(i\,k_{x0}\,x) \sin(j\,k_{z0}\,z) e^{-\imath \omega t}, \tag{3.113}$$

a result that could have been derived directly from the boundary value problem (3.112) and (3.111) using the method of separation of variables.

In (3.113) the discrete horizontal and vertical wavenumbers are integer multiples of $k_{x0} = \pi/L$ and $k_{z0} = \pi/H$, respectively, and $\mathcal{A}_{ij}$ is the (possibly complex) streamfunction amplitude of the i–jth mode. The actual disturbance is the real part of the double sum in (3.113).

The frequency of each mode is determined from the dispersion relation. Setting $k_x = i\pi/L$ and $k_z = j\pi/H$ in (3.36), the frequency can be expressed in terms of the tank dimensions and mode number, i–j:

$$\omega = \omega_{ij} = N_0 \left[1 + \left(\frac{jL}{iH} \right)^2 \right]^{-1/2}, \quad i,j = 1,2,\dots \tag{3.114}$$

Any small-amplitude disturbance in the container can be represented by (3.113). In particular, if the initial disturbance at time $t = 0$ is the real-valued function $\psi_0(x,z)$ with $\psi_t(x,z,0) = 0$, then $\mathcal{A}_{ij} = A_{ij}$ is real-valued and is given by the double Fourier sine series coefficient

$$A_{ij} = \frac{4}{\pi^2} \int_0^L \int_0^H \psi_0(x,z) \sin(ik_{x0}x) \sin(jk_{z0}z) \, dz \, dx, \quad i,j = 1,2,\dots \tag{3.115}$$

When modelling internal waves in a basin, usually only the low-order modes are of interest. Each of these is referred to as the 'i–j mode', in which i is the horizontal and j the vertical mode number. In the case of the rectangular box, the lowest is the '1–1' mode corresponding to internal waves with half the horizontal and vertical wavelength spanning the length and height of the container, respectively. The structures of this and the 2–1 and 1–2 modes are illustrated in Figure 3.10.

3.5.2 Modes in non-uniform stratification

Above we assumed the stratification was uniform, a circumstance often studied in laboratory experiments, but which is not representative of geophysical fluids. In the ocean, for example, the stratification is stronger at the thermocline and weaker in the abyss. Nonetheless, through a change of co-ordinate it is possible to transform the structures of sufficiently short waves in non-uniformly stratified fluid to those in uniformly stratified fluid. This procedure is known as 'WKB renormalization' or

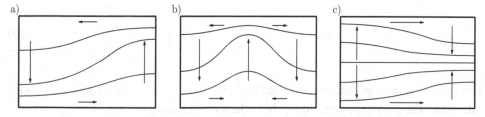

Fig. 3.10. Structure of a) 1–1, b) 2–1 and c) 1–2 modes at one phase of their oscillation. Solid lines show the position of isopycnal surfaces and arrows indicate the direction of horizontal and vertical displacements.

'WKB stretching'. It is a particular application of ray theory, discussed in detail in Section 6.3.

For simplicity, we discuss this method here for two-dimensional Boussinesq internal waves in non-rotating fluid, which are bounded above and below by horizontal boundaries, but which are periodic in horizontal space and time with horizontal wavenumber k_x and frequency ω. Replacing $N_0{}^2$ with $N^2(z)$ in equation (3.112) and assuming $\psi = \hat{\psi}(z)\exp[\iota(k_x x - \omega t)]$, the vertical structure of the streamfunction amplitude is found to satisfy

$$\frac{d^2\hat{\psi}}{dz^2} + \gamma^2\hat{\psi} = 0, \tag{3.116}$$

in which

$$\gamma^2(z) = k_x{}^2\left[\frac{N^2(z)}{\omega^2} - 1\right]. \tag{3.117}$$

If $N^2(z) = N_0{}^2$, a constant, then γ is a constant and is identified with the vertical wavenumber. In particular, with boundary conditions $\hat{\psi}(0) = \hat{\psi}(-H) = 0$, we find that $\hat{\psi}$ is represented by the series of vertical modes

$$\hat{\psi}_j(z) \propto \sin(\gamma_j z) \quad \text{with} \quad \gamma = \gamma_j \equiv j\pi/H \quad \text{for } j = 1, 2, \ldots \tag{3.118}$$

Even if N^2 is non-constant, we can arrive at an analogous result if we assume the waves in which we are interested have a vertical wavelength much smaller than the scale of vertical variations in N^2. This is the essential assumption of WKB theory. Approximate solutions of (3.117) are given by assuming $\hat{\psi}$ is the product of a slowly varying real-valued amplitude, A_ψ, times an oscillatory function with real-valued phase, φ: $\hat{\psi} = A_\psi(z)\exp(\pm\iota\varphi(z))$. Substituting this into (3.117), taking real and imaginary parts and neglecting second derivatives of $A_\psi(z)$ gives

$$\hat{\psi} \propto \gamma^{-1/2}\exp\left(\pm\iota\int\gamma\,dz\right). \tag{3.119}$$

The pair of solutions can be combined to write $\hat{\psi}$ as a sequence of vertical modes satisfying the zero upper and lower boundary conditions.

In particular, if we consider hydrostatic waves, for which $\omega \ll N$, (3.117) simplifies to $\gamma \simeq k_x N/\omega$. Substituting this into (3.119) and comparing the result to

(3.118) inspires us to define a new co-ordinate Z by

$$Z \equiv \frac{1}{N_\star} \int_z^0 N(\tilde{z})\, d\tilde{z}, \qquad (3.120)$$

in which $N_\star$ is a characteristic value of N, typically taken to be its maximum value. In terms of this new 'WKB-stretched' co-ordinate, the streamfunction of the jth mode is

$$\hat{\psi}_j \propto \sqrt{\frac{N_\star}{N}} \sin(\gamma_j Z), \qquad (3.121)$$

in which

$$\gamma_j = j\pi N_\star \left[\int_{-H}^0 N\, dz \right]^{-1}, \ \text{for } j = 1,2,\ldots \qquad (3.122)$$

This procedure is called 'WKB renormalization'.

For example, Figure 3.11 shows the result of transforming vertical wave modes in uniform stratification to waves in non-uniform stratification. In uniform stratification, modes 2 and 3 have vertical structure given by (3.118) with $j = 2$ and 3. The streamfunction has one and two zero-crossings, respectively, within the domain. In non-uniform stratification their structure is given by (3.121) with respect to the stretched co-ordinate defined in terms of z by (3.120). When plotted against z the effect is to have waves with the same number of zero-crossings but larger streamfunction amplitude where N is large.

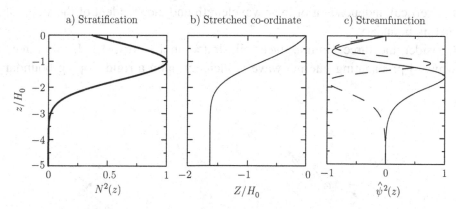

Fig. 3.11. a) Non-uniform stratification given by $N^2 = -N_0^2\,(z/H_0)$ $\exp\left(-(z - H_0)^2/2H_0^2\right)$ with $H_0 = 1$, b) transformation (3.120) from z to stretched co-ordinate Z, and c) normalized streamfunction of mode-2 (solid) and mode-3 (dashed) waves calculated using (3.121) and (3.122) and recast in terms of z.

Given the form of the streamfunction, we can take derivatives to find the velocity field. Again assuming that variations in N are slow, to leading order we find

$$\hat{u}_j \propto N^{1/2} \cos(\gamma_j Z), \quad \hat{w}_j \propto N^{-1/2} \sin(\gamma_j Z). \tag{3.123}$$

Likewise we can use the polarization relations to determine other fields of interest. This analysis shows that the energy of hydrostatic modes at any depth z is dominated by the square of the horizontal velocity field and is proportional to N.

The process of WKB renormalization is particularly useful in the interpretation of oceanographic data: horizontal oscillations may be larger in the thermocline than in the abyss because of variations in the vertical structure of the mode, not necessarily because energy has dissipated going from the surface downwards.

The method is easily extended to include rotation and three-dimensional effects. The hydrostatic approximation is convenient because the transformation becomes independent of the wave frequency, but this too can be relaxed. Though we have performed this analysis for modes, Section 6.3 generally explains how WKB theory can be used to describe the evolution of propagating wavepackets in non-uniform media.

3.5.3 Internal waves near slopes

If the basin is not rectangular-shaped, the pattern of waves it contains can be so complicated as to make analytic mode-like solutions impractical. However, internal waves possess a peculiar property that can organize their structure in a manner revealed by attractors, not modes, through a process called 'geometric focusing'. The property in question is that waves with fixed frequency propagate at fixed angles to the vertical, independent of their wavelength and independent of the slope of a reflecting boundary.

Consider the three circumstances illustrated in Figure 3.12. In all three, a downward-propagating internal wave is incident upon a rigid sloping boundary.

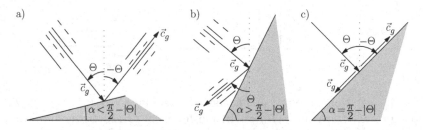

Fig. 3.12. Schematic illustration of the reflection of an internal wave incident upon a) a shallow slope, b) a steep slope and c) a critical slope.

The waves have frequency $\omega < N_0$ such that constant-phase lines form an angle $\Theta = \pm\cos^{-1}(\omega/N_0)$ to the vertical. The slope forms an angle α with the horizontal.

It is well known that when light reflects from a boundary, the angle of reflection equals the angle of incidence. Not so here. Upon reflection the frequency does not change, so neither does $|\Theta|$; the constant-phase lines of internal waves still form an angle with the vertical whose magnitude $|\Theta|$ is independent of α. In Figure 3.12a, $|\alpha| < \pi/2 - |\Theta|$ so the waves reflect upwards. If the slope is so large that $\pi/2 \geq |\alpha| > \pi/2 - \Theta$, then as Fig 3.12b illustrates, the waves reflect downwards but their phase-lines still form an angle $|\Theta|$ to the vertical, independent of α. The shallow-slope case is referred to as 'subcritical reflection' and the steep-slope case as 'supercritical reflection'.

In the critical case, illustrated in Fig 3.12c, $|\alpha| = \pi/2 - |\Theta|$ and the waves propagate both upslope and downslope. This last case provides a mechanism for dissipation of the waves if we relax the free-slip boundary conditions and allow viscous damping to erode the waves whose motions are large along the sloping boundary.

Now consider waves propagating in a wedge-shaped domain with a sloping lower boundary and horizontal upper boundary, as shown in Figure 3.13a. Subcritical waves reflect alternately from the sloping bottom and horizontal surface thus

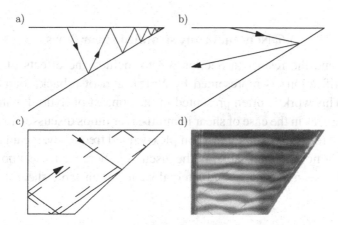

Fig. 3.13. a) Propagation of subcritical internal waves in a wedge towards a point-attractor. b) Propagation of supercritical internal waves away from the point-attractor. c) Propagation of internal waves towards a parallelogram-shaped attractor in a trapezoidal basin. d) Snapshot from a laboratory experiment showing dye-line displacements are largest near the predicted attractor (dashed line) for internal waves generated by vertically oscillating the trapezoidal-shaped tank at a fixed frequency. [The image in d) is reproduced, with permission of the Nature Publishing Group, from Figure 3a of Maas *et al.*, Nature, **388**, 557–561 (1997).]

focusing the waves upon the corner of the wedge where presumably the waves dissipate. The corner acts as a point-attractor for the waves.

If the incident waves have sufficiently small frequency that they are supercritical then, as shown in Figure 3.13b, they reflect downwards away from the wedge and back into the open basin.

If instead the domain is enclosed both laterally and vertically, the waves must continue to reflect around the basin. In this circumstance even supercritical waves can be drawn towards an attractor that forms a closed path. For example, Figure 3.13c shows that waves in a trapezoidal domain which are supercritical with respect to the sloping side are drawn towards a parallelogram-shaped attractor. Thus internal waves in a non-rectangular box are manifest as a cyclic beam rather than a mode.

The manifestation of such an attractor has been demonstrated in laboratory experiments as shown in Figure 3.13d. Here a trapezoidal tank filled with uniformly stratified fluid with buoyancy frequency N_0 oscillates vertically at a fixed frequency, $\omega < N_0$. Over time dyed isopycnal surfaces reveal that the largest disturbances occur along the lines of a parallelogram whose sides form angles to the vertical of $|\Theta| = \cos^{-1}(\omega/N_0)$.

For waves of fixed frequency, the attractor is unique provided at least one side of the basin is sloping. In a rectangular box, the attractor is not unique, which is why one sees modes with uniform amplitude everywhere in the domain rather than a cyclical beam of waves.

3.6 Non-uniformly stratified shear flows

Here we extend the results in Section 3.3 to include the effects of a vertically varying stratification, as represented by $N^2(z)$, and of a background horizontal flow $\bar{U}(z)$. This work is often presented in the context of hydrodynamic stability theory because, as in the case of shear in multi-layer fluids discussed in Section 2.6, the dispersion relation can involve a complex-valued frequency, meaning that some stratified shear flows are unstable. In the discussion here we focus upon drawing a connection between interfacial and internal waves in unstable shear flows.

3.6.1 Equations of motion

We restrict our attention to two-dimensional, non-rotating, Boussinesq flow. This is a 'parallel flow', meaning that background fields are functions of z alone. In particular, streamlines of the background flow are horizontal and so lie parallel to each other. It is mathematically convenient here to use density rather than potential temperature as a basic state variable. But, as we saw in Section 3.3.2, the conclusions we draw apply equally well to disturbances in a stratified Boussinesq gas as well as in a stratified liquid.

The fully nonlinear momentum and continuity equations (3.37) are modified by replacing u with $\bar{U} + u$, the sum of the background and fluctuation horizontal velocity. Linearizing these equations by assuming small fluctuations, we have

$$\frac{\partial u}{\partial t} + \bar{U}\frac{\partial u}{\partial x} + w\frac{d\bar{U}}{dz} = -\frac{1}{\rho_0}\frac{\partial p}{\partial x}, \tag{3.124}$$

$$\frac{\partial w}{\partial t} + \bar{U}\frac{\partial w}{\partial x} = -\frac{1}{\rho_0}\frac{\partial p}{\partial z} - \frac{1}{\rho_0}\rho g, \tag{3.125}$$

$$\frac{\partial \rho}{\partial t} + \bar{U}\frac{\partial \rho}{\partial x} = w\frac{\varrho_0}{g}N^2, \tag{3.126}$$

and

$$\frac{\partial u}{\partial x} + \frac{\partial w}{\partial z} = 0. \tag{3.127}$$

In (3.126) we have recast the background density gradient in terms of the Boussinesq form of the squared buoyancy frequency (3.8).

Together these form four linear partial differential equations in the four unknown perturbation fields. Unlike the corresponding matrix equation for uniformly stratified fluid considered in Section 3.3, here the coefficients are not constant but involve the z-dependent functions N^2 and $\bar{U}$. It is impractical therefore to write the equations in matrix form because the z-derivative operator, ∂_z, does not commute with functions of z: $\partial_z(\bar{U}b) \neq \bar{U}\partial_z b$ for some arbitrary function b.

Furthermore, the symmetry-breaking in z means that we can represent only the x- and t-dependence of the basic state fields as complex exponentials. Explicitly, we can write each basic state field b, as

$$b(x,z,t) = \hat{b}(z)e^{\iota(k_x x - \omega t)}, \tag{3.128}$$

in which b can be any of u, w, ρ or p. The z-dependent amplitude $\hat{b}$ can be a complex-valued function, and it is understood that the actual field is the real part of the right-hand side of (3.128).

Combining the four coupled equations forms a single partial differential equation for one field. Because the fluid is incompressible we can write u and w in terms of derivatives of the streamfunction ψ by

$$(u,w) = \left(-\frac{\partial \psi}{\partial z}, \frac{\partial \psi}{\partial x}\right). \tag{3.129}$$

We eliminate pressure in the momentum equations by taking the curl to get an equation for the y-component of vorticity $\zeta = \partial_z u - \partial_x w = -\nabla^2 \psi$. Using (3.128), this simplifies to give

$$-(\bar{U} - c)\left(-k_x^2 + \frac{d^2}{dz^2}\right)\hat{\psi} + \frac{d^2\bar{U}}{dz^2}\,\hat{\psi} = \frac{g}{\varrho_0}\hat{\rho}, \qquad (3.130)$$

in which $c \equiv \omega/k_x$ is the horizontal phase speed c_{Px} as would be recorded from a horizontal time series as in (1.116).

Equation (3.126) with (3.128) gives a second equation in $\hat{\psi}$ and $\hat{\rho}$:

$$(\bar{U} - c)\hat{\rho} = \frac{\varrho_0}{g}N^2\,\hat{\psi}. \qquad (3.131)$$

Using this to eliminate $\hat{\rho}$ in (3.130) gives a single equation in $\hat{\psi}$:

$$\frac{d^2\hat{\psi}}{dz^2} + \left[\frac{N^2}{(\bar{U} - c)^2} - \frac{\bar{U}''}{(\bar{U} - c)} - k_x^2\right]\hat{\psi} = 0. \qquad (3.132)$$

This is the Taylor–Goldstein equation, named after the scientists who first derived and analysed the formula in the early 1930s.

Alternately, we can define the z-dependent extrinsic frequency

$$\bar{\Omega}(z) = \omega - k_x\bar{U}(z), \qquad (3.133)$$

and so rewrite (3.132) in terms of Ω:

$$\frac{d^2\hat{\psi}}{dz^2} + k_x^2\left[\frac{N^2}{\bar{\Omega}^2} - \frac{\bar{\Omega}''}{k_x^2\bar{\Omega}} - 1\right]\hat{\psi} = 0. \qquad (3.134)$$

For given background profiles of $N^2(z)$ and $\bar{U}(z)$ and for given k_x, the Taylor–Goldstein equation forms a differential eigenvalue problem. Nontrivial disturbances exist for eigenvalues $c(k_x)$ which gives the dispersion relation $\omega(k_x) = k_xc(k_x)$. The vertical structure of the disturbances is given by the corresponding eigenfunction $\hat{\psi}$ which, together with (3.128), completely describes the evolution of the small-amplitude disturbance. Such solutions are sometimes called 'normal modes'. Just as we found in our examination of instability of multi-layer fluids, if $\omega = \omega_r + \iota\sigma$ is complex-valued for some k_x, then the streamfunction

$$\psi(x, z, t) = \hat{\psi}(z)e^{\iota(k_xx - \omega_rt)}e^{\sigma t} \qquad (3.135)$$

grows exponentially in time. This is called an 'unstable normal mode'.

3.6.2 Singularities

The Taylor–Goldstein equation is singular in the sense that the coefficients of the ordinary differential equation (3.132) become infinitely large at levels $z \equiv z_c$ where the horizontal phase speed matches the background flow speed: $c = \bar{U}(z_c)$. Equivalently, this is where the extrinsic frequency is zero: $\Omega(z_c) = 0$. The height where this occurs is called a 'critical level'.

A Frobenius expansion reveals the behaviour of solutions near critical levels. For mathematical simplicity, we assume that $z_c = 0$ and that the flow about z_c is approximately uniformly stratified, $N \simeq N_0$, and has nearly uniform shear s_0 so that $\bar{U} \simeq c + s_0 z$.

With these approximations, the Taylor–Goldstein equation becomes

$$\frac{d^2\hat{\psi}}{dz^2} + \left[\frac{N_0^2}{s_0^2 z^2} - k_x^2\right]\hat{\psi} \simeq 0, \quad \text{for } z \simeq z_c = 0. \tag{3.136}$$

In forming a Frobenius expansion, we assume that $\hat{\psi}$ can be written

$$\hat{\psi} = z^p(a_0 + a_1 z + \ldots). \tag{3.137}$$

This is similar to a power series, except that p may be a fractional exponent.

The value of p and the coefficients a_0, a_1, etc. are determined by substituting (3.137) into (3.136) and matching terms with equal powers of z. Thus to leading order $p(p-1) = -N_0^2/s_0^2$. Explicitly, we find the leading power is

$$p = p_\pm \equiv \frac{1}{2} \pm \sqrt{\frac{1}{4} - \text{Ri}}. \tag{3.138}$$

Here $\text{Ri} \equiv N_0^2/s_0^2$ defines the bulk Richardson number for shear flows in continuously stratified fluid having characteristic shear s_0 and characteristic buoyancy frequency N_0. More generally, we define

$$\text{Ri}_g(z) \equiv \frac{N^2}{(\bar{U}')^2} \tag{3.139}$$

to be the gradient Richardson number. This is a function of z and measures the ratio of the squared buoyancy frequency to the squared shear at each height in the flow. In (3.138), $\text{Ri} = \text{Ri}_g(0)$.

If $\text{Ri} \le 1/4$ at a critical level, (3.138) shows that

$$\hat{\psi} \simeq C_+ z^{p+} + C_- z^{p-} \quad \text{for } z \simeq z_c = 0, \tag{3.140}$$

in which $C_\pm$ are constants and the exponents hold real values between 0 and 1. In particular, if the fluid is unstratified at the critical level, then $\psi = C_+ z + C_-$.

The exponents are complex-valued if $\text{Ri} > 1/4$. In this circumstance we can recast (3.140) in terms of real functions using the relation $z^{a+\iota b} = z^a \exp(\iota b \ln|z|)$. Thus we can write

$$\hat{\psi} = z^{1/2}\left[C_+ \cos\left(\sqrt{\text{Ri}-1/4}\,\ln|z|\right) + C_- \sin\left(\sqrt{\text{Ri}-1/4}\,\ln|z|\right)\right]. \tag{3.141}$$

Although its magnitude decreases to zero as $|z| \to 0$, it also exhibits rapid oscillatory behaviour about the critical level.

The change in the structure of $\hat{\psi}$ with Ri is one indication that the value $\text{Ri} = 1/4$ is special in the consideration of stratified shear flows. Indeed, the next section demonstrates that a necessary condition for instability is that the gradient Richardson number is less than $1/4$ for some z. If $\text{Ri}_g(z) \geq 1/4$ for all z, the flow is stable.

3.6.3 Stability criteria

A stratified shear flow becomes unstable if disturbances can overcome the stabilizing influence of the background stratification by drawing kinetic energy from the mean flow. A fluid parcel argument can be used to gain some intuition into these dynamics.

Suppose the flow about some arbitrary level has an approximately constant buoyancy frequency and shear. Moving in a frame of reference with the speed of the background flow at $z = 0$, we have $\bar{\rho} \simeq \varrho_0(1 - z/H)$ and $\bar{U} \simeq s_0 z$. We wish to exchange two parcels located above and below the origin at $z = \pm \delta$. In (3.80) we found that the total available potential energy associated with the exchange is

$$\Delta E_P = 4\varrho_0\delta^2 g/H. \tag{3.142}$$

In the process of exchanging positions, the parcels can change speed to values between their initial speeds $-s_0\delta$ and $s_0\delta$. When changing speeds momentum must be conserved. Working in the Boussinesq approximation, we can assume that the density difference negligibly affects the momentum so really the sum of the speeds of both parcels does not change: the total speed must always be zero. The maximum loss in kinetic energy occurs if the speed of both parcels after the exchange is zero. Therefore the change in kinetic energy due to the exchange is

$$\Delta E_K = -\frac{1}{2}\varrho_0[(-s_0\delta)^2 + (s_0\delta)^2] = -\varrho_0(s_0\delta)^2. \tag{3.143}$$

If the parcels gain available potential energy given by (3.142), at the expense of kinetic energy given by (3.143), the sum must be negative, meaning that no more energy was required for the exchange than was already available. So, combining (3.142) and (3.143) and dividing through by $\varrho_0\delta^2$, we require $4g/H - s_0^2 < 0$. Finally, if we recognize that g/H is just the local value of the squared buoyancy frequency about $z = 0$, we have the condition

$$\text{Ri}_g(0) = \frac{N_0^2}{s_0^2} < 1/4. \tag{3.144}$$

Here we have used the definition (3.139) for Ri_g. And so we see that fluid parcels can draw energy from the mean flow to overcome restoring buoyancy forces provided the gradient Richardson number is less than $1/4$.

More generally, the necessary condition for instability is

$$Ri_g(z) < 1/4 \qquad (3.145)$$

for some z. However, the arguments leading to this result do not comprise a proof because we have assumed the ambient fluid is not affected by the exchange of the two parcels. In fact the fluid must move as a continuous whole.

To prove (3.145) is indeed a necessary condition for the existence of unstable normal modes, we need to manipulate the Taylor–Goldstein equation so as to establish under what conditions c and hence $\omega = ck_x$ is complex-valued.

The approach is to substitute $\hat{\psi} = (\bar{U} - c)^{1/2}\phi$ in (3.132) for which $\phi(z)$ is possibly a complex-valued function. Rearranging the resulting equation gives

$$\frac{d}{dz}\left((\bar{U}-c)\frac{d\phi}{dz}\right) + \left[\frac{N^2}{\bar{U}-c} - \frac{1}{2}\frac{d^2\bar{U}}{dz^2} - k_x^2(\bar{U}-c) - \frac{1}{4}\frac{1}{\bar{U}-c}\left(\frac{d\bar{U}}{dz}\right)^2\right]\phi = 0.$$
$$(3.146)$$

Now we multiply through by the complex conjugate function $\phi^\star$ and integrate over z within the domain. The vertical velocity is zero at the extremes of the domain, whether bounded or unbounded and so, integrating by parts, the first term on the left-hand side becomes $\int(\bar{U}-c)|\phi'|^2\,dz$, in which the prime denotes a z-derivative. Taking the imaginary part of the entire resulting equation gives

$$c_i\int\left|\phi'\right|^2 + \left[\frac{N^2 - (\bar{U}')^2/4}{|\bar{U}-c|^2} + k_x^2\right]|\phi|^2\,dz = 0.$$

Here $c_i \equiv \omega_i/k_x$ is the imaginary part of c in which the complex frequency is represented by $\omega = \omega_r + \iota\omega_i$. Assuming the flow is unstable so that $|c_i| > 0$, the integral must be zero. This can occur only if $N^2 - (\bar{U}')^2/4$ is negative over some range of z, that is if $Ri_g(z) < 1/4$ for some z, in which Ri_g is given by (3.139).

3.6.4 Piecewise-linear theory

Whether or not the flow is unstable, the Taylor–Goldstein equation (3.132) can be used to predict the structure and dispersion relation of internal waves in a stratified shear flow. In special cases analytic solutions of the equation can be found. In particular, if $\bar{U}$ is piecewise-linear and the fluid is unstratified where $\bar{U}$ is not constant, then the coefficient of $\hat{\psi}$ in the square brackets of (3.132) is constant. So the equation becomes the simple ordinary differential equation

$$\frac{d^2\hat{\psi}}{dz^2} + \gamma^2\hat{\psi} = 0, \tag{3.147}$$

in which $\gamma^2 = -k_x^2$ in regions where $N = 0$ and $\gamma^2 = N_0^2/(U_0 - c)^2 - k_x^2$ in regions where $N = N_0$ and $\bar{U} = U_0$ are constant. Thus solutions of (3.147) for $\hat{\psi}$ are either oscillatory or exponential in z depending on whether γ^2 is positive or negative, respectively.

The solutions found in each unstratified constant-shear and in each uniformly-stratified constant-flow region are matched across discontinuities of the background profiles by requiring continuity of vertical velocity and pressure. These are the same conditions given by (2.134) and (2.135), which we used to examine the effects of shear in multi-layer fluids. Explicitly, we require continuity of

$$\frac{1}{\bar{U} - c}\hat{\psi}, \tag{3.148}$$

and

$$\bar{\rho}\left\{(\bar{U} - c)\frac{d\hat{\psi}}{dz} - \bar{U}'\hat{\psi} - \frac{g}{\bar{U} - c}\hat{\psi}\right\}, \tag{3.149}$$

or approximations thereof, as listed in Table 2.2.

Boundary conditions must also be imposed at the vertical extremes of the domain. If the domain is bounded by a rigid plane, then the condition for no normal flow is that $\hat{\psi}$ is constant at the boundary; usually one takes $\hat{\psi} = 0$. In an unbounded domain two possibilities exist. If $\gamma^2 < 0$ so that the general solution of (3.147) gives increasing and decreasing exponentials, we keep only the solution which is bounded. This circumstance occurs, for example, if the fluid is unstratified in the far field. If the fluid is stratified and unsheared so that $\gamma^2 > 0$ in the far field, then (3.147) gives upward- and downward-propagating wave solutions. Invoking causality, we keep only the solution that transports energy outward: $\gamma \equiv k_z < 0$ as $z \to \infty$ and $\gamma \equiv k_z > 0$ as $z \to -\infty$.

The matching and boundary conditions together constitute $2n$ equations in the $2n$ unknown coefficients of the general solutions in each of the n layers. The differential eigenvalue problem is thus reduced to a matrix eigenvalue problem.

For any prescribed $\bar{U}$ and N, solutions are found numerically by integrating (3.132) for given k_x and a guess at the possibly complex eigenvalues $c(k_x)$. By iteration, the correct eigenvalue is found as those values c for which $\hat{\psi}$ satisfies both the upper and lower boundary conditions. If c is complex, the flow is unstable for that value of k_x and the growth rate is $\sigma = |\Im(c)k_x|$.

3.6.5 Waves in uniform flow with uniform stratification

As a simple starting point, we first consider the solution of the Taylor–Goldstein equation in the special case of uniform flow, $\bar{U}(z) = U_0$, in uniformly stratified fluid, $N^2(z) = N_0{}^2$.

Equation (3.147) gives two solutions for $\hat{\psi}(z)$: $C_+ \exp(\iota\gamma z)$ and $C_- \exp(-\iota\gamma z)$ in which

$$\gamma = k_x \sqrt{\frac{N_0{}^2}{\Omega_0{}^2} - 1}. \tag{3.150}$$

If the discriminant of the square root is non-negative, γ plays the role of the magnitude of the vertical wavenumber k_z.

Comparing this result with (3.59), we see that the effect of introducing a constant background flow is to Doppler-shift the frequency from its intrinsic value ω (the value measured by a stationary observer) to its extrinsic value $\Omega_0 = \omega - U_0 k_x$, which is the frequency of the waves measured with respect to the background flow. Rearranging (3.150) with $\gamma = |k_z|$ gives the dispersion relation for internal waves in uniform flow:

$$\Omega_0{}^2 = N_0{}^2 \frac{k_x{}^2}{k_x{}^2 + k_z{}^2}. \tag{3.151}$$

So in this circumstance the Taylor–Goldstein equation shows that the flow supports both upward- and downward-propagating plane waves provided that $\Omega_0 < N_0$. Recalling that the vertical group velocity is opposite in sign to the vertical component of the phase velocity, $k_z = +\gamma$ corresponds to downward-propagating waves and $k_z = -\gamma$ corresponds to upward-propagating waves.

If the extrinsic frequency exceeds the buoyancy frequency, then γ in (3.150) must be a pure imaginary number, which corresponds to exponential solutions for $\hat{\psi}$ in (3.147). But a wave whose amplitude increases or decreases exponentially with height cannot exist in a Boussinesq fluid with no boundaries or interfaces. In this unbounded domain, we must have $\Omega_0 \leq N_0$ so that γ, and hence k_z, is real-valued.

3.6.6 Trapped internal gravity waves

We now consider the moderately more complicated case of waves in a stationary fluid whose stratification is prescribed by

$$N^2 = \begin{cases} N_0{}^2 & |z| \leq H \\ 0 & z > H. \end{cases} \tag{3.152}$$

Such a finite-depth layer of strongly stratified fluid is sometimes called a 'duct' and internal waves with vertical scale comparable to the width of the duct are said to be 'trapped' or 'ducted'. The more complicated problem of the resonant energy transfer by internal waves in a system of two ducts is considered in Section 6.6.1.

We assume that solutions to (3.147) have the form

$$\hat{\psi} = \begin{cases} \mathcal{A}e^{-k_x z} & z > H \\ \mathcal{B}_1 \cos(\gamma z) + \mathcal{B}_2 \sin(\gamma z) & |z| \leq H \\ \mathcal{C}e^{k_x z} & z < -H, \end{cases} \qquad (3.153)$$

in which γ is defined in (3.150) with $\Omega_0 = \omega$. Seeking wave-like solutions, we assume $\omega < N_0$ so that γ is real.

The interface conditions (3.148) and (3.149) reduce to continuity of $\hat{\psi}$ and its derivative (see Table 2.2). Therefore nontrivial eigenvalues of the problem are given implicitly as a function of k_x by values of γ satisfying

$$\gamma \tan(\gamma H) = k_x, \qquad (3.154)$$

in which case $\mathcal{B}_2 = 0$, and

$$k_x \tan(\gamma H) = -\gamma, \qquad (3.155)$$

in which case $\mathcal{B}_1 = 0$.

The structure of the streamfunction amplitude $\hat{\psi}$ corresponding to the dispersion relation (3.154) is an even function in z. These waves are thus referred to as 'even modes' or 'sinuous waves'. The streamfunction amplitude corresponding to solutions of (3.155) are odd functions and so the waves are called 'odd modes' or 'varicose waves'.

The naming convention is analogous to that for interfacial waves in a three-layer fluid, as discussed in Section 2.4. In that case there was a single dispersion relation corresponding to the even mode and another corresponding to the odd mode. In the case of trapped waves in a stratified layer, there is an infinite but countable family of dispersion relations for the even and odd modes. Given k_x and H, the sequence of γ (and hence ω) can be identified as the successive intersection points of the graphs of $\cot(\gamma H)$ and γ/k_x for even modes and the intersection points of the graphs of $\tan(\gamma H)$ and $-\gamma/k_x$ for odd modes. The lowest even mode has no zero-crossings within the stratified layer, the next has two zero-crossings, the next has four, and so on. The lowest odd mode has one zero-crossing point within the layer, the next has three, and so on.

The dispersion relations and structure for the lowest two even modes computed numerically from (3.154) are shown as the solid lines in Figure 3.14. The lowest two odd modes have dispersion relations given by (3.155) and are shown as the dashed lines in this figure.

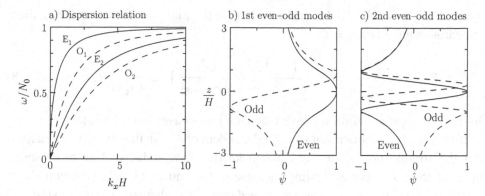

Fig. 3.14. a) Dispersion relation of first two even (solid lines) and odd (dashed lines) modes for trapped waves in a stratified layer of depth $2H$ and buoyancy frequency N_0 as given by (3.152). b) Streamfunction amplitude, $\hat{\psi}(z)$, of the lowest even and odd modes, E1 and O1 and c) of the next lowest even and odd modes, E2 and O2. In b) and c) $\hat{\psi}$ is calculated for the case $k_x H = 1$ and is normalized to have maximum amplitude $\|\hat{\psi}\| = 1$.

In the limit $k_x H \to 0$ the roots of the lowest even modes have $\gamma \sim (k_x/H)^{1/2}$. Using (3.150) and rearranging, we find the dispersion relation for long ducted waves:

$$\omega^2 \sim \frac{k_x H}{1 + k_x H} N_0^{\,2} \simeq \frac{1}{2} g' k_x. \tag{3.156}$$

Here $\omega = \Omega_0$ because there is no background flow. The result (3.156) is indeed the dispersion relation (2.85) for an interfacial wave in a two-layer fluid with infinite upper- and lower-layer depths and with the density difference between the two layers given in terms of the reduced gravity $g' = (2H)N_0^{\,2} = g(\Delta\rho)/\rho_0$.

The horizontal velocity amplitude $\hat{u} = -d\hat{\psi}/dz$ corresponding to the even mode is

$$\hat{u} = A_\xi \omega \begin{cases} \cos(\gamma H)\exp[k_x H(1 - z/H)] & z > H \\ (\gamma/k_x)\sin(\gamma z) & |z| \le H \\ -\cos(\gamma H)\exp[k_x H(1 + z/H)] & z < H, \end{cases} \tag{3.157}$$

in which A_ξ is the maximum vertical displacement at $z = 0$.

Therefore long waves introduce shear across the stratified layer. It is natural to ask under what circumstances the shear will be so strong that the layer can become dynamically unstable. To this end we compute the maximum value of the gradient Richardson number at $z = 0$:

$$\text{Ri}_g(0) = \frac{N_0{}^2}{|\hat{u}'(0)|^2} = \frac{1}{(A_\xi k_x)^2} \frac{N_0{}^2}{\omega^2} \left(\frac{\omega^2}{N_0{}^2 - \omega^2}\right)^2 \simeq \frac{k_x H}{(A_\xi k_x)^2}. \qquad (3.158)$$

In this last approximation we have taken the long-wave limit (3.156).

Immediately we notice that Ri_g is independent of $N_0{}^2$ in this limit. Thus increasing the stratification alone does not make the wave-induced flow more stable because there is a corresponding increase in the square of the maximum shear. For fixed $k_x H$, the only parameter that determines whether or not the stratified shear layer is stable is the amplitude relative to the horizontal wavelength. Using (3.145), a necessary condition for the flow to be unstable is $A_\xi k_x > 2(k_x H)^{1/2}$.

3.6.7 Shear instability in stratified fluid

In Section 2.6.3 we showed that an unstratified piecewise-linear shear layer of half-depth H is unstable to disturbances with relative horizontal wavenumbers $k_x H$ lying in the range $0 < k_x H \lesssim 0.64$. This is the phenomenon of Kelvin–Helmholtz instability. The most unstable mode, being the disturbance with the fastest growth rate, has a relative wavenumber $k_x H \simeq 0.398$ and growth rate $\sigma \simeq 0.20 s_0$, in which s_0 is the strength of the shear.

In that analysis, we showed that the flow is unstable even if the shear is flanked by density interfaces. However, if the fluid is continuously stratified, then density variations acting across the shear can act to stabilize the flow. In particular, we know from (3.145) that the flow is guaranteed to be stable to normal mode disturbances if the gradient Richardson number everywhere exceeds $1/4$.

Here we consider the stability of a uniformly stratified piecewise-linear shear layer (2.142). Numerically computed marginal stability curves (where the growth rate becomes zero) and the growth rates of the most unstable modes are shown in Figure 3.15. This illustrates the stability characteristics in terms of the minimum gradient Richardson number $\min(\text{Ri}_g)$, which measures the relative strength of the stratification to the maximum shear $\bar{U}'(0)$. In unstratified fluid, $\min(\text{Ri}_g) = 0$ and we recover the result found in Section 2.6.3. In weakly stratified fluid, the growth rate of the most unstable mode is smaller than that in unstratified fluid and the corresponding horizontal wavenumber is moderately larger. The range of unstable wavenumbers decreases as $\min(\text{Ri}_g)$ increases, as indicated by the solid curve in the figure. As expected the marginal stability curve lies below the line $\min(\text{Ri}_g) = 1/4$.

Figure 3.16 shows the characteristics of the most unstable mode of the piecewise-linear shear layer (2.142), with uniform stratification prescribed by

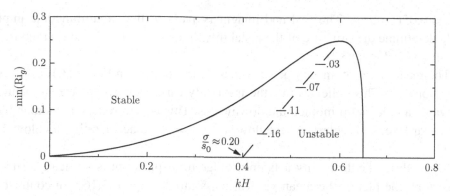

Fig. 3.15. Stability of piecewise-linear shear in uniform stratification. Non-dimensional growth rates, σ/s_0, of the most unstable mode are indicated at $\min(\mathrm{Ri}_g) = 0, 0.05, 0.10, 0.15$ and 0.20.

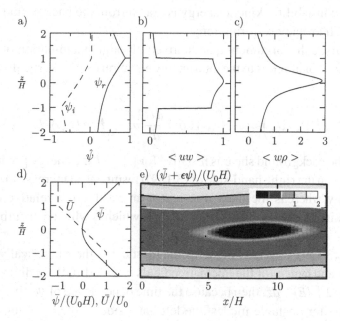

Fig. 3.16. a) Normalized perturbation streamfunction amplitude (real part – solid line; imaginary part – dashed line) and b) corresponding normalized momentum flux per unit mass and c) mass flux computed for the most unstable mode of the piecewise-linear shear layer $\overline{U}(z)$ given by (2.142) in uniform stratification with $N^2 = 0.05(U_0/H)^2$ (a bulk Richardson number of $\mathrm{Ri} = 0.05$). d) The background flow (dashed) and the corresponding background streamfunction (solid). e) Contours of the total streamfunction computed with the superimposed perturbation at a relatively small amplitude in its growth phase: $\epsilon = 0.1$.

$N^2 = 0.05(U_0/H)^2$. The real and imaginary parts of the streamfunction amplitude determine the structure of the instability. In this case, the mode is sinuous, as illustrated in Figure 3.16e.

The mean momentum flux per unit mass $\langle uw \rangle$, plotted in Figure 3.16b, is positive for all z. This indicates that the instability carries the rightward momentum upwards and leftward momentum downwards. This acts to reduce the strength of the shear layer by accelerating the flow above $z = 0$ and decelerating it below this level.

Meanwhile, the instability acts on average to displace mass vertically. This is shown by the plot of the mean vertical mass flux in Figure 3.16c. In contrast to internal waves in uniformly stratified fluid, here $\langle w\rho \rangle$ is non-zero, having its largest value where the shear is strongest. Positive density fluctuations move upwards and negative fluctuations move downwards. Consequently, the stratification in the shear layer weakens.

These diagnostics are consistent with the fluid parcel argument we used to examine shear flow instability: kinetic energy is drawn from the background shear to do work against buoyancy restoring forces.

We can pursue this argument quantitatively through examination of the kinetic energy equation, analogous to the total energy equation (3.92). Vertically integrating gives

$$\frac{\partial}{\partial t}\left(\int \langle E_K \rangle \, dz \right) = - \int \varrho_0 \langle uw \rangle \frac{d\bar{U}}{dz} \, dz - \int g \langle w\rho \rangle \, dz. \qquad (3.159)$$

Because the background shear is negative for $|z| \leq H$ the energy production term (the first term on the right-hand-side of (3.159)) results in an increase in disturbance kinetic energy whereas the mass transport term results in a relative decrease in disturbance kinetic energy. The former is larger which is why the disturbance grows in time.

Equation (3.159) provides a useful diagnostic for the numerically integrated profiles of $\hat{\psi}$ and hence for the fluctuation fields u, w and ρ. If we divide both sides of (3.159) by $2 \int \langle E_K \rangle \, dz$, then because the time dependence of $\hat{\psi}$ given by (3.135) is exponential for unstable modes, the left-hand side of the resulting equation is the growth rate σ. And so computing the ratio of the right-hand side of (3.159) to twice the total kinetic energy should likewise give the growth rate. Thus σ can be determined from integrating correlations of the eigenfunctions in (3.159). The result should be the same as the imaginary part of the eigenvalue $\omega = ck$.

In nature and in laboratory experiments, shear is often large where density gradients are large, and so there has been great interest in the stability of non-uniformly stratified shear flows. In the symmetric circumstance in which the stratification is greatest where the shear is greatest, unstable modes are qualitatively similar to the Kelvin–Helmholtz modes discussed above. In the asymmetric circumstance in which the peak in N^2 does not coincide with the peak shear, we also see the appearance of unstable Holmboe modes. These are the continuous-stratification analogue of the Holmboe modes in a two-layer fluid, examined in Section 2.6.4.

3.7 Non-Boussinesq internal waves

Thus far we have considered the characteristics of small-amplitude Boussinesq internal waves. To model internal waves that propagate vertically over distances larger than the density scale height, H_ϱ, it is necessary to relax the constraint that density variations are significant only in the buoyancy term of the equations of motion.

So that we may more intuitively compare the dynamics of anelastic and Boussinesq internal waves, we first examine the non-Boussinesq equations for a liquid in which we imagine its density increases substantially with depth. This might occur if a dissolved salt with high solubility, like sodium iodide, rapidly increases concentration with depth. Such solutions are used in laboratory experiments to examine non-Boussinesq effects. Non-Boussinesq equations for liquids may also describe disturbances in a particle-bearing fluid, in which the effective density increases substantially due to the increasing concentration of suspended particles.

Following the analysis of non-Boussinesq waves in a liquid, we turn to the solution of the anelastic equations describing the propagation of internal waves in an adiabatic gas. These equations are typically employed in studies of atmospheric internal waves.

3.7.1 Non-Boussinesq waves in a liquid

The non-Boussinesq equations were derived in Section 1.12.3. Here we restrict ourselves to two dimensions, we neglect Coriolis forces and we assume the background flow is stationary. The resulting fully nonlinear equations describing the conservation of momentum, internal energy and mass for an incompressible liquid are

$$\bar{\rho}\frac{Du}{Dt} = -\frac{\partial p}{\partial x},$$
$$\bar{\rho}\frac{Dw}{Dt} = -\frac{\partial p}{\partial z} - g\rho,$$
$$\frac{D\rho}{Dt} = -\frac{d\bar{\rho}}{dz}w,$$
$$\nabla \cdot \vec{u} = 0.$$

(3.160)

This differs from the Boussinesq equations (3.41) in that ϱ_0 is replaced with the prescribed background density profile $\overline{\rho}(z)$ in the x- and z-momentum equations.

For small-amplitude waves, the advection terms in the material derivatives of (3.160) are negligibly small and so the equations simplify to give the following system of four coupled linear partial differential equations:

$$\overline{\rho}\frac{\partial u}{\partial t} = -\frac{\partial p}{\partial x},$$

$$\overline{\rho}\frac{\partial w}{\partial t} = -\frac{\partial p}{\partial z} - g\rho,$$

$$\frac{\partial \rho}{\partial t} = -\frac{d\overline{\rho}}{dz}w, \tag{3.161}$$

$$\frac{\partial u}{\partial x} + \frac{\partial w}{\partial z} = 0.$$

The appearance of the z-dependent coefficients, $\overline{\rho}(z)$, in the momentum equations poses an additional level of algebraic complexity in solving the coupled system of partial differential equations. In particular, the operator ∂_z and the function $\overline{\rho}(z)$ are not commutative. For this reason, we do not write the formulae in matrix form because taking a determinant does not straightforwardly give a single differential equation, even for a uniformly stratified fluid.

Instead we proceed by successively reducing the number of equations and variables from four to one. We get an equation in p and ρ by eliminating w in the internal energy and vertical momentum equation:

$$-\frac{\overline{\rho}}{\overline{\rho}'}\frac{\partial^2 \rho}{\partial t^2} = -\frac{\partial p}{\partial z} - g\rho, \tag{3.162}$$

in which $\overline{\rho}'$ is the background density gradient. A second equation in p and ρ is determined by taking the divergence of the momentum equations and using $\nabla \cdot \vec{u} = 0$. Also using the internal energy equation to eliminate w, we find

$$\nabla^2 p + \left(g\frac{\partial}{\partial z} - \frac{\partial^2}{\partial t^2}\right)\rho = 0. \tag{3.163}$$

Finally, eliminating p we have a single partial differential equation in ρ:

$$\nabla^2\left(\frac{1}{N^2}\frac{\partial^2 \rho}{\partial t^2} + \rho\right) - \frac{\partial^2 \rho}{\partial z^2} + \frac{1}{g}\frac{\partial^3 \rho}{\partial t^2 \partial z} = 0. \tag{3.164}$$

Here the background density gradient is written in terms of the squared buoyancy frequency using (3.7).

At this point, we have made no assumptions about the background density profile; N^2 is an arbitrary function of z. If the fluid is uniformly stratified, meaning that $\overline{\rho}$ decreases exponentially with height so that $N = N_0$ is constant, the equation simplifies further. Using (3.31) with $\overline{\rho} = \varrho_0 \exp(-z/H_\varrho)$, the corresponding squared buoyancy frequency is $N_0{}^2 = g/H_\varrho$, in which H_ϱ is the density scale height. Thus (3.164) becomes

$$\frac{\partial^2}{\partial t^2} \nabla^2 \rho + \frac{1}{H_\varrho} \frac{\partial^3 \rho}{\partial t^2 \partial z} + N_0{}^2 \frac{\partial^2 \rho}{\partial x^2} = 0. \tag{3.165}$$

In the limit as H_ϱ becomes large the second term on the left-hand side of the equation is negligibly small and the equation reduces to the corresponding Boussinesq equation (3.42). Note that the last term on the left-hand side of (3.165) does not vanish in the large H_ϱ limit due to the relatively large magnitude of g which acts to keep $N_0{}^2 = g/H_\varrho$ finite.

Although $\overline{\rho}$ is an exponential function of z, the coefficients of (3.165) are constant. Therefore we may assume that solutions have the form

$$\rho(x, z, t) = A_\rho e^{\iota(k_x x + \gamma z - \omega t)}, \tag{3.166}$$

in which k_x, γ and ω are constants. In particular, k_x and ω can be taken as real constants, but we will see that $\gamma = \gamma_r + \iota \gamma_i$ must be complex-valued.

Substituting (3.166) into (3.165), we find the waves satisfy the dispersion relation

$$\omega^2 = N_0{}^2 \frac{k_x{}^2}{k_x{}^2 + \gamma^2 - \iota \frac{N_0{}^2}{g} \gamma}. \tag{3.167}$$

The background flow is uniform and time invariant, so ω must be real-valued. Likewise there is no shear to act as an energy source that would lead to spatial instability. So k_x is real. The imaginary part of $\gamma^2 - \iota(N_0{}^2/g)\gamma$ must therefore be zero. This condition poses no restriction on γ_r, but it fixes the imaginary part to be

$$\gamma_i = \frac{N_0{}^2}{2g} = \frac{1}{2H_\varrho}. \tag{3.168}$$

Putting this into the last two terms in the denominator of (3.167) gives $\gamma^2 - \iota(N_0{}^2/g)\gamma = \gamma_r{}^2 + 1/(4H_\varrho^2)$.

Finally, defining $k_z \equiv \gamma_r$ and $\vec{k} = (k_x, k_z)$ we have found that two-dimensional non-Boussinesq waves satisfy the dispersion relation

$$\omega^2 = N_0^2 \frac{k_x^2}{|\vec{k}|^2 + 1/(4H_\varrho^2)}. \tag{3.169}$$

Using (3.166) with $\gamma = k_z + i/(2H_\varrho)$, their structure as represented by the fluctuation density field is given by

$$\rho(x, z, t) = A_\rho e^{i(k_x x + k_z z - \omega t)} e^{-z/2H_\varrho}. \tag{3.170}$$

As expected, in the limit $H_\varrho \to \infty$ (3.169) reduces to the dispersion relation (3.54) for Boussinesq waves. Likewise, the structure prescribed by (3.170) reduces to that for plane waves if $z \ll H_\varrho$. This is why we can use the Boussinesq approximation for waves in a non-Boussinesq fluid that propagate over distances z much smaller than H_ϱ.

Going back to the equations of motion, we can represent the amplitude of the other fields in terms of A_ρ. In particular, the vertical displacement field may be found from the vertical momentum equation together with the definition $w = \partial \xi / \partial t$. Thus

$$\xi(x, z, t) = A_\xi e^{i(k_x x + k_z z - \omega t)} e^{z/2H_\varrho}, \tag{3.171}$$

in which $A_\xi = A_\rho H_\varrho / \varrho_0$. Although the density fluctuation field decreases with height, the vertical displacement of non-Boussinesq waves increases with height.

Likewise, velocity fluctuations increase with height according to

$$\vec{u}(x, z, t) = \imath \omega A_\xi \left(\left[\frac{k_z}{k_x} + \frac{\imath}{2k_x H_\varrho} \right], -1 \right) e^{\imath(k_x x + k_z z - \omega t)} e^{z/2H_\varrho}, \tag{3.172}$$

but the dynamic pressure decreases with height as a result of the exponentially decreasing background pressure:

$$p(x, z, t) = \imath \varrho_0 \frac{\omega^2}{k_x^2} \left[k_z + \frac{\imath}{2H_\varrho} \right] A_\xi e^{\imath(k_x x + k_z z - \omega t)} e^{-z/2H_\varrho}. \tag{3.173}$$

The polarization relations (3.171–3.173) show that non-Boussinesq waves significantly change amplitude with height if they propagate over vertical distances comparable to $2H_\varrho$. This vertical symmetry-breaking is a distinguishing characteristic of non-Boussinesq waves. In the Boussinesq approximation there is no such structural difference between upward- and downward-propagating waves.

At first it appears that the exponential growth with height of ξ and $\vec{u}$ contradicts the small-amplitude assumption used to derive the linear partial differential equations (3.161). To be consistent, we must imagine the solution corresponds to the behaviour of a vertically localized small-amplitude wavepacket. Equation (3.171) predicts that this wavepacket grows exponentially in amplitude as it propagates upwards provided the amplitude remains sufficiently small for linear theory to be valid.

In reality, upward-propagating waves do not grow in amplitude indefinitely. Weakly nonlinear effects modify the structure of the waves and ultimately non-Boussinesq internal waves grow to such amplitude that they overturn and break, as discussed in Sections 4.5 and 4.6.

3.7.2 Anelastic waves in a gas

The essential dynamics of non-Boussinesq internal waves in a liquid extend to those of internal waves in air, although the governing equations are somewhat modified to account for the thermodynamics of a gas. These are called 'anelastic internal waves'. Whereas the mass, momentum and internal energy equations for a non-Boussinesq liquid are written in terms of density, the evolution of a dry adiabatic gas is given in terms of the potential temperature. We will show, however, that these can be manipulated for small-amplitude waves to yield a governing equation equivalent to that for waves in a non-Boussinesq liquid.

The fully nonlinear equations of motion for an anelastic gas were derived in Section 1.12.3. Neglecting the effects of rotation and background winds, these are given in the x–z plane by

$$\frac{Du}{Dt} = -\frac{\partial}{\partial x}\left(\frac{p}{\overline{\rho}}\right),$$

$$\frac{Dw}{Dt} = -\frac{\partial}{\partial z}\left(\frac{p}{\overline{\rho}}\right) + \frac{g}{\overline{\theta}}\theta,$$

$$\frac{D\theta}{Dt} = -w\frac{d\overline{\theta}}{dz},$$

$$\nabla \cdot (\overline{\rho}\vec{u}) = 0.$$

(3.174)

The equations differ from those for a non-Boussinesq liquid (3.160) in several ways. Here the key thermodynamic variable is the fluctuation potential temperature θ. Thus the internal energy equation now involves θ and $\overline{\theta}$ instead of ρ and $\overline{p}$. Likewise, the buoyancy term in the vertical momentum equation is now expressed in terms of θ as in (3.21). The manipulation of the momentum equations has resulted in the background density $\overline{\rho}$ now appearing inside the pressure gradient terms. Finally, the continuity equation has been approximated by the divergence of the background density times the velocity field. This differs from the continuity equation for an incompressible fluid (1.30), but still serves to filter vertically propagating sound waves. The approximation to the continuity equation is less restrictive than $\nabla \cdot \vec{u} = 0$ and, to emphasize this, the waves are sometimes said to be compressible.

As usual, the equations for small-amplitude waves are linearized through the removal of the advection terms in the material derivative. Thus we have

$$
\frac{\partial u}{\partial t} = -\frac{\partial}{\partial x}\left(\frac{p}{\rho}\right),
$$

$$
\frac{\partial w}{\partial t} = -\frac{\partial}{\partial z}\left(\frac{p}{\rho}\right) + \frac{g}{\overline{\theta}}\theta,
$$

$$
\frac{\partial \theta}{\partial t} = -\frac{d\overline{\theta}}{dz}w,
$$

$$
\frac{\partial \overline{\rho}u}{\partial x} + \frac{\partial \overline{\rho}w}{\partial z} = 0.
$$

(3.175)

Taking a different approach from our treatment of a non-Boussinesq liquid, we will reduce the coupled equations in (3.175) to a single equation first by implicitly defining a 'mass streamfunction' Ψ by

$$
(u, w) = \left(-\frac{1}{\overline{\rho}}\frac{\partial \Psi}{\partial z}, \frac{1}{\overline{\rho}}\frac{\partial \Psi}{\partial x}\right).
$$

(3.176)

The fourth equation of (3.175) guarantees that Ψ can be defined in this way.

Taking the curl of the momentum equations eliminates pressure and gives the vorticity equation

$$
\frac{\partial \zeta}{\partial t} = -\frac{g}{\overline{\theta}}\frac{\partial \theta}{\partial x},
$$

(3.177)

in which

$$
\zeta = (\nabla \times \vec{u}) \cdot \hat{y} = \frac{\partial}{\partial z}\left(-\frac{1}{\overline{\rho}}\frac{\partial \Psi}{\partial z}\right) - \frac{\partial}{\partial x}\left(\frac{1}{\overline{\rho}}\frac{\partial \Psi}{\partial x}\right) = -\frac{1}{\overline{\rho}}\left(\nabla^2 \Psi - \frac{\overline{\rho}'}{\overline{\rho}}\frac{\partial \Psi}{\partial z}\right).
$$

(3.178)

Likewise, we can write the internal energy equation explicitly in terms of Ψ and θ:

$$
\frac{\partial \theta}{\partial t} = -\frac{\overline{\theta}'}{\overline{\rho}}\frac{\partial \Psi}{\partial x},
$$

(3.179)

in which we have used a prime to denote the background potential temperature gradient.

Eliminating θ from (3.177) and (3.179), and using (3.178) gives

$$
\frac{\partial^2}{\partial t^2}\left(\nabla^2 \Psi\right) + \frac{1}{H_\varrho}\frac{\partial^3 \Psi}{\partial z \partial t^2} + N^2 \frac{\partial^2 \Psi}{\partial x^2} = 0,
$$

(3.180)

in which the squared buoyancy frequency is defined for an anelastic gas using (3.24), and we have defined the density scale height by $H_\varrho \equiv \left(-\overline{\rho}'/\overline{\rho}\right)^{-1}$, as in (1.75).

Until this point no assumptions have been made about the form of $\overline{\theta}$ and $\overline{\rho}$ and, correspondingly, N^2 and H_ϱ. These background fields are interrelated through (3.27). They can simultaneously be taken as constant if c_s, given by (1.59), is constant. This would be the case for an isothermal gas. Even if the temperature changes with height, variations in c_s are so small that H_ϱ can be taken as approximately constant in a uniformly stratified gas.

Remarkably, if both N and H_ϱ are constant, then (3.180) is identical to the differential equation (3.165) that describes the evolution of non-Boussinesq internal waves in a liquid. The distinction is that (3.180) describes the evolution of the mass streamfunction instead of the fluctuation density.

By analogy, we have solutions of the form

$$\Psi(x,z,t) = A_\Psi e^{i(k_x x + k_z z - \omega t)} e^{-z/2H_\varrho}, \tag{3.181}$$

in which ω satisfies the same dispersion relation as (3.169):

$$\omega^2 = N_0^2 \frac{k_x^2}{|\vec{k}|^2 + 1/(4H_\varrho^2)}. \tag{3.182}$$

The other relevant fields are related to A_Ψ through the polarization relations. In particular, using the second equation in (3.176) and relating the vertical displacement field to the vertical velocity through $\partial\xi/\partial t = w$, we find

$$\xi(x,z,t) = A_\xi e^{i(k_x x + k_z z - \omega t)} e^{z/2H_\varrho}, \tag{3.183}$$

where

$$A_\xi = -\frac{1}{\varrho_0} \frac{k_x}{\omega} A_\Psi. \tag{3.184}$$

The relationships between the velocity fields and A_ξ are then found to be the same as in (3.172). These and other polarization relations are listed in Table 3.2.

Using the fourth equation of (3.174) together with the horizontal momentum equation, the background density can be brought inside the derivatives of the advection terms when converting the equation to flux-form. Horizontally averaging thereby gives the momentum transport equation:

$$\frac{\partial \overline{\rho} \langle u \rangle}{\partial t} = -\frac{\partial}{\partial z} (\overline{\rho} \langle uw \rangle). \tag{3.185}$$

That is, the change in momentum is given directly by the divergence of the mean momentum flux

$$\langle \mathcal{F}_M \rangle \equiv \overline{\rho} \langle uw \rangle. \tag{3.186}$$

Like ξ, the amplitudes of the fluctuation velocity fields increase exponentially with height as $\exp(z/2H_\varrho)$. Because the background density decreases exponentially as $\overline{\rho} = \varrho_0 e^{-z/H_\varrho}$, the momentum flux given by (3.186) is independent of z.

Table 3.2. *Polarization relations and correlations between basic state fields for small-amplitude anelastic internal waves in a gas. The fluid is assumed to be uniformly stratified with $\overline{p} = \varrho_0 e^{-z/H_\varrho}$ and $N_0{}^2 = g\overline{\theta}'/\overline{\theta} = g/H_\theta$ constant. Each field is related to the vertical displacement amplitude A_ξ which is assumed to be real and positive. The middle column gives the values of the (complex) amplitudes at the reference height $z = 0$. The rightmost column gives the real fields in terms of the phase angle $\varphi = k_x x + k_z z - \omega t$. Amplitudes are represented in terms of time and length scales $N_0{}^{-1}$ and $k_x{}^{-1}$, respectively. The relative vertical wavenumber is represented by $\Theta = \tan^{-1}(k_z/k_x)$ and the frequency ω is given in terms of N_0, k_x and Θ through the dispersion relation given at the top of the table. The fluctuation potential temperature is related to the vertical displacement through the length scale $H_\star$, defined by $H_\star{}^{-1} = 1/(2H_\varrho) + 1/H_\theta$. For an isothermal gas, $H_\star = (14/11)H_\varrho$.*

Dispersion relation: $\quad \omega = N_0 k_x/\sqrt{|\vec{k}|^2 + 1/(2H_\varrho)^2} = N_0 \cos\Theta[1 + \cos^2\Theta/(2k_x H_\varrho)^2]^{-1/2}$

Vertical group speed: $\quad c_{gz} = -\dfrac{N_0}{k_x}\cos^2\Theta\sin\Theta[1 + \cos^2\Theta/(2k_x H_\varrho)^2]^{-3/2}$

Defining formula	Relationship to vertical displacement	
ξ	A_ξ	$\xi = A_\xi\, e^{z/2H_\varrho}\,\cos\varphi$
$\theta = -\dfrac{d\overline{\theta}}{dz}\xi$	$\mathcal{A}_\theta = -\theta_0 N_0{}^2/g\, A_\xi$	$\theta = -\theta_0\dfrac{1}{H_\theta}A_\xi e^{z/H_\star}\cos\varphi$
$\rho = -\dfrac{\overline{\rho}}{\overline{\theta}}\theta$	$\mathcal{A}_\rho = \rho_0 N_0{}^2/g\, A_\xi$	$\rho = \rho_0\dfrac{1}{H_\varrho}A_\xi e^{-z/2H_\varrho}\cos\varphi$
$\vec{u} = \dfrac{1}{\overline{\rho}}\nabla\times(\Psi\hat{y})$	$\mathcal{A}_\Psi = -\rho_0\dfrac{\omega}{k_x}A_\xi$	$\Psi = -\rho_0\dfrac{\omega}{k_x}A_\xi\, e^{-z/2H_\varrho}\cos\varphi$
$w = \dfrac{\partial\xi}{\partial t}$	$\mathcal{A}_w = -\imath\omega A_\xi$	$w = \omega A_\xi\, e^{z/2H_\varrho}\sin\varphi$
$u = -\dfrac{1}{\overline{\rho}}\dfrac{\partial\Psi}{\partial z}$	$\mathcal{A}_u = \dfrac{\omega}{k_x}(\imath k_z - 1/2H_\varrho)\, A_\xi$	$u = -\omega\tan\Theta\, A_\xi\, e^{z/2H_\varrho}\sin\varphi$ $\times[1 + \cot\varphi\cot\Theta/(k_x H_\varrho)]$
$\zeta = \partial_z u - \partial_x w$	$\mathcal{A}_\zeta = -\dfrac{k_x}{\omega}N_0{}^2 A_\xi$	$\zeta = -\dfrac{k_x}{\omega}N_0{}^2\, A_\xi\, e^{z/2H_\varrho}\,\cos\varphi$
$\rho_0\dfrac{\partial u}{\partial t} = -\dfrac{\partial p}{\partial x}$	$\mathcal{A}_p = \rho_0\dfrac{\omega^2}{k_x{}^2}(\imath k_z - 1/2H_R ho)A_\xi$	$p = \rho_0\dfrac{\omega^2}{k_x}\tan\Theta\, A_\xi e^{-z/2H_\varrho}\sin\varphi$ $\times[1 + \cot\varphi\cot\Theta/(k_x H_\varrho)]$

Correlations

$$\langle u\rangle_L = -\langle\xi\zeta\rangle = \frac{1}{2}\frac{k_x}{\omega}N_0{}^2 A_\xi{}^2 e^{z/H_\varrho}$$

$$\langle\mathcal{F}_{Mz}\rangle = \overline{\rho}\,\langle uw\rangle = -\frac{1}{2}\varrho_0\omega^2\tan\Theta A_\xi{}^2$$

$$\langle E\rangle = \langle E_K + E_P\rangle = \frac{1}{2}\varrho_0 N_0{}^2 A_\xi{}^2$$

$$\langle\mathcal{F}_{Ez}\rangle = \langle pw\rangle = -\frac{1}{2}\varrho_0\frac{\omega^3}{k_x{}^3}\tan\Theta A_\xi{}^2 = c_{gz}\langle E\rangle$$

Therefore the reason non-Boussinesq waves grow exponentially with height is a consequence of momentum conservation. As internal waves propagate upwards, the fluctuation velocity must increase as the ambient fluid density decreases so that the momentum transport remains constant. This is why the amplitude of the velocity components grows as the inverse square root of the background density. The same conclusion follows from energy conservation.

The wave-induced mean flow for anelastic waves is conveniently represented, as for Boussinesq waves, by the correlations between the vertical displacement and vorticity fields: $\langle u \rangle_L = -\langle \xi \zeta \rangle$. From the polarization relations, this is given explicitly by

$$\langle u \rangle_L = \frac{1}{2} \frac{k_x}{\omega} N_0{}^2 A_\xi{}^2 e^{z/H_\varrho}. \tag{3.187}$$

Unlike Boussinesq waves, this formula shows that the wave-induced mean flow increases exponentially as the waves propagate upwards. However, the momentum associated with the flow $\overline{\rho} \langle u \rangle_L$ remains constant. It is straightforward to check that (3.187) is equivalent to $\langle uw \rangle / c_{gz}$, consistent with the interpretation that the wave-induced mean flow results from vertical variations in the transport of momentum per unit mass by wavepackets moving at the vertical group velocity.

We can also derive the mean energy equation through manipulation of the linearized momentum and internal energy equations in (3.175). As with Boussinesq waves, we find

$$\frac{\partial}{\partial t} (\langle E_K \rangle + \langle E_P \rangle) = -\frac{\partial}{\partial z} \langle \mathcal{F}_E \rangle, \tag{3.188}$$

in which

$$\langle E_K \rangle = \frac{1}{2} \overline{\rho} \langle u^2 + w^2 \rangle \tag{3.189}$$

is the kinetic energy density,

$$\langle E_P \rangle = \frac{1}{2} \overline{\rho} N_0{}^2 \langle \xi^2 \rangle \tag{3.190}$$

is the available potential energy, and

$$\langle \mathcal{F}_E \rangle = \langle pw \rangle \tag{3.191}$$

is the vertical energy flux. Explicit formulae for these correlations are provided in Table 3.2. As expected, the kinetic and available potential energy are in equipartition and the mean energy flux is equal to the vertical group velocity times the mean energy. Like the momentum flux, although the velocity and displacement amplitudes grow exponentially with height, the mean energy and flux are constant as a result of multiplying correlations of velocity and displacement by the background density, which decreases with height.

3.8 Internal waves influenced by rotation

Here we examine how Coriolis forces influence the dynamics of internal waves. This discussion is limited to waves on the f-plane, meaning that we assume the inertial frequency, $f_0 = 2\Omega_e \sin\phi_0$, can be treated as a constant. In Section 2.7.2 we referred to surface and interfacial waves on the f-plane as inertial waves. Here we distinguish internal waves on the f-plane from inertial waves by referring to them as 'inertia gravity waves'. Some texts also describe them as 'inertio-gravity waves'.

Because Coriolis forces act to deflect horizontal motions, we need to examine their motion in fully three-dimensional space. For mathematical simplicity we restrict our attention here to small-amplitude Boussinesq waves in stationary, uniformly stratified fluid.

The addition of Coriolis forces on the f-plane amounts to extending the Boussinesq equations of motion (3.41) to $\vec{x} = (x,y,z)$ space and, from (1.54), adding $-f_0 v$ and $f_0 u$ to the left-hand sides of the x- and y-momentum equations, respectively. The resulting equations represented in matrix form are

$$
\begin{pmatrix}
\partial_t & -f_0 & 0 & 0 & +\frac{1}{\rho_0}\partial_x \\
f_0 & \partial_t & 0 & 0 & +\frac{1}{\rho_0}\partial_y \\
0 & 0 & \partial_t & +\frac{g}{\rho_0} & +\frac{1}{\rho_0}\partial_z \\
0 & 0 & -\frac{\rho_0}{g}N_0^2 & \partial_t & 0 \\
\partial_x & \partial_y & \partial_z & 0 & 0
\end{pmatrix}
\begin{pmatrix} u \\ v \\ w \\ \rho \\ p \end{pmatrix} = 0.
\tag{3.192}
$$

The equations have constant coefficients and the domain is unbounded. Therefore we may assume there exist plane wave solutions of the form

$$
b(x,y,z,t) = A_b e^{\iota(\vec{k}\cdot\vec{x}-\omega t)},
\tag{3.193}
$$

in which $\vec{k} = (k_x, k_y, k_z)$ and b represents any of the basic state fields, u, v, w, ρ or p.

As in the solution of the equations for Boussinesq internal waves, we may substitute (3.193) into (3.192) to give an algebraic eigenvalue problem from which the matrix determinant gives the dispersion relation, or we may first take the determinant to get a partial differential equation from which substituting (3.193) gives the dispersion relation. In either case we find

$$
\omega^2 = \frac{N_0^2 k_h^2 + f_0^2 k_z^2}{|\vec{k}|^2} = N_0^2\cos^2\Theta + f_0^2\sin^2\Theta,
\tag{3.194}
$$

in which $k_h \equiv |(k_x, k_y)|$ is the magnitude of the horizontal component of the wavenumber vector and Θ is defined as in (3.35) so that, implicitly,

$$\tan \Theta = \frac{k_z}{|\vec{k_h}|}. \tag{3.195}$$

The evolution equation given in terms of the vertical displacement field ξ is

$$\frac{\partial^2}{\partial t^2} \nabla^2 \xi + \left(N_0{}^2 \nabla_h{}^2 + f_0{}^2 \partial_z{}^2 \right) \xi = 0, \tag{3.196}$$

in which $\nabla = (\partial_x, \partial_y, \partial_z)$ is the gradient operator in three dimensions and $\nabla_h = (\partial_x, \partial_y)$ is the horizontal gradient operator. This equation could be derived directly from (3.194) by transforming the wavenumber components and frequency in the dispersion relation to spatial and temporal derivatives, as in (1.127).

In the case $f_0 = 0$, the dispersion relation (3.194) reduces to the three-dimensional analogue of that for Boussinesq internal waves (3.54). Likewise (3.196) reduces to the evolution equation for three-dimensional Boussinesq internal waves.

Just as N_0 prescribes an upper bound on propagating internal wave frequencies, (3.194) shows that f_0 is a low-frequency cut-off: internal waves cannot propagate with frequencies lower than f_0. Internal waves with frequency close to but larger than f_0 are sometimes specifically referred to as inertia gravity waves.

Because the Coriolis frequency is much smaller than characteristic values of the buoyancy frequency, inertia gravity waves have much smaller frequency than N_0 and so are hydrostatic. Consequently, as discussed in Section 3.3.6, the phase lines of inertia gravity waves are almost horizontal: $|\Theta| \sim 90°$. By combining (3.194) and (3.195), the magnitude of the slope of lines of constant phase is

$$|k_h/k_z| = \cot|\Theta| = \sqrt{\frac{\omega^2 - f_0{}^2}{N_0{}^2 - \omega^2}}. \tag{3.197}$$

This depends upon frequency alone and not on the spatial scale of the waves.

Assuming $f_0 \lesssim \omega \ll N_0$ so that $|\vec{k}| \simeq |k_z|$, the dispersion relation (3.194) is sometimes written in the approximate form for inertia gravity waves as

$$\omega^2 = \frac{N_0{}^2}{k_z{}^2} k_h{}^2 + f_0{}^2. \tag{3.198}$$

This should be compared with the dispersion relation (2.160) for inertial waves in a shallow one-layer fluid. The equations are identical if one replaces the shallow water wave speed c with $N_0/|k_z|$. Indeed, this heuristic generally provides a convenient way to extend the results of shallow water theory to hydrostatic internal waves in continuously stratified fluid.

The structure of the waves can be determined from the eigenvectors of (3.192), which gives the polarization relations for inertia gravity waves. In particular, for waves with no structure in the y-direction ($k_y = 0$), the velocity field is given in terms of the real-valued vertical displacement amplitude A_ξ by

$$\mathcal{A}_u = \iota\omega\frac{k_z}{k_x}A_\xi \;\Rightarrow\; u = -\omega\tan\Theta\, A_\xi\sin(k_x x + k_z z - \omega t) \qquad (3.199)$$

$$\mathcal{A}_v = f_0\frac{k_z}{k_x}A_\xi \;\Rightarrow\; v = f_0\tan\Theta\, A_\xi\cos(k_x x + k_z z - \omega t) \qquad (3.200)$$

$$\mathcal{A}_w = -\iota\omega A_\xi \;\Rightarrow\; w = \omega A_\xi\sin(k_x x + k_z z - \omega t), \qquad (3.201)$$

in which we have used (3.195) in evaluating the actual value of the fields to the right. For inertia gravity waves, we may use (3.198) to approximate $|\tan\Theta| = |k_h/k_z| \simeq N_0/\sqrt{\omega^2 - f_0^2}$.

The relationship between the velocity fields for downward- and upward-propagating waves is illustrated in Figure 3.17. The u and v fields are 90° out of phase, implying that fluid parcels follow near-horizontal elliptical paths during the passage of the waves. The motion becomes circular in the horizontal as the wave frequency ω approaches the Coriolis frequency f_0. Also in this limit, $|\Theta| \to 90°$ which means $\tan|\Theta| \gg 1$. So the motion is predominately horizontal with vertical excursions that are smaller by a factor $\cot|\Theta|$. This result is consistent with the dynamics of hydrostatic waves, for which buoyancy forces are exactly balanced by vertical pressure gradient forces.

The horizontal orbital motion of inertia gravity waves together with their small vertical motion is similar to that for inertial (shallow water) waves. However, comparing the polarization relations (2.165) and (2.166) to (3.199) and (3.200), respectively, we see some crucial differences.

First, for inertia gravity waves u is out of phase, not in phase, with the vertical displacement field. It is the spanwise velocity v that is in phase with A_ξ. Second, whereas the motion of fluid parcels associated with inertial waves is synchronized over the depth of the fluid, inertia gravity waves have vertically varying motion that depends upon the sign of $\Theta = \tan^{-1}(k_z/k_x)$ as well as the sign of f_0.

At any position, fluid parcels for both types of waves undergo anticyclonic orbits in time, as illustrated in Figure 3.17b and d. That is, fluid moves along tilted elliptical paths opposite to the direction of the background rotation given by the sign of f_0. In the northern hemisphere, this motion is clockwise seen from above. At a snapshot in time the velocity field changes its orientation with increasing depth so that it rotates anticyclonically as seen from above if the crests move upwards and the group velocity is downwards. However, if the crests move downwards and the

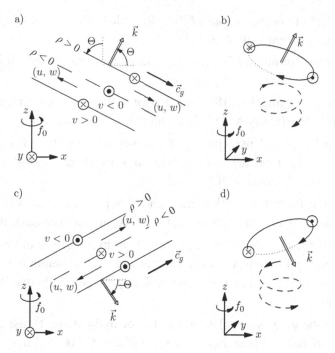

Fig. 3.17. Schematic showing polarization relations for rightward a),b) upward-moving phase (downward group velocity) and c), d) downward-moving phase (upward group velocity) inertia gravity waves. All figures are shown with $f_0 > 0$, consistent with motion in the northern hemisphere. The velocity field at a snapshot in time is shown in a) and c) for which the y-axis is directed into the page. The parcel motion is illustrated in a perspective view in b) and d). At a fixed vertical level seen from above, the motion (solid line) is anticyclonic whereas the sense of rotation with increasing depth (dashed line) is b) clockwise if $k_z > 0$ and $c_{gz} < 0$ and d) anti-clockwise if $k_z < 0$ and $c_{gz} > 0$.

group velocity is upwards, the sense of rotation with depth at a snapshot in time is cyclonic. This property of inertia gravity waves proves useful for oceanographers in diagnosing, based upon the rotation of the velocity vector with depth, whether the observed waves originated from surface or bottom processes.

Exercises

3.1 Derive the force equation (3.22) for a vertically displaced parcel of an ideal gas.

3.2 Using a fluid parcel argument and the definition for potential density (1.21), derive a formula for the squared buoyancy frequency of a non-Boussinesq liquid of great depth.

3.3   Use (3.15) to determine an explicit formula for $\bar{\rho}$ in terms of a prescribed background temperature, $\bar{T}$.

3.4   Derive the formula (3.26) for the squared buoyancy frequency of an ideal gas given in terms of temperature.

3.5   Derive the formula (3.18), which relates the potential temperature scale height to the density scale height and pressure scale height.

3.6   The background density profile of a non-Boussinesq liquid is given by $\bar{\rho} = \varrho_0(1 - z/H)$. Find the squared buoyancy frequency and show that N^2 is approximately constant if $|z| \ll H$.

3.7   Write the formulae for the phase and group velocity of two-dimensional Boussinesq waves in terms of N_0, Θ, $k \equiv |\vec{k}|$ and, if necessary, the sign of k_x.

3.8   Write explicit formulae for the phase and group velocity of three-dimensional Boussinesq internal waves characterized by wavenumber vector $\vec{k} = (k_x, k_y, k_z)$. Show that the group velocity is perpendicular to the phase velocity.

3.9   Derive the dispersion relation, (3.71), for hydrostatic, Boussinesq internal waves in the x–z plane. For what angle $\Theta \equiv \tan^{-1}(k_z/k_x)$ does the resulting frequency differ from that for non-hydrostatic waves by 1%? For waves at this angle, what is their frequency relative to N_0 and what slope is formed by lines of constant phase?

3.10   Derive the mass conservation equation, given for stationary fluid by (3.76), by including the existence of a mean background horizontal flow $\overline{U}(z)$ that varies in the vertical.

3.11   For Boussinesq internal waves in unsheared, non-rotating continuously stratified fluid, show that the mass flux $\rho\bar{u}$ averages to zero.

3.12   Use the polarization for internal waves and the period-averaged forms of (3.79) and (3.81) to calculate the internal wave kinetic and available potential energy, respectively, and show that they are in equipartition.

3.13   Find the period-averaged formulae for the wave action $\langle A \rangle$ and its flux $\langle \mathcal{F}_A \rangle$. Give your results in terms of ϱ_0, N_0, k_x and $\Theta = \tan^{-1}(k_z/k_x)$. What are the corresponding formulae for the small-amplitude form of the pseudoenergy?

3.14   Show that the discrete spectrum of internal waves in a rectangular box is given by (3.114).

3.15   The Garrett–Munk spectrum represents the stratification in the ocean at mid-latitudes by $N(z) = N_* \exp(z/H_\sigma)$. Under the procedure of WKB renormalization, write an explicit formula for the stretched vertical co-ordinate, Z, and determine the vertical structure of the jth hydrostatic mode in terms of the horizontal and vertical velocity. In this problem assume the ocean depth, H, is much larger than the e-folding depth H_σ.

3.16 A non-rectangular, two-dimensional domain has boundaries at $z = 0$, $z = 1$, $x = 0$ and along the line $z = x - 1$. Sketch the attractor for internal waves in this domain in which the fluid is a uniformly stratified fluid with buoyancy frequency N_0 and the waves are excited with relative frequency a) $\omega/N_0 = 1/2$, b) $\omega/N_0 = \sqrt{3}/2$, c) $\omega/N_0 \gtrsim 1/\sqrt{2}$ and d) $\omega/N_0 \lesssim 1/\sqrt{2}$.

3.17 Derive the mean energy equation (3.92) for internal waves in the presence of a mean background shear flow. Give your results in terms of ϱ_0, N_0, k_x and Θ.

3.18 Derive the analogue to the Taylor–Goldstein equation for a non-Boussinesq liquid.

3.19 For the uniformly stratified jet flow prescribed by $\bar{U}(z) = U_0 \text{sech}^2(z/L)$ and $N = N_0$, use the Frobenius expansion to find the leading-order power law behaviour of $\hat{\psi}$ about the critical level at $z = 0$ for waves with phase speed $c = U_0$.

3.20 A fluid has a continuous background density profile given by

$$\bar{\rho}(z) = \begin{cases} \rho_1 & z > H/2 \\ -\frac{\rho_2 - \rho_1}{H} z + \frac{\rho_2 + \rho_1}{2} & |z| \leq H/2 \\ \rho_2 & z < -H/2. \end{cases}$$

(a) Use piecewise-linear theory to find the corresponding dispersion relation, $\omega = \omega(k, m)$, for two-dimensional waves in x–z space.
(b) Examine your solution in the limits $mH \ll 1$ and $mH \gg 1$ and compare your result with the dispersion relation for interfacial waves and waves in a continuously stratified fluid.

3.21 The kinetic energy density associated with anelastic internal waves is given by

$$E_K = \frac{1}{2}\bar{\rho}(u^2 + w^2).$$

(a) For waves in a uniformly stratified fluid, explicitly determine $\langle E_K \rangle$ in terms of the vertical displacement amplitude A_ξ.
(b) Derive the flux-form of the total energy equation for anelastic internal waves.
(c) Explicitly evaluate the mean vertical flux of total energy, $\langle F_{Ez} \rangle$, and show that this equals the energy times the vertical group velocity.

3.22 Using (3.192) and (3.193), convert the differential matrix eigenvalue problem to an algebraic matrix eigenvalue problem and so derive the dispersion relation (3.194) for inertia gravity waves.

3.23 Explicitly determine the polarization relations for inertia gravity waves from the governing equations (3.192). Specifically, find how the amplitude of the velocity field components u, v, and w are related to the vertical displacement amplitude A_ξ.

3.24 Determine how the phase of the horizontal velocity field of inertia gravity waves changes with increasing depth as a function of f_0 and k_z. In the northern hemisphere, does the velocity vector associated with waves having downward group velocity rotate clockwise or counter-clockwise with depth?

4

Nonlinear considerations

4.1 Introduction

The linear theory calculations performed so far have the advantage that analytic solutions can be found for a broad range of circumstances. However, the solutions are strictly valid only in the limit of infinitesimally small amplitude waves. Waves are said to be weakly nonlinear when their amplitude is sufficiently large that nonlinear effects arising from the advection terms in the material derivative begin to play an important role. Using perturbation theory, simplified equations can be derived that capture the leading-order nonlinear effects, referred to as 'weak nonlinearity'. Analysis of such equations not only reveals how nonlinear effects change the evolution of the waves, but they serve to establish bounds on the wave amplitude for which linear theory is valid.

As well as modifying the structure of small-amplitude waves, nonlinear effects can give rise to new classes of steady waves, such as solitary waves, and they can result in the breakdown of steady waves through mechanisms such as modulational instability and parametric subharmonic instability.

In this chapter we focus primarily upon the weakly nonlinear dynamics of interfacial and internal waves in otherwise stationary fluid; there is no ambient mean flow. For those not familiar with perturbation theory for differential eigenvalue problems, we begin with a brief review of the mathematics necessary for the treatment of weakly nonlinear waves. This analysis shows that frequency is a function of amplitude as well as wavenumber. We then examine how weakly nonlinear effects modify the evolution of interfacial and internal waves. The chapter closes with a discussion of instabilities associated with finite-amplitude internal waves.

4.2 Weakly nonlinear theory

Waves are said to be 'weakly nonlinear' if the nonlinear terms in the evolution equations have a small but not insignificant effect upon their dynamics: their amplitude is finite, but not too large. In this case, perturbation methods may be employed to determine approximate analytic representations of their structure and dispersion characteristics.

We begin by working through a simple example from mechanics, which serves to illustrate the mathematical technique. We then describe the generic approach to finding the weakly nonlinear dispersion relation and the structure of one-dimensional waves that are periodic in x and t. Using this result, we go on to derive the nonlinear Schrödinger equation that describes the evolution of the amplitude envelope of finite-amplitude wavepackets.

4.2.1 The nonlinear pendulum

The formula describing the motion of a point mass at the end of a weightless rigid string of length ℓ is given by

$$\phi'' + \frac{g}{\ell}\sin\phi = 0, \tag{4.1}$$

in which g is the acceleration due to gravity and $\phi(t)$ is the angle formed between the string and the vertical. If the oscillations are small, then we can approximate $\sin\phi \simeq \phi$, in which case (4.1) reduces to a spring equation describing oscillatory motion with frequency $(g/\ell)^{1/2}$.

We wish to examine how the motion and frequency change when oscillations are moderately large. That is, we seek solutions of

$$\phi'' + \frac{g}{\ell}\left(\phi - \frac{1}{6}\phi^3\right) = 0, \tag{4.2}$$

in which we have kept the order ϕ^3 term in the Taylor-series expansion of $\sin\phi$.

The standard though (as it turns out here) incomplete perturbation theory assumption is to write

$$\phi = A_0\left(\phi_0 + A_0\phi_1 + A_0^2\phi_2 + \ldots\right). \tag{4.3}$$

Here A_0 represents the small but finite amplitude of the pendulum and ϕ_0, ϕ_1, ... are order unity functions of time which are yet to be determined.

Substituting (4.3) into (4.2) and extracting leading-order terms, we recover the linearized problem of (4.1) in which $\sin\phi \simeq \phi$. The solution gives the squared frequency

$$\omega_0^2 = g/\ell, \tag{4.4}$$

and the time-dependent motion is represented by the order-unity function

$$\phi_0 = \cos(\omega_0 t). \tag{4.5}$$

Here we have arbitrarily set the phase so that the linear pendulum reaches maximum amplitude at $t = 0$.

At next order in A_0 (found by extracting terms of order $A_0{}^2$) we have

$$\phi_1'' + \omega_0{}^2 \phi_1 = 0. \tag{4.6}$$

This result is independent of ϕ_0. Because higher-order terms in the perturbation expansion (4.3) must depend on the terms before them we insist that $\phi_1 = 0$. Another way to reach this conclusion is to see that (4.6) is the same as (4.4) and so we can combine the first two terms of the expansion into the leading-order term.

At next order in A_0 we have

$$\phi_2'' + \omega_0{}^2 \phi_2 = \frac{1}{6}\omega_0{}^2 \cos^3(\omega_0 t) = \frac{1}{24}\omega_0{}^2 \cos(3\omega_0 t) + \frac{1}{8}\omega_0{}^2 \cos(\omega_0 t). \tag{4.7}$$

On the right-hand side of this equation we have applied a trigonometric identity to the cosine-cubed term.

We ignore the homogeneous solutions, just as we did for solutions of (4.6), and we focus upon finding the particular solution of (4.7). This gives ϕ_2 as it depends upon forcing by lower-order terms.

However, the last term on the right-hand side of (4.7) poses a problem. Because $\cos \omega_0 t$ is a solution of the linear operator on the left-hand side of (4.7) its presence means that ϕ_2 involves terms proportional to $t \cos \omega_0 t$ and $t \sin \omega_0 t$. The culprit is called a 'secular term', which gives rise to particular solutions that exhibit unbounded growth of the amplitude in time. This would occur if a pendulum was externally forced at its resonant frequency. But there is no external force in this problem. Because no external energy source exists to give rise to a temporally increasing amplitude the secular term leads to an unphysical result.

The problem is rectified if we assume that finite-amplitude effects additionally modify the frequency ω of the oscillations, an insight attributed to Poincaré and Lindstedt. Specifically, we suppose that the frequency can be written as a power series in A_0 so that

$$\omega^2 = \omega_0{}^2 \left(1 + A_0{}^2 \sigma_2 + \ldots \right). \tag{4.8}$$

Because (4.2) involves two time derivatives, we have formulated the expansion in ω^2 and we have neglected the $A_0 \sigma_1$ term because secular terms did not arise in (4.6). Alternately, one could argue that (4.8) is the appropriate perturbation expansion on the basis of symmetry: the results should not depend upon the sign of ω or A_0.

Using (4.8), the equation for ϕ_2 becomes

$$\phi_2'' + \omega_0{}^2 \phi_2 + (\omega_0{}^2 \sigma_2) \cos(\omega_0 t) = \frac{1}{24}\omega_0{}^2 \cos(3\omega_0 t) + \frac{1}{8}\omega_0{}^2 \cos(\omega_0 t). \tag{4.9}$$

We choose the value of σ_2 precisely so that the secular terms, the ones proportional to $\cos(\omega_0 t)$, vanish. This is done by setting $\sigma_2 = 1/8$. The remaining forcing term on the right-hand side of (4.9) is not resonant and so we find a solution for ϕ_2 that does not grow with time. Explicitly, the motion of the pendulum is given by

$$\phi \simeq A_0 \cos(\omega t) - \frac{1}{192}A_0{}^3 \cos(3\omega t), \qquad (4.10)$$

in which the frequency of oscillation is modified according to

$$\omega^2 \simeq \frac{g}{\ell}\left(1 + \frac{1}{8}A_0{}^2\right). \qquad (4.11)$$

Thus we have found that the fundamental frequency of the oscillation increases as the square of the amplitude and that weakly nonlinear effects introduce superharmonic modulations to the periodic motion, though these would be barely discernible even if $A_0 \simeq 1$.

4.2.2 Weakly nonlinear theory for partial differential equations

A fully nonlinear system of equations can be represented generally by

$$Lf = N(f), \qquad (4.12)$$

in which L is a linear operator acting on f, N is a nonlinear operator involving f and its derivatives, and f itself is a function that represents the wave structure, such as its vertical displacement, velocity potential or streamfunction. We are looking for time-periodic solutions and so we write $L \equiv L(\omega)$, a linear operator involving the frequency ω and spatial derivatives of f but not time derivatives.

As above, we will use the wave amplitude A_0 as our perturbation parameter. Saying A_0 is small would make more sense if we defined it nondimensionally as the product of the amplitude and horizontal wavenumber. However, here we will work with dimensional variables in order to establish a stronger link with the theory for small-amplitude waves, developed in earlier chapters in this book.

The usual assumption of perturbation theory is that f can be represented by a power series in A_0 so that

$$f = A_0\left(f_0 + A_0 f_1 + A_0{}^2 f_2 + \ldots\right). \qquad (4.13)$$

Here the sequence of functions, $f_0, f_1, \ldots$ are assumed to be of order unity. For example, this could be done by assuming the maximum value $\|f_i\|$ is of order unity.

As in the nonlinear pendulum example, we will see in the description of weakly nonlinear interfacial and internal waves that the dispersion relation assumes a power series of the form (4.8):

$$\omega^2 = \omega_0{}^2 \left(1 + A_0{}^2 \sigma_2 + \dots\right). \tag{4.14}$$

The expansion is in terms of ω^2 because of the temporal symmetry typically associated with interfacial and internal waves: there is no dynamical difference between waves that move forwards or backwards in time. Symmetry also dictates that the terms with odd powers of A_0 vanish.

With these assumptions, we now expand the linear operator, L, and the nonlinear operator, N, as power series in A_0:

$$L(\omega) = L_0(\omega_0) + A_0{}^2 L_2(\omega_0, \sigma_2) + \dots \tag{4.15}$$

and

$$\frac{1}{A_0} N(f) = A_0 N_1(f_0) + A_0{}^2 N_2(f_0, f_1) + \dots \tag{4.16}$$

Here $L_0, L_2, \dots$ are linear operators that do not depend upon A_0. Likewise, $N_1, N_2, \dots$ are independent of A_0 and we have made the assumption that, to leading order, the nonlinearity in N is quadratic in f. That is, we assume N involves pairwise products of f and its derivatives. This is the case, for example, in the advection terms of the material derivative.

Substituting the perturbation expansions (4.13), (4.15) and (4.16) into (4.12), a sequence of linear partial differential equations are extracted at successive powers in A_0.

At leading order we have

$$L_0 f_0 = 0. \tag{4.17}$$

This eigenvalue problem (4.17) gives the linear dispersion relation, $\omega_0(\vec{k})$, and the corresponding structure of small-amplitude waves as represented by f_0, a result that could have been found using linear theory.

The weakly nonlinear equations involve products of fields and so it is necessary to represent f explicitly as the real part by (1.107) rather than by (1.106). Thus, for horizontally periodic small-amplitude waves,

$$A_0 f_0 \propto \frac{1}{2} \mathcal{A}_0 e^{i\varphi} + \text{cc}, \tag{4.18}$$

in which, for one-dimensional waves, $\varphi = kx - \omega t$ represents the relative phase and cc represents the complex conjugate. The absolute phase of f_0 is set by the argument of the complex amplitude $\mathcal{A}_0$. The factor $1/2$ is included on the right-hand side so that the magnitude $A_0 = |\mathcal{A}_0|$ represents the half peak-to-peak amplitude.

At next order in A_0 we have

$$L_0 f_1 = N_1(f_0). \tag{4.19}$$

This equation reveals how the finite-amplitude structure at order A_0 depends upon nonlinear interactions between small-amplitude waves. Because N_1 is quadratic in f_0 and its derivatives, the right-hand side of (4.19) is proportional to $\exp(\pm 2\iota\,\varphi)$ and a constant; explicitly, $\exp(0\iota\,\varphi)$. We ignore the homogeneous solution of (4.19), effectively incorporating it into the leading-order linear solution. Thus f_1 is a linear combination of $\exp(\pm 2\iota\varphi)$ and $\exp(0\iota\varphi)$. That is, nonlinear wave–wave interactions can excite superharmonic waves with double the wavenumber, $2k$, and they can induce a mean disturbance (of zero wavenumber).

Importantly, the quadratic form of N_1 means that $N_1(f_0)$ does not have a term proportional to $\exp(\pm\iota\,\varphi)$, which is a homogeneous solution of the linear differential equation (4.17): f_1 is not forced resonantly by the nonlinearity. So secular terms do not arise at this order of the perturbation expansion, further justifying why the $A_0\sigma_1$ term was not included in the expansion of ω^2 in (4.14).

At next highest order we have

$$L_0(\omega_0)f_2 + L_2(\omega_0,\sigma_2)f_0 = N_2(f_0,f_1), \qquad (4.20)$$

an equation that involves the unknown constant σ_2 and the unknown function f_2. Now quadratic interactions between f_0 and f_1 excite waves with structure proportional to $\exp(\pm\iota\varphi)$ and $\exp(\pm 3\iota\varphi)$. The waves with oscillatory behaviour given by $\exp(\pm\iota\varphi)$ resonate with the fundamental waves through the operator L_0. If not rectified, the solutions of (4.20) would have the form $\varphi\exp(\pm\iota\varphi)$. As in the pendulum problem, the secular terms unphysically result in the unbounded growth in time and space of the wave amplitude.

This is why we have introduced the perturbation expansion (4.14) in terms of even powers of A_0. The value of σ_2 is chosen precisely so as to cancel the secular terms. Once σ_2 is determined and the resonant forcing terms on the right-hand side of (4.20) removed we can solve (4.20) for the particular solution f_2.

The procedure then repeats itself by determining (4.12) at successively higher powers in A_0 and solving for successive values of σ_i and f_i. This is easier said than done because the algebra can quickly become unwieldy. It advisable to use a symbolic-algebra program like 'MAPLE' to provide a check on the work.

4.2.3 The nonlinear Schrödinger equation

The above procedure allows us to determine how the dispersion relation is modified by weakly nonlinear effects. For interfacial and internal waves the resulting formula for ω may be written as

$$\omega = \omega_0 + A_0^2\,\omega_2 + \ldots, \qquad (4.21)$$

in which, using (4.14), the leading-order nonlinear correction to the frequency is represented by the coefficient

$$\omega_2 = \frac{1}{2}\omega_0\sigma_2. \tag{4.22}$$

Knowing how finite-amplitude effects modify the dispersion relation as in (4.21), we can formulate a differential equation that describes the evolution of a finite-amplitude wavepacket. This is known as a 'nonlinear Schrödinger' or 'NLS' equation.

The derivation of the linear Schrödinger equation was discussed in Section 1.15.6. Here this is extended to include finite-amplitude effects. Again we represent the structure in terms of η as a Fourier transform

$$\eta(x,t) = \int_{-\infty}^{\infty} \hat{\eta}(k)e^{\iota(kx-\omega t)}\,dk, \tag{4.23}$$

in which $\hat{\eta}(-k) = \hat{\eta}^{\star}(k)$ ensures the integral is real-valued. Its initial structure is prescribed by the real part of

$$\eta(x,0) = \mathcal{A}(x,0)\,e^{\iota k_0 x}. \tag{4.24}$$

The characteristic amplitude of the waves is typically taken to be the maximum value $A_0 = ||\mathcal{A}||$.

Assuming the wavepacket is quasi-monochromatic with peak wavenumber k_0 and corresponding frequency $\omega(k_0)$, the waves at later times can be represented by

$$\eta(x,t) = \mathcal{A}(x,t)e^{\iota[k_0 x - \omega(k_0)t]}, \tag{4.25}$$

in which the amplitude envelope, $\mathcal{A}(x,t)$, evolves slowly in space and time compared with the wavelength and period of the waves. Note that $\mathcal{A}$ is generally complex valued, so the relative phase as well as the amplitude of the waves varies in space and time.

As justified in Section 1.15.6, we Taylor-expand the dispersion relation (4.21), which now includes the finite-amplitude correction:

$$\omega(k) \simeq \omega_0(k_0) + \omega_0'(k_0)(k - k_0) + \frac{1}{2}\omega_0''(k_0)(k - k_0)^2 + \omega_2(k_0)|A_0|^2. \tag{4.26}$$

Substituting (4.25) into (4.23), using (4.26), taking x- and t-derivatives and comparing terms, we arrive at the following approximate evolution equation for the amplitude envelope $\mathcal{A}$:

$$\frac{\partial \mathcal{A}}{\partial t} \simeq -\omega_0'\frac{\partial \mathcal{A}}{\partial x} + \iota\frac{1}{2}\omega_0''\frac{\partial^2 \mathcal{A}}{\partial x^2} - \iota\omega_2|\mathcal{A}|^2\mathcal{A}. \tag{4.27}$$

The last two terms on the right-hand side of (4.27) represent the leading-order competing effects of linear and nonlinear dispersion, respectively. To focus upon these dispersion effects, it is convenient to transform the equation into a co-ordinate system that moves with the wavepacket at the group velocity. This is done by defining a translating spatial co-ordinate $X = x - c_g t$. Transforming from (x,t) to (X,t) has the effect of combining the first term on the right-hand side of (4.27) with the time-derivative on the left-hand side. Rearranging the result gives the standard form of the one-dimensional nonlinear Schrödinger equation:

$$i\mathcal{A}_t = -\frac{1}{2}\omega_0''(k_0)\,\mathcal{A}_{XX} + \omega_2(k_0)\,|\mathcal{A}|^2\mathcal{A}. \tag{4.28}$$

For small-amplitude waves, the last term on the right-hand side, which is proportional to amplitude-cubed, is negligibly small compared to the linear terms and we recover (1.136).

While (4.28) describes the evolution of the amplitude and phase of weakly nonlinear wavepackets, the following section provides fundamental insights into the qualitative behaviour of its solutions.

4.2.4 Modulational stability theory

AM radio works by sending a signal at fixed frequency but varying the amplitude of superimposed waves. For FM radio, the frequency varies slightly about a mean value. As such, the respective signals are said to be amplitude and frequency modulated.

Radio waves, which are just low frequency electromagnetic (light) waves, are small-amplitude and are nondispersive in uniform media: their dispersion relation is expressed simply by $\omega = \pm ck$, in which c is a k-independent constant. So pulses formed by a superposition of these nondispersive waves maintain their form as the wavepacket propagates. However, if the waves in question are dispersive, linear theory predicts that the pulses spread out. Here we show that nonlinear effects can sometimes cause the pulses to narrow and grow in amplitude. This development of the equations draws from ray theory, which is discussed in more detail in Section 6.3 regarding the propagation of waves in non-uniform media.

The propagation of plane waves can be represented by an amplitude multiplying the complex exponential $\exp[i(kx - \omega t)]$. Defining the phase to be $\varphi = kx - \omega t$, we have $k = \partial\varphi/\partial x$ and $\omega = -\partial\varphi/\partial t$. Eliminating φ from these equations gives an evolution equation for k as it depends upon spatial variations of the frequency:

$$\frac{\partial k}{\partial t} = -\frac{\partial \omega}{\partial x}. \tag{4.29}$$

Using the definition of the group speed, (4.29) can be rewritten entirely in terms of an advection equation for k:

$$\frac{\partial k}{\partial t} + c_g \frac{\partial k}{\partial x} = 0. \tag{4.30}$$

Separately, we develop an equation for the evolution of the amplitude of the waves. If the waves are sufficiently small amplitude and propagate in the absence of background shear, the energy flux is just $c_g \langle E \rangle$, in which $\langle E \rangle$ is the average energy of the waves. In Sections 2.2.6 and 3.4.2 we showed that wavepackets transport energy according to

$$\frac{\partial}{\partial t} \langle E \rangle = -\frac{\partial}{\partial x} \left(c_g \langle E \rangle \right). \tag{4.31}$$

For small-amplitude waves, $\langle E \rangle$ can generally be written as a function of k times the squared amplitude, $|\mathcal{A}|^2$. Substituting this into (4.31) and using (4.30), we therefore find

$$\frac{\partial}{\partial t} |\mathcal{A}|^2 = -\frac{\partial}{\partial x} \left(c_g |\mathcal{A}|^2 \right). \tag{4.32}$$

If the waves are finite-amplitude, then (4.29) and (4.32) are coupled through the weakly nonlinear dispersion relation of the form (4.21). Here this is given to second order in amplitude by

$$\omega \simeq \omega_0(k) + |\mathcal{A}|^2 \omega_2(k). \tag{4.33}$$

Substituting (4.33) into (4.29), keeping only the leading-order terms, gives the following evolution equation for k and $|\mathcal{A}|^2$:

$$\frac{\partial}{\partial t} \begin{pmatrix} k \\ |\mathcal{A}|^2 \end{pmatrix} = - \begin{bmatrix} \partial_k \omega_0 & \omega_2 \\ |\mathcal{A}|^2 \partial_{kk} \omega_0 & \partial_k \omega_0 \end{bmatrix} \frac{\partial}{\partial x} \begin{pmatrix} k \\ |\mathcal{A}|^2 \end{pmatrix}. \tag{4.34}$$

The eigenvalues of the matrix on the right-hand side of (4.34) predict the speed at which amplitude or wavenumber variations propagate. Explicitly, we find the leading-order amplitude correction to the group speed, $c_g \equiv \partial_k \omega_0$, is

$$c_g{}^A = c_g \pm |\mathcal{A}| \sqrt{\omega_2 \frac{\partial^2 \omega_0}{\partial k^2}}. \tag{4.35}$$

The terms discarded in formulating (4.34) contribute only to corrections of $c_g{}^A$ of the order of $|\mathcal{A}|^2$.

If the term inside the square root of (4.35) is positive, then a finite-amplitude dispersive wavepacket will spread as pulses moving at moderately different speeds. Such waves are said to be 'modulationally stable'. This behaviour is different from that predicted for small-amplitude waves in that the spreading is faster than that due to linear dispersion alone.

On the other hand, if the term inside the square root is negative, then $c_g{}^A$ is complex-valued, meaning that weakly nonlinear effects cause the wavepacket envelope to grow exponentially in time. Such waves are said to be 'modulationally unstable'.

The formula (4.35) is used to give a mathematical definition for modulational stability:

$$\text{modulationally stable waves:} \quad \omega_2 \frac{\partial^2 \omega_0}{\partial k^2} > 0$$

$$\text{modulationally unstable waves:} \quad \omega_2 \frac{\partial^2 \omega_0}{\partial k^2} < 0. \tag{4.36}$$

The criterion depends upon the rate of change with the wavenumber of the group velocity predicted by linear theory and upon the order-amplitude-squared correction to the dispersion relation.

Modulational instability does not imply wave breaking. It means only that weakly nonlinear effects act initially within a wavepacket to increase the maximum value of its amplitude envelope. The consequent evolution depends upon the fully nonlinear dynamics of the waves in question. For example, depending upon the dispersion relation and the initial wave amplitude, finite-amplitude waves can transfer energy back and forth between different frequencies, through what is known as the 'Fermi–Pasta–Ulam' recurrence phenomenon or, for deep water waves in particular, 'Benjamin–Feir' instability.

4.3 Weakly nonlinear interfacial waves

In Section 2.3.2 we examined the structure and dispersion relation associated with small-amplitude waves at the interface of a two-layer fluid with infinitely deep upper and lower layers. Here we will apply the general approach outlined above to examine how the structure and dispersion relation is modified when the wave amplitude is not negligibly small.

4.3.1 Theory for interfacial waves in infinitely deep fluid

As in Section 2.3.2, we suppose the background density profile is given by

$$\bar{\rho}(z) = \begin{cases} \rho_1 & z \geq 0 \\ \rho_2 & z < 0, \end{cases} \tag{4.37}$$

and we let φ and ϕ represent the velocity potentials in the upper and lower layers, respectively, as shown in Figure 4.1. The vertical displacement of the interface is represented by η.

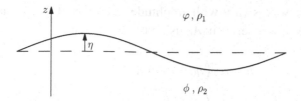

Fig. 4.1. Schematic illustrating the displacement η of the interface between two infinitely deep layers of fluid. The upper fluid has density ρ_1 with a velocity potential ϕ_1, and the lower fluid has density ρ_2 with a velocity potential ϕ_2.

The motion within each layer due to the passage of finite-amplitude interfacial waves is governed by Laplace's equations (2.72). We seek solutions in the x–z plane that are horizontally periodic with wavenumber k and which are bounded as $z \to \pm\infty$. Although the partial differential equations are linear, finite amplitude effects are captured by the nonlinear interface conditions (2.73) and (2.74), here with ϕ_1 replaced by φ and ϕ_2 replaced by ϕ.

Isolating η in the first of these conditions gives

$$g(\rho_2 - rho_1)\eta = \left[\rho_1 \left(\frac{\partial \varphi}{\partial t} + \frac{1}{2}|\nabla\varphi|^2 \right) - \rho_2 \left(\frac{\partial \phi}{\partial t} + \frac{1}{2}|\nabla\phi|^2 \right) \right]\Bigg|_{z=\eta}. \qquad (4.38)$$

The remaining interface condition gives two equations:

$$\frac{\partial \varphi}{\partial z}\Bigg|_{z=\eta} = \frac{\partial \phi}{\partial z}\Bigg|_{z=\eta} \qquad (4.39)$$

and

$$\frac{\partial \phi}{\partial z}\Bigg|_{z=\eta} = \left(\frac{\partial}{\partial t} + \nabla\phi \cdot \nabla \right) \eta \Bigg|_{z=\eta}. \qquad (4.40)$$

Another condition could be posed by replacing ϕ with φ in (4.40), but this adds no new information.

Assuming that η is small but not negligibly so, we perform a Taylor-series expansion about $\eta = 0$ for each of the interface conditions (4.38), (4.39) and (4.40). Generally, for a function $f(x,z,t)$ which is smooth near $z = 0$ we can write

$$f|_{z=\eta} \simeq f|_{z=0} + \eta \frac{\partial f}{\partial z}\Bigg|_{z=0} + \frac{1}{2}\eta^2 \frac{\partial^2 f}{\partial z^2}\Bigg|_{z=0} + \cdots \qquad (4.41)$$

In particular,

$$\phi(x,\eta,t) \simeq \phi(x,0,t) + \eta \phi_z(x,0,t) + \cdots \qquad (4.42)$$

Following the methodology described in Section 4.2.2, we now use perturbation theory for differential eigenvalue problems to determine how the structure and frequency of the waves vary with amplitude. We expand η, φ, ϕ and the squared frequency ω^2 as power series in A_0 using (4.13) so that

$$\eta = A_0 \left(\eta_0 + A_0 \eta_1 + A_0{}^2 \eta_2 + \ldots, \right),$$

$$\varphi = A_0 \left(\varphi_0 + A_0 \varphi_1 + A_0{}^2 \varphi_2 + \ldots, \right),$$

$$\phi = A_0 \left(\phi_0 + A_0 \phi_1 + A_0{}^2 \phi_2 + \ldots, \right),$$

$$\omega^2 = \omega_0{}^2 \left(1 + A_0{}^2 \sigma_2 + \ldots \right). \tag{4.43}$$

Arbitrarily, we can treat A_0 as the vertical displacement amplitude. Though it would be more rigorous to use the nondimensional amplitude $\alpha = k A_0$ as our perturbation parameter, it is sufficient here to use A_0 and then extract terms in successive powers of A_0, as done in Section 4.2.2. The functions η_i, φ_i and ϕ_i are defined to be independent of A_0 so that these functions and their derivatives are of order unity.

Substituting the expansions for φ and ϕ into Laplace's equation for the upper and lower layers and extracting terms at successive powers of A_0 gives the sequence of partial differential equations $\nabla^2 \varphi_i = 0$ and $\nabla^2 \phi_i = 0$. The subscript $i = 0, 1, 2, \ldots$ corresponds to successive orders in the perturbation expansion.

At leading order, for which terms are of order A_0, we seek horizontally periodic, vertically bounded solutions of $\nabla^2 \varphi_0 = 0$ and $\nabla^2 \phi_0 = 0$, subject to the leading-order terms of the interface conditions (4.38), (4.39) and (4.40). Eliminating η_0 from these gives a matrix operator equation in φ_0 and ϕ_0 alone:

$$L_0 \vec{\phi}_0 \Big|_{z=0} = 0, \tag{4.44}$$

in which

$$L_0 \equiv \begin{pmatrix} \partial_z & -\partial_z \\ \rho_1 \omega_0{}^2 & \rho_2 (g' \partial_z - \omega_0{}^2) \end{pmatrix} \quad \text{and} \quad \vec{\phi}_0 \equiv \begin{pmatrix} \varphi_0 \\ \phi_0 \end{pmatrix}. \tag{4.45}$$

The first row comes from (4.39) after expanding φ and ϕ using (4.41). The second row comes from combining the linearized forms of (4.38) and (4.40) to eliminate η. The second time derivatives in the result are replaced with $-\omega^2$ and this is expanded using the last expression in (4.43).

From (4.38), the leading-order equation defining the vertical displacement in terms of $\vec{\phi}_0$ is

$$\eta_0 = -\frac{1}{\rho_2 g'} \left[\rho_1 \frac{\partial \varphi_0}{\partial t} - \rho_2 \frac{\partial \phi_0}{\partial t} \right] \Big|_{z=0}. \tag{4.46}$$

Just as we found in Section 2.3, the eigensolutions for small-amplitude interfacial waves are given by

$$\eta_0 = \cos(kx - \omega t)$$

$$\varphi_0 = -\frac{\omega}{k} e^{-kz} \sin(kx - \omega t), \quad z > 0 \tag{4.47}$$

$$\phi_0 = \frac{\omega}{k} e^{kz} \sin(kx - \omega t), \quad z < 0,$$

in which

$$\omega^2 = \omega_0{}^2 = \frac{\rho_2 - \rho_1}{\rho_1 + \rho_2} gk. \tag{4.48}$$

Here the structure of the waves is represented by the actual values rather than as complex exponentials and the phase has been chosen so that $\eta = A_0 \eta_0$ has its maximum value A_0 at $x = t = 0$.

At the next highest order the interface conditions can be written

$$L_0 \vec{\phi}_1 \big|_{z=0} = \vec{N}_1(\varphi_0, \phi_0, \eta_0) \big|_{z=0}, \tag{4.49}$$

in which L_0 is defined as in (4.45) and the forcing due to quadratic interactions between the leading-order fields is given by

$$\vec{N}_1 = \begin{pmatrix} -\eta_0 \partial_z [\partial_z \varphi_0 - \partial_z \phi_0] \\ -\eta_0 \partial_z [\rho_1 \omega_0{}^2 \varphi_0 + \rho_2 (g' \partial_z - \omega_0{}^2) \phi_0] + \\ \rho_1 \nabla \varphi_0 \cdot \nabla \varphi_{0t} - \rho_2 \nabla \phi_0 \cdot \nabla \phi_{0t} + \rho_2 g' \nabla \phi_0 \cdot \nabla \eta_0 \end{pmatrix}. \tag{4.50}$$

From (4.44) and (4.45), it follows immediately that both terms within square brackets in (4.50) are zero. Furthermore, with substitution of the leading-order solutions of (4.48) into the remaining nonlinear terms in (4.50), we find that both expressions in the vector are identically zero. Therefore, $\vec{\phi}_1$ is independent of $\vec{\phi}_0$, and so we set

$$\varphi_1 = \phi_1 = 0. \tag{4.51}$$

However, η_1 is not independent of $\vec{\phi}_0$. At this order of A_0, the power series expansion of (4.38) gives

$$\rho_2 g' \eta_1 = \left[\rho_1 \left(\eta_0 \partial_{zt} \varphi_0 + \frac{1}{2} |\nabla \varphi_0|^2 \right) - \rho_2 \left(\eta_0 \partial_{zt} \phi_0 + \frac{1}{2} |\nabla \phi_0|^2 \right) \right]\bigg|_{z=0}.$$

Using (4.47), we find

$$\eta_1 = \frac{1}{2} k \frac{\rho_2 - \rho_1}{\rho_1 + \rho_2} \cos 2(kx - \omega t). \tag{4.52}$$

The next order in the perturbation expansion introduces the order $A_0{}^2$ correction to the squared frequency. Thus, by analogy with (4.20), the interface conditions are written

$$\left[L_0\vec{\phi}_2 + L_2\vec{\phi}_0\right]\Big|_{z=0} = \vec{N}_2(\varphi_0, \phi_0, \eta_0, \eta_1)\Big|_{z=0}, \qquad (4.53)$$

in which

$$L_2(\sigma_2) \equiv \begin{pmatrix} 0 & 0 \\ \rho_1\omega_0{}^2\sigma_2 & -\rho_2\omega_0{}^2\sigma_2 \end{pmatrix}. \qquad (4.54)$$

The nonlinear terms in $\vec{N}_2$ are either cubic in the leading-order terms or involve products of η_1 with the leading-order terms. As in the simplification of (4.50), some of the terms can immediately be set to zero. After extensive algebra the remaining terms can be written as a superposition of functions involving $\sin(kx - \omega t)$, $\cos(kx - \omega t)$, $\sin 3(kx - \omega t)$ and $\cos 3(kx - \omega t)$.

The terms proportional to $\sin(kx - \omega t)$ and $\cos(kx - \omega t)$ are resonant with the linear term $L_0\phi_2$ and would result in the unphysical growth of the waves over time. But these secular terms can be eliminated with the appropriate choice of σ_2. Explicitly, we find the correct choice that gets rid of the secular terms is

$$\sigma_2 = \frac{\rho_1{}^2 + \rho_2{}^2}{(\rho_1 + \rho_2)^2} A_0{}^2 k^2. \qquad (4.55)$$

Therefore, to this order accuracy the dispersion relation is

$$\omega^2 = gk\frac{\rho_2 - \rho_1}{\rho_2 + \rho_1}\left(1 + \frac{\rho_1{}^2 + \rho_2{}^2}{(\rho_1 + \rho_2)^2} A_0{}^2 k^2\right). \qquad (4.56)$$

We next compute the correction to φ, ϕ and η at this order. In particular, together with (4.47) and (4.52), the interfacial displacement accurate to amplitude-cubed is

$$\eta = A_0\cos(kx - \omega t) + \frac{1}{2}A_0{}^2 k\frac{\rho_2 - \rho_1}{\rho_1 + \rho_2}\cos 2(kx - \omega t)$$

$$+ \frac{3}{8}A_0{}^3 k^2\left(\frac{(\rho_2 - \rho_1)^2 - 4\rho_1\rho_2/3}{(\rho_1 + \rho_2)^2}\right)\cos 3(kx - \omega t). \qquad (4.57)$$

The following sections examine the structure and dispersion of weakly nonlinear interfacial waves in the limit of deep surface waves and of Boussinesq interfacial waves.

4.3.2 Deep water waves

Because the density of air is much less than that of water, the dispersion relation and structure of deep water waves is found by taking the limit $\rho_1 \to 0$ in (4.56) and

(4.57), respectively. Explicitly, their dispersion relation is given by

$$\omega^2 = gk\left(1 + A_0^2 k^2\right).$$

(4.58)

As for small-amplitude waves, it is understood that $k > 0$ in (4.58) and that the directionality of the waves is set by the sign of ω. Without loss of generality, we assume $\omega > 0$ corresponding to rightward-propagating waves. Explicitly, $\omega \simeq \sqrt{gk}\left(1 + A_0^2 k^2/2\right)$.

If $A_0 k \ll 1$, (4.58) reduces to the dispersion relation of small amplitude deep water waves as given by (2.18). The frequency and phase speed of the waves increases with amplitude. The group velocity is modified according to (4.35) in which $\omega_0 = (gk)^{1/2}$ and $\omega_2 = (gk^5)^{1/2}/2$. But computing the term inside the square root on the right-hand side of (4.35), we find

$$\omega_2 \omega_0'' = -\frac{1}{8}gk.$$

(4.59)

Because k is strictly positive this expression is always negative. Therefore deep water waves of all wavenumbers are modulationally unstable. Even infinitesimally small amplitude wavepackets will initially grow in amplitude, though at a very small growth rate compared to the frequency.

Taking the first two terms in (4.57), the surface displacement of moderately large amplitude deep surface waves is

$$\eta \simeq A_0\left[\cos(kx - \omega t) + \frac{1}{2}A_0 k \cos 2(kx - \omega t)\right].$$

(4.60)

At crests the two terms superimpose to increase the surface displacement. Conversely, at troughs they superimpose so that the downward displacement is not so great. That is, finite-amplitude effects act to flatten the troughs and sharpen the crests, as illustrated in Figure 4.2. The top two graphs show the structure of waves with small and moderately large amplitude.

As the amplitude A_0 becomes larger, it is necessary to compute increasingly higher-order terms in the Fourier cosine series representation for A_0. The amplitude is limited, however. As first shown by Stokes, at a critical value of A_0 the wave crests develop sharp peaks which form an angle of $120°$ across the cusp, as shown in Figure 4.2c (see Exercises). The crest-to-trough distance in this case is approximately $0.44 k^{-1} \simeq 0.071\lambda$. For still larger amplitude waves, the crest spills and the wave breaks.

In idealized circumstances, the weakly nonlinear structure of a train of deep water waves is also modified through development of periodic cusping along the span of the waves. Such dynamics have been neglected here but are discussed in references cited in Appendix A.

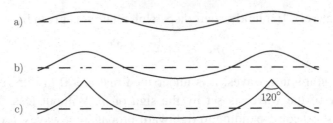

Fig. 4.2. Comparison of the surface displacement structure of a deep water wave whose amplitude is a) small b) moderate and c) at the point of breaking. The structure of the small- and moderate-amplitude waves is given by (4.60). The large-amplitude wave has a crest-to-trough distance of $0.44 k^{-1} \simeq 0.071 \lambda$ and the cusps at their crests form a 120° angle. Note, the aspect ratios of the plots are not to scale.

4.3.3 Deep interfacial plane waves

We now consider the motion of the interface between two fluids of comparable density for which the upper and lower layers are infinitely deep. Taking the Boussinesq limit $\rho_1 \to \rho_2^-$ in (4.56), the dispersion relation becomes

$$\omega^2 \simeq \frac{1}{2} g' k \left(1 + \frac{1}{2} A_0^2 k^2 \right), \qquad (4.61)$$

in which the reduced gravity g' is given by

$$g' \equiv g \frac{\rho_2 - \rho_1}{\rho_2}. \qquad (4.62)$$

As for deep water waves, finite-amplitude effects act to increase their frequency and phase speed, and the group velocity (4.35) is modified by an imaginary term proportional to amplitude, meaning that the waves are modulationally unstable.

The structure of the waves is not purely sinusoidal when they grow to large amplitude. Taking the Boussinesq limit of (4.57), the surface displacement is

$$\eta = A_0 \left[\cos(kx - \omega t) - \tfrac{1}{8} A_0^2 k^2 \cos 3(kx - \omega t) \right]. \qquad (4.63)$$

Note that the order A_0^2 term vanishes in the Boussinesq limit so that, unlike the formula (4.60) for deep water waves, the leading-order finite-amplitude correction to the small-amplitude wave structure is of order A_0^3. This has a number of interesting implications.

First, there is no bias for the waves to have sharp crests and shallow troughs, as was the case for deep water waves. Instead, both the crests and troughs of finite-amplitude interfacial Boussinesq waves are flatter than their small-amplitude sinusoidal counterparts, as shown in Figure 4.3. This symmetry is anticipated

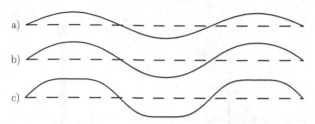

Fig. 4.3. Comparison of the structure of interfacial Boussinesq waves at a) small, b) moderate and c) large amplitude.

because in the Boussinesq approximation the equations of motion are invariant upon reflection in z.

On the other hand, because the finite-amplitude correction is of order A_0^3, in comparison with deep surface waves interfacial waves would have to get to quite large amplitude before the flattening of crests and troughs becomes evident. In reality, they are unlikely to occur to any significant degree because of other dynamics that occur at finite amplitude. In Section 2.3.2 we found that the fluid moves leftwards above and rightwards below the interface of a rightward-advancing wave crest. Thus the mechanism for interfacial wave breaking is different than that for deep surface waves: their amplitude is limited not by the development of cusped peaks but by the development of small-scale shear instabilities as discussed, for example, in Section 2.6.3.

4.3.4 Deep interfacial wavepackets

The evolution of the amplitude envelope $A(x,t)$ of a quasi-monochromatic wavepacket whose wavenumber spectrum is peaked about $k = k_0$ is given by (4.27). Using (4.61), the equation for the interfacial displacement is given explicitly by

$$\iota(A_t - c_g A_x) = \frac{1}{4}\omega_0(k_0)\left[+\frac{1}{2}k_0^{-2}A_{xx} + k_0^2|A|^2A\right], \tag{4.64}$$

in which x- and t-subscripts denote space and time derivatives, $\omega_0(k_0) = (g'k_0/2)^{1/2}$, and the group velocity of small-amplitude waves is $c_g = (g'/8k_0)^{1/2} = \frac{1}{2}\omega_0/k_0$.

Equation (4.64) has been written in a way that is straightforwardly converted into nondimensional form. Defining $\tilde{T} = \omega_0 t$, $\tilde{X} = k_0(x - c_g t)$ and $\tilde{A} = k_0 A$ gives the nonlinear Schrödinger equation

$$\iota\tilde{A}_{\tilde{T}} = \frac{1}{8}\tilde{A}_{\tilde{X}\tilde{X}} + \frac{1}{4}|\tilde{A}|^2\tilde{A}. \tag{4.65}$$

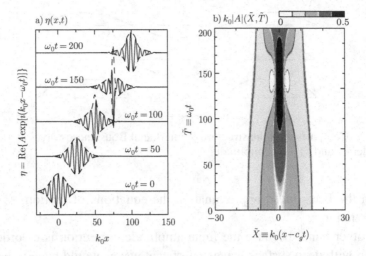

Fig. 4.4. Weakly nonlinear evolution of a Gaussian interfacial wavepacket pre-scribed initially with amplitude $A_0 k_0 = 0.3$ and width $\sigma k_0 = 10$. The solid lines in a) show a vertically shifted sequence of graphs of the vertical displacement field, η (solid line) with the amplitude envelope, $|A|$ (dashed line), at the times indicated. The time series of the amplitude envelope $|A|$ are shown by the greyscale contours in b). This is plotted in a frame of reference moving with the wavepacket.

Note that the advection term $c_g A_x$ on the left-hand side of (4.64) has disappeared as a consequence of defining $\tilde{X}$ to be a spatial co-ordinate in the frame of reference moving with the wavepacket at the group velocity.

Figure 4.4 shows the predicted evolution of a weakly nonlinear deep interfacial wavepacket whose vertical displacement field is given initially by

$$\eta = A_0 e^{-x^2/2\sigma^2} \cos(k_0 x - \omega_0 t).$$

Solving (4.64) in the case with $A_0 k_0 = 0.3$ and $\sigma k_0 = 10$ shows that the wavepacket narrows and its maximum amplitude more than doubles at times around $t \simeq 150 \omega_0^{-1}$. Thereafter the wavepacket broadens and the peak amplitude decreases again.

The growth in amplitude is the result of modulational instability and the narrowing then broadening of the wavepacket is an example of the Fermi–Pasta–Ulam recurrence phenomenon.

4.3.5 Interfacial waves in finite-depth fluid

Up until now we have ignored the presence of solid horizontal boundaries situated above or below the interface. Although the mathematical method to determine the weakly nonlinear behaviour of finite-depth interfacial waves is the same as that above, the algebra is more cumbersome and only the salient results will be given here.

We first consider a two-layer Boussinesq fluid with an unbounded upper layer and a lower layer bounded below by a rigid horizontal boundary at depth H below the mean-depth of the interface. This circumstance might crudely model an atmospheric inversion in the absence of a mean wind. By symmetry, the result can be flipped vertically to describe waves at an interface beneath a finite-depth upper layer. This might describe a model thermocline in the ocean or a lake, for which the surface can be treated as rigid.

The dispersion relation in both circumstances is the same:

$$\omega^2 \simeq g'k \, \frac{1}{\coth kH + 1} \times \left[1 + \frac{9 - 22\tanh kH + 13\tanh^2 kH + 4\tanh^3 kH}{8\tanh kH} (A_0 k)^2 \right].$$
(4.66)

As expected, in the limit $H \to \infty$, this reduces to the dispersion relation (4.61) for interfacial waves in infinitely deep fluid.

The polynomial in $\tanh kH$ in the numerator of the fraction in (4.66) is always positive. As a consequence these waves, like those discussed above, are modulationally unstable for all k.

At second order in amplitude, the weakly nonlinear structure of the waves above a finite-depth lower layer is

$$\eta = A_0 \left[\cos(kx - \omega t) + \frac{3}{4} \frac{1 - \tanh kH}{\tanh^2 kH} (A_0 k) \cos 2(kx - \omega t) \right].$$
(4.67)

Unlike the case of interfacial waves in infinite-depth fluid, but similar to the weakly nonlinear behaviour of surface waves, here we find the wave crests sharpen and the troughs flatten as a consequence of finite-amplitude effects. Specifically, the waves form peaks in the direction oriented towards the deeper fluid, as shown in Figure 4.5.

A second class of weakly nonlinear waves exists in semi-infinite fluids in the near-shallow water limit. These are solitary waves. Because the development and solution of these equations are distinct from those discussed in this section, we defer discussion of solitary waves to Section 4.4.

Beforehand we consider the special case of a finite-depth Boussinesq fluid in which the upper- and lower-layer fluids have approximately the same depth, $H_1 \simeq H_2 \equiv H$. Though geophysically irrelevant, it is a symmetric geometry that has been singled out for study in laboratory experiments.

As in the case of interfacial waves in a two-layer fluid that is unbounded above and below, we find the interfacial waves in a bounded, equal-depth, two-layer fluid

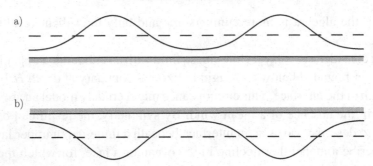

Fig. 4.5. The structure of weakly nonlinear interfacial waves in a fluid with a) a shallow lower layer and b) a shallow upper layer.

do not form cusps but flatten with respect to the sinusoidal structure of small-amplitude waves at both the crests and troughs. For long waves, an expansion in the form (4.13) is possible provided $[(A_0k)/(kH)^2]^2$ is small. In this case, the dispersion relation is

$$\omega^2 \simeq \frac{1}{2}g'k\left[1+\left(1-\frac{1}{2}\coth^2 kH\right)(A_0k)^2\right]. \tag{4.68}$$

If the amplitude is sufficiently large, the structure of interfacial waves in finite-depth fluid depends sensitively upon the departure from symmetry. The crests and troughs are smooth if $(H_2/H_1)^2 < \rho_2/\rho_1$ and they form downward cusps if $(H_2/H_1)^2 > \rho_2/\rho_1$. The transition occurs when the lower layer is deeper than the upper layer by an amount $H_1(\Delta\rho/\rho_1)/2$.

4.4 Solitary waves

Originally, solitary waves referred to finite-amplitude and isolated (hence solitary) disturbances of permanent form. In fluid dynamics these are hump-shaped waves which, for example, have a crest but no trough. The waves exist in finite-depth fluid which typically has one or two layers and the horizontal extent of the wave is long, but not too long, compared with the depth of the fluid.

Despite being large-amplitude, the waves maintain their structure through a balance between dispersion, which tends to broaden the crest, and weakly nonlinear effects, which tend to steepen it. Whereas fluid beneath a small-amplitude shallow water wave oscillates back and forth but experiences no net displacement, fluid beneath a solitary wave is permanently displaced a finite distance which is comparable to the horizontal extent of the wave, as required by continuity of mass. This is shown in Figure 4.6.

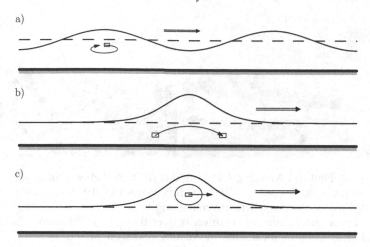

Fig. 4.6. Schematics showing the motion of fluid parcels beneath surface waves: a) orbiting motion with no net displacement beneath a small amplitude wave, b) finite displacement due to a passing solitary wave and c) long distance transport for fluid within a closed streamline in a 'closed-core' nonlinear wave.

Also shown in this figure is the case of a solitary wave of such large amplitude that it traps fluid within the hump and transports it long distances. This is sometimes called a closed-core solitary wave. Because the theory for such waves is not so well established, we will restrict our attention here to moderately large-amplitude solitary waves.

Surprisingly, if two solitary waves collide, they are able to pass through each other and maintain their shape after the interaction. Thus they act as though they obey the superposition principle of linear theory, even though the waves are nonlinear. This intriguing property, and the challenge (and successes) of finding solutions of the nonlinear equations has resulted in wide ranging research in mathematics, physics and engineering.

The definition of solitary waves is sometimes extended to include any finite-amplitude wave which can be represented as a solution of the weakly nonlinear shallow water equations. Such formulae have been used to describe the dramatic occurrence of internal solitary wave trains generated by thunderstorm outflows interacting with an atmospheric inversion, as shown in Figure 4.7.

4.4.1 Nonlinear shallow water equations

If a wave's horizontal wavelength is much longer than the characteristic fluid depth $\mathcal{H}$, it is called a shallow water wave. As shown in Chapter 2, these waves are non-dispersive with z-independent horizontal velocity and linear variation with z

a) Morning Glory cloud　　　　　　　　　b) Internal bore

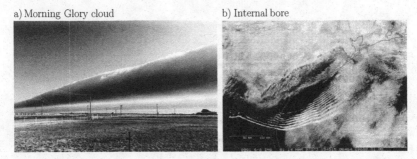

Fig. 4.7. a) Photo of Morning Glory clouds over Burketown, in north-eastern Australia taken on August 24, 1986. b) Image from GOES-8 satellite showing internal waves at an atmospheric inversion originating from a storm complex over central Texas and emanating southwards over the Gulf of Mexico. The image was taken at 9:45 am local time in Austin, Texas on March 14, 1997. Wave crests are separated by approximately 10 km and a single crest is as long as 400 km. [The image in a) is reproduced with permission of Noel Henderson (see http://www.burkeshirecouncil.com/morningglory.htm). The image in b) is reproduced, with permission ©American Meteorological Society, from Figure 1 of Clarke, Mon. Wea. Rev., **126**, 1098–1100 (1998).]

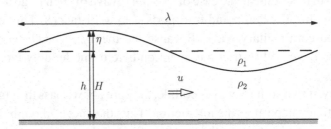

Fig. 4.8. The variables used to describe a rightward-propagating wave in shallow water. For a shallow water wave the horizontal wavelength, λ, is much greater than the mean depth, H, of the layer.

of vertical velocity. This structure may be applied to the momentum and continuity equations for an incompressible two-layer fluid, and from this we can derive a simplified set of equations that describe the motion of shallow water waves.

Here we will focus upon waves in a two-layer fluid with infinite upper-layer depth and with a lower layer of mean depth H, as illustrated in Figure 4.8. Letting the upper- and lower-layer fluid densities be ρ_1 and ρ_2, respectively, the fully nonlinear shallow water equations become

$$\frac{\partial u}{\partial t} + u\frac{\partial u}{\partial x} + g'\frac{\partial h}{\partial x} = 0 \tag{4.69}$$

and

$$\frac{\partial h}{\partial t} + \frac{\partial\, hu}{\partial x} = 0, \tag{4.70}$$

in which $h = H + \eta$ is the total depth of the fluid between the bottom boundary and the interface, and $g' = g(\rho_2 - \rho_1)/\rho_2$ is the reduced gravity. These form two coupled equations in u and h. Explicitly, they can be recast in the vector form of an advection equation:

$$\frac{\partial}{\partial t}\begin{pmatrix} u \\ h \end{pmatrix} + \begin{pmatrix} u & g' \\ h & u \end{pmatrix}\frac{\partial}{\partial x}\begin{pmatrix} u \\ h \end{pmatrix} = 0. \tag{4.71}$$

For the linear scalar advection equation $v_t + c_0 v_x = 0$, we know the solutions are $v(x,t) = v(x - c_0 t, 0)$: v is a constant in a frame of reference moving at constant speed c_0. For this reason c_0 is called the characteristic velocity and the family of lines, $x - c_0 t = \text{constant}$, over which v is constant, are called 'characteristic curves'. More generally, the classic wave equation $v_{tt} - c_0^2 v_{xx} = 0$ can be written as two coupled advection equations describing the motion of rightward-propagating waves with characteristic velocity $c_0 > 0$ moving along characteristic curves $\Gamma_+ = x - c_0 t = \text{constant}$ and leftward-propagating waves with characteristic velocity $-c_0$ moving along characteristic curves $\Gamma_- = x + c_0 t = \text{constant}$, as illustrated in Figure 4.9. Transforming into a co-ordinate system $(\xi_+, \xi_-) = (x - c_0 t, x + c_0 t)$ aligned with these curves the wave equation takes the canonical form $v_{\xi_+ \xi_-} = 0$. This is straightforwardly integrated to give $v = f(\xi_+) + g(\xi_-)$ for functions f and g that are determined by initial conditions. Focusing upon rightward-propagating disturbances, $g = 0$ and f is prescribed by the value of $v(x, 0)$, which in the transformed co-ordinates is given by values evaluated along the ξ_--axis.

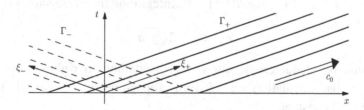

Fig. 4.9. Characteristics of rightward-propagating disturbances $\Gamma_+ = x - c_0 t$ (solid lines) and leftward-propagating disturbances $\Gamma_- = x + c_0 t$ (dashed lines) that evolve according to the classic wave equation. The rightward-propagating disturbances are defined initially by values on Γ_- at $t = 0$. Thereafter they move at characteristic speed c_0 along the curves $\Gamma_+ = \text{constant}$. The equations describing the motion are simplified when transformed into a co-ordinate system (ξ_+, ξ_-) with axes parallel to the characteristic curves.

These ideas can be generalized to the nonlinear matrix equation given by (4.71). The plan is to transform the matrix equation into a co-ordinate system that effectively decouples the two scalar equations. The transformation is found in terms of the eigenvalues and eigenvectors of the matrix on the left-hand side of (4.71). The eigenvalues are the characteristic velocities

$$c_\pm = u \pm \sqrt{g'h}. \tag{4.72}$$

The result shows that the shallow water wave speed, $\sqrt{g'H}$, increases at a crest where $h = H + \eta > H$ and it is Doppler-shifted as a result of the perturbed flow, u, associated with the waves. The two speeds correspond to waves that move upstream and downstream with respect to this flow. The family of curves Γ_+ for which $dx/dt = c_+$ are the characteristic curves for rightward-propagating waves.

The corresponding eigenvectors of the matrix in (4.71) dictate how u and h are related along a characteristic curve. Letting ξ be the along-curve direction for rightward-propagating waves, we have

$$\begin{pmatrix} u - c_+ & g' \\ h & u - c_+ \end{pmatrix} \frac{\partial}{\partial \xi} \begin{pmatrix} u \\ h \end{pmatrix} = 0,$$

from which it follows that $u_\xi = (\sqrt{g'/h}) h_\xi$.

This result inspires an explicit way of transforming the coupled nonlinear equations into a single equation describing the evolution of rightward-propagating disturbances along the characteristic curves. Adding the momentum equation (4.69) to $(g'/h)^{1/2}$ times the continuity equation (4.70) gives

$$(\partial_t u + c_+ \partial_x u) + \sqrt{\frac{g'}{h}} (\partial_t h + c_+ \partial_x h) = 0. \tag{4.73}$$

On Γ_+ the condition $dx/dt = c_+$ means (4.73) can be written as the differential $du + (g'/h)^{1/2} dh = d \left(u + 2(g'h)^{1/2} \right) = 0$. Integrating therefore shows that

$$r_+ \equiv u + 2\sqrt{g'h} \tag{4.74}$$

is a constant on characteristic curves corresponding to rightward-propagating waves. A similar calculation for leftward-propagating waves reveals that $r_- \equiv u - 2(g'h)^{1/2}$ is constant.

The quantities $r_\pm$, called 'Riemann invariants', can be determined from initial conditions. Focusing upon rightward-propagating waves, we use the r_- invariants to initialize the system, assuming that $u = 0$ where the interface is undisplaced ($h = H$). Therefore

$$u = 2\sqrt{g'h} - 2\sqrt{g'H} \tag{4.75}$$

along characteristic curves Γ_+.

Substituting (4.75) into (4.70) and recasting the problem in terms of η using $h = H + \eta$ gives the nonlinear shallow water equation for rightward-propagating waves:

$$\frac{\partial \eta}{\partial t} + \left(3\sqrt{g'(H + \eta)} - 2\sqrt{g'H}\right)\frac{\partial \eta}{\partial x} = 0. \tag{4.76}$$

In the limit $\eta \ll H$, (4.76) reduces to the shallow water equation for small-amplitude rightward-propagating waves: $\eta_t + c_0 \eta_x = 0$, in which $c_0 = \sqrt{g'H}$ is the shallow water wave speed. If η is not negligibly small, (4.76) shows that the wave moves relatively more quickly where $\eta > 0$ and more slowly where $\eta < 0$. One consequence of such motion is that crests tend to catch up to leading troughs. That is, nonlinear effects lead to wave steepening and, barring the development of other retarding factors, the waves will ultimately overturn and break.

4.4.2 Moderately long weakly nonlinear interfacial waves

Here we assume that η is small but not too small and that the horizontal extent is long but not too long compared to the fluid depth H. The first assumption introduces the weakly nonlinear effects of wave steepening. The second assumption allows for linear dispersion, which acts to broaden the waves. We will see that the two effects can be balanced, resulting in a new class of steady waves known as solitary waves.

We are concerned only with moderately large amplitude effects, so we expand the leading square root in (4.76) for small η and discard terms of order $(\eta/H)^2$ and higher. Thus we find

$$\eta_t + c_0 \eta_x + \frac{3}{2H}\eta \eta_x = 0, \tag{4.77}$$

in which $c_0 = \sqrt{g'H}$. The first two terms on the left-hand side of (4.77) correspond to the linear equation for rightward-propagating, non-dispersive waves moving at the shallow water wave speed.

To splice in the effect of weak dispersion, we consider the small kH limit of the dispersion relation for infinitesimally small amplitude waves in finite-depth fluid. From (2.92), we have for rightward-propagating waves

$$\omega = \sqrt{g'k \frac{\tanh kH}{1 + (\rho_1/\rho_2)\tanh kH}}$$

$$\simeq c_0 k \left[1 - \frac{1}{2}\frac{\rho_1}{\rho_2}(|k|H) - \frac{1}{6}\left(1 + \frac{3}{4}\left(\frac{\rho_1}{\rho_2}\right)^2\right)(kH)^2\right]. \tag{4.78}$$

The absolute value of k in the second term in square brackets ensures that the fastest wave speed is c_0 even in the case $\omega < 0$ and $k < 0$.

Our plan is to convert (4.78) into a partial differential equation using the methodology introduced in Section 1.15.6. Note that in the long-wave limit ($|kH| \ll 1$), (4.78) reduces to the shallow water dispersion relation $\omega = c_0 k$, which has the corresponding partial differential equation $\partial_t \eta + c_0 \partial_x \eta = 0$. This advection equation we recognize as the first two terms on the left-hand side of (4.77). For somewhat shorter waves, the other terms of (4.78) cannot be neglected. When transformed into higher-order derivatives they give extra linear terms that can be included in (4.77).

This is done here in two special circumstances.

4.4.2.1 Korteweg–de Vries equation

First, suppose $\rho_1 \ll \rho_2$ as would be the case for surface waves. The dispersion relation becomes $\omega = c_0 k - c_0 H^2 k^3/6$ corresponding to the differential equation

$$\eta_t + c_0 \eta_x + \frac{1}{6} c_0 H^2 \eta_{xxx} = 0. \tag{4.79}$$

Combining this result with (4.77), we have an equation that includes the effects both of weak nonlinearity and weak dispersion:

$$\eta_t + c_0 \left(1 + \frac{3}{2H} \eta \right) \eta_x + \frac{1}{6} c_0 H^2 \eta_{xxx} = 0. \tag{4.80}$$

This is the 'Korteweg–de Vries equation', often referred to simply as the 'KdV equation'.

Sometimes this equation is transformed to a frame moving at speed c_0. Defining $X = x - c_0 t$ gives

$$\eta_t + \beta \eta \eta_X + \gamma \eta_{XXX} = 0, \tag{4.81}$$

in which $\beta \equiv 3c_0/2H$ and $\gamma \equiv c_0 H^2/6$ are constants.

Remarkably, analytical solutions of the KdV equation can be found for a broad range of circumstances. In the most straightforward case solutions can be found for waves of permanent form. Thus we can write $\eta(x,t) = \xi(x - Ut)$, in which U is the constant translation speed of the wave.

Substituting this into (4.80), integrating, multiplying by ξ' and integrating again gives the nonlinear, first-order ordinary differential equation

$$(\xi')^2 + \frac{3}{H^3} \xi^3 - \frac{6}{H^2} \left(\frac{U}{c_0} - 1 \right) \xi^2 + C_1 \xi + C_0 = 0, \tag{4.82}$$

in which C_0 and C_1 are integration constants.

A hump-shaped solitary wave is the solution found by setting the integration constants C_0 and C_1 to zero – a consequence of requiring ξ and its derivative to be

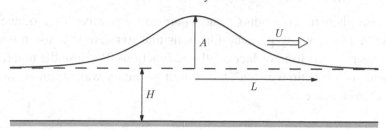

Fig. 4.10. A solitary wave solution of equation (4.80). The horizontal extent L is a function of the amplitude A and the equilibrium depth of the layer H. The interface displacement is given by (4.83).

zero at infinity. Substituting in (4.82) gives

$$\eta = A \; \text{sech}^2 \left(\sqrt{\frac{3A}{4H^3}} (x - Ut) \right), \tag{4.83}$$

which is plotted in Figure 4.10. The solitary wave speed is

$$U = c_0 \left(1 + \frac{1}{2} \frac{A}{H} \right). \tag{4.84}$$

This is a distinguishing feature of solitary waves: they propagate faster than the fastest speed, c_0, associated with long small-amplitude waves.

The solitary wave amplitude, A, is similar to the half peak-to-peak amplitude of sinusoidal waves in that it represents the maximum vertical displacement above the equilibrium fluid depth. But it is wrong to call A the half peak-to-peak amplitude because there is only a crest and no trough associated with a solitary wave. This fact leads to a second distinguishing feature of solitary waves. Whereas fluid parcels beneath small-amplitude sinusoidal waves oscillate around an equilibrium position, during the passage of a solitary wave fluid parcels are displaced a finite horizontal distance in the direction of the wave propagation. This is a direct consequence of mass conservation.

Although one cannot attribute a wavelength to a solitary wave, one can characterize its horizontal extent, L. Integrating (4.83) over all x at fixed t and dividing by A gives the estimate

$$\frac{L}{H} = \frac{4}{\sqrt{3}} \sqrt{\frac{H}{A}}. \tag{4.85}$$

Unlike small-amplitude waves, the horizontal extent of a solitary wave is not independent of the amplitude, but increases as the inverse square root of A both taken relative to the upstream depth H of the fluid.

In the case where the constants C_0 and C_1 are nonzero, waves of permanent form can be written as spatially periodic functions that translate at constant speed U. These waves have the form of Jacobi-elliptic functions. Because the notation used to represent such functions is 'cn', the resulting solitary wave train is sometimes called a 'cnoidal wave'.

4.4.2.2 Benjamin–Ono equation

Next, we will assume the fluid is Boussinesq so that $\rho_2 - \rho_1 \ll \rho_2$. In this case the second term in the square brackets of (4.78) gives the leading-order dispersion term, and so the partial differential equation describing the motion of waves involves second-order rather than third-order spatial derivatives. The formula is complicated by the presence of the absolute value of k in the second term. Together with the nonlinear term in (4.77) and transforming to a frame of reference moving at speed c_0, we have

$$\eta_t + c_0 \frac{3}{2H} \eta \eta_X + \frac{1}{2} c_0 H \frac{\partial^2}{\partial X^2} \fint \frac{\eta(\tilde{X}, t)}{\tilde{X} - X} d\tilde{X} = 0, \qquad (4.86)$$

in which $X = x - c_0 t$. The third term on the left-hand side involves the Cauchy principal-value integral. This integral operator acting on η is similar in effect to the Dirac δ-function in that it pulls out the value of η at x. But whereas the δ-function is even, the integral operator is odd. Thus the two x-derivatives acting on the integral results in the absolute value of k.

The result (4.86) is called the 'Benjamin–Ono equation'. Sometimes this equation is represented in a somewhat simpler form that allows for leftward and rightward wave propagation in the linearized equations of motion. The canonical form of this equation is

$$\eta_{tt} + \beta \eta \eta_x + \gamma \eta_{xxxx} = 0, \qquad (4.87)$$

in which β and γ are constants distinct from those in the canonical form of the Korteweg–de Vries equation (4.81).

Analytic solutions of the Benjamin–Ono equation include solitary wave trains and solitary waves written in terms of Jacobian elliptical and hyperbolic functions. However, their structure is substantially more complicated than the simple analytic form (4.83) for a Korteweg–de Vries solitary wave.

4.5 Weakly nonlinear internal waves

We now consider the evolution of finite-amplitude internal waves in uniformly stratified fluid. As with interfacial waves, the background flow is taken to be stationary and the waves are assumed to be two-dimensional, having structure only in the x and z directions. As will be shown, the theory for weakly nonlinear internal waves is significantly different from that for weakly nonlinear interfacial waves.

In particular, because the formula for plane (monochromatic) internal waves exactly solves the fully nonlinear equations of motion the perturbation approach laid out in Section 4.2.2 cannot be applied. Instead, one must assume at the outset that the waves have the form of a quasi-monochromatic wavepacket. The resulting analysis will show that the dominant weakly nonlinear interaction for horizontally periodic wavepackets of finite vertical extent is through interactions between the waves and the wave-induced mean flow. This lies in contrast with the interfacial wave dynamics for which weakly nonlinear effects involved interactions between the waves and their superharmonics.

4.5.1 Perturbation theory

For mathematical convenience, here we consider internal waves in the x–z plane propagating in stationary uniformly stratified, non-rotating, Boussinesq fluid. The fully nonlinear equations of motion are

$$
\begin{aligned}
\varrho_0 \frac{Du}{Dt} &= -\frac{\partial p}{\partial x}, \\
\varrho_0 \frac{Dw}{Dt} &= -\frac{\partial p}{\partial z} - g\rho, \\
\frac{D\rho}{Dt} &= \frac{\varrho_0}{g} N_0{}^2 w, \\
\nabla \cdot \vec{u} &= 0,
\end{aligned}
\tag{4.88}
$$

in which $N_0{}^2$ is the constant squared buoyancy frequency defined in the Boussinesq approximation by (3.2).

The system of equations can be represented in a more succinct form by writing them in terms of the streamfunction ψ and the vertical displacement field

$$
\xi = -\rho/\bar{\rho}'.
\tag{4.89}
$$

Because $\bar{\rho}$ varies linearly with z in a uniformly stratified Boussinesq fluid, the definition (4.89) applies even for large-amplitude disturbances. Taking the curl of the momentum equations and relating the vorticity to ψ by $\zeta = u_z - w_x = -\nabla^2\psi$ gives

$$
\frac{D}{Dt}\nabla^2\psi = -N_0{}^2\frac{\partial \xi}{\partial x}.
\tag{4.90}
$$

The internal energy equation for the time evolution of density perturbations becomes

$$\frac{D\xi}{Dt} = \frac{\partial \psi}{\partial x}. \tag{4.91}$$

These equations can be combined further to yield a single nonlinear differential equation in ψ which can be written as a linear differential operator acting on the left-hand side and the divergence of nonlinear terms on the right-hand side:

$$L\psi = \nabla \cdot \vec{F}, \tag{4.92}$$

in which

$$L(\partial_x, \partial_z, \partial_t) \equiv \partial_t^2 \left(\partial_x^2 + \partial_z^2 \right) + N_0^2 \partial_x^2 \tag{4.93}$$

and

$$\vec{F} \equiv \partial_t(\vec{u}\zeta) + N_0^2 \partial_x(\vec{u}\xi). \tag{4.94}$$

Given this result, we would like to follow the perturbation theory approach taken for interfacial waves to determine the weakly nonlinear dispersion relation for internal waves in uniformly stratified fluid. However, it turns out that this approach is not straightforward because plane waves are exact solutions of the fully nonlinear equations of motion. This can be seen by using the polarization relations for internal waves to relate $\vec{u}$ and ζ to ξ. Substituting these expressions in (4.94) gives $\nabla \cdot \vec{F} = 0$. Even in the presence of background rotation and in a three-dimensional domain, plane internal waves are exact solutions of the fully nonlinear equations because the condition $\nabla \cdot \vec{u} = 0$ requires the advection operator $\vec{u} \cdot \nabla$ to be zero (see Exercises).

Instead it is necessary to consider the evolution of quasi-monochromatic wavepackets in the perturbation theory analysis of (4.92). We will do this here for wavepackets that are horizontally periodic but have limited vertical extent. At leading order in the perturbation expansion the streamfunction can be written as

$$A_{\psi_0}\psi_0 = \Re\{A_\psi \exp[\imath\,(k_x x + k_z z - \omega_0 t)]\}, \tag{4.95}$$

in which $\|\psi_0\|$ is order unity and the amplitude envelope A_ψ, with characteristic amplitude $A_{\psi 0}$, varies slowly in z and t. Explicitly, if σ_z is the characteristic width of the wavepacket, the slow scales can be defined in terms of $\epsilon = 1/(k_x \sigma_z) \ll 1$. Thus $A_\psi \equiv A_\psi(Z, T)$ in which $Z = \epsilon z$ and $T = \epsilon t$. Likewise we define $\alpha = A_{\psi 0} k_x^2 / N_0$ to be a nondimensional measure of the amplitude.

Taking the perturbation theory approach laid out in Section 4.2.2, we assume in general that finite-amplitude waves can be represented by a power series:

$$\psi = A_{\psi 0} \left(\psi_0 + A_{\psi 0}\psi_1 + A_{\psi 0}^2 \psi_2 + \ldots \right), \tag{4.96}$$

in which $\|\psi_i\|$ is order unity. The corresponding weakly nonlinear dispersion relation is assumed to take the form

$$\omega^2 = \omega_0{}^2\left(1 + A_\psi{}_0{}^2\sigma_2 + \dots\right). \tag{4.97}$$

We will substitute these expansions into (4.92) and extract terms in successive orders of ϵ and α, which are assumed to be of comparable magnitude.

At leading order, we have $L_0\psi_0 = 0$ in which

$$L_0 \equiv L(\imath k_x, \imath k_z, -\imath\omega_0), \tag{4.98}$$

with L defined by (4.93). The space and time derivatives acting on the amplitude envelope are order ϵ and higher and so are neglected. Thus we recover the linear dispersion relation for Boussinesq internal waves:

$$\omega_0{}^2 = N_0{}^2\frac{k_x{}^2}{k_x{}^2 + k_z{}^2}. \tag{4.99}$$

At the next order in perturbation theory, the nonlinear terms on the right-hand side of (4.92) vanish. This again is ultimately a consequence of plane internal waves being a solution of the fully nonlinear equations. Thus ψ_1 can be taken to be zero. Although the streamfunction associated with waves is unaffected at this order in perturbation theory, we have already seen in Section 3.4.5 that internal waves self-interact to produce a wave-induced mean flow $\langle u \rangle_L$ whose magnitude is of the order of the amplitude squared. This is given in terms of the vertical displacement amplitude by (3.107). In terms of A_ψ we have

$$\langle u \rangle_L = \frac{1}{2}\frac{|\vec{k}|^3}{N}|A_\psi|^2. \tag{4.100}$$

At next order we have the equation

$$L_0\psi_2 + L_2(\sigma_2)\psi_0 = N_2. \tag{4.101}$$

From (4.93) with $\partial_{tt} \to -\omega^2$ and using (4.97), the linear operator at this order is

$$L_2(\sigma_2) = |\vec{k}|^2\omega_0{}^2\sigma_2. \tag{4.102}$$

Within N_2 the terms that are resonant with the linear operator arise from the wave-induced mean flow interacting with the waves. Explicitly, replacing $\vec{u}$ with $\langle u \rangle_L \hat{x}$ in (4.94), $\nabla \cdot \vec{F}$ becomes

$$\langle u \rangle_L\left(\partial_{xt}\zeta_0 + N_0{}^2\partial_{xx}\xi_0\right).$$

At leading order the amplitude envelope of the vorticity field is $|\vec{k}|^2 A_\psi$ and the amplitude envelope of the vertical displacement field is $(-k_x/\omega_0)A_\psi$. Together with (4.100), this determines the coefficient of the resonant part of the nonlinear forcing term. This secular term is removed with the appropriate choice of σ_2. Thus we find

$$\sigma_2 = |\vec{k}|^4/N_0{}^2. \tag{4.103}$$

Thus we have found the leading nonlinear correction to the dispersion relation (4.97) with σ_2 given by (4.103). For comparison with the results for weakly nonlinear interfacial waves, we recast the result in terms of the vertical displacement amplitude, A_0, using the polarization relation $A_\psi = -(\omega/k_x)A_\xi$. Thus we find

$$\omega^2 = \omega_0{}^2 \left(1 + A_0{}^2|\vec{k}|^2 + \dots\right). \tag{4.104}$$

Taking the square root of both sides and Taylor-series expanding the right-hand side gives $\omega \simeq \omega_0 + A_0{}^2\omega_2$ in which

$$\omega_2 = \omega_0|\vec{k}|^2/2. \tag{4.105}$$

4.5.2 Modulational stability

As in (4.36), the modulational stability of the wavepacket is determined by the sign of $\omega_2 \partial_{k_z k_z} \omega_0$. Using the dispersion relation (4.99), we find

$$\frac{\partial^2 \omega_0}{\partial k_z{}^2} = -\frac{Nk_x}{|\vec{k}|^3}\left[\frac{k_x{}^2 - 2k_z{}^2}{|\vec{k}|^2}\right]. \tag{4.106}$$

Whereas ω_2 in (4.105) is always positive, the right-hand side of (4.106) is negative if $|k_z| < 2^{-1/2}|k_x|$ and positive otherwise. Thus horizontally periodic internal wavepackets of finite vertical extent are modulationally unstable if $|k_z| < 2^{-1/2}|k_x|$ and are modulationally stable otherwise. That is to say, the envelope of these weakly nonlinear internal waves is expected to narrow and increase in amplitude if the frequency of the waves exceeds $(2/3)^{1/2}N_0 \simeq 0.82N_0$. For lower-frequency weakly nonlinear wavepackets the envelope is expected to broaden faster than linear dispersion predicts. From (4.106), the transition from modulational instability to stability occurs when the frequency of the wavepacket corresponds to that for internal waves moving with the fastest vertical group velocity.

The evolution of modulationally stable and unstable quasi-monochromatic internal wavepackets is illustrated in Figure 4.11, which shows the results of fully nonlinear numerical simulations of moderately large-amplitude horizontally periodic and vertically localized wavepackets. In Figures 4.11a, b and c the ratio

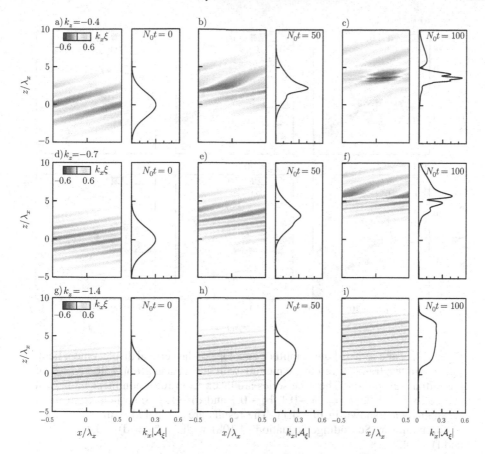

Fig. 4.11. Results of a fully nonlinear numerical simulation showing the evolution of a Gaussian wavepacket with initial amplitude $k_x A_0 = 0.3$, width $k_x \sigma_z = 10$ and vertical wavenumber $k_z = -0.4 k_x$ shown at times a) $N_0 t = 0$, b) 50 and c) 100. For each time, the greyscale contours to the left show the vertical displacement field, ξ, whereas the plot to the right shows the amplitude envelope $|\xi|$. This wavepacket is modulationally unstable. The corresponding plots in d), e) and f) show the evolution of a Gaussian wavepacket with the same amplitude and width but with vertical wavenumber $k_z = -0.7 k_x$. This wavepacket is marginally modulationally unstable. Likewise, the plots in g), h) and i) show the evolution of a wavepacket having vertical wavenumber $k_z = -1.4 k_x$. This wavepacket is modulationally stable.

$|k_z/k_x| = 0.4 < 2^{-1/2}$ is sufficiently small and the amplitude sufficiently large that the wavepacket is modulationally unstable. As expected, the simulation shows that the maximum amplitude of the wavepacket initially increases in time as the wavepacket as a whole moves upwards.

In contrast, in the simulation with $|k_z/k_x| = 1.4 > 2^{-1/2}$ the amplitude decreases in time as a result of the wavepacket being modulationally stable. This is illustrated in Figures 4.11g, h and i. The simulation with $|k_z/k_x| = 0.7$, illustrated in

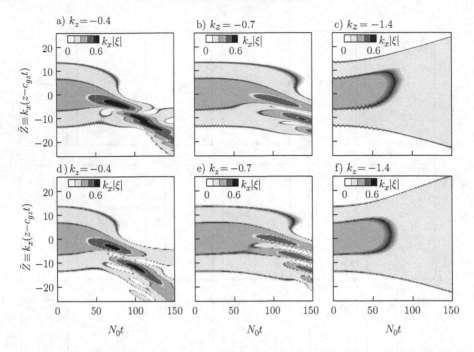

Fig. 4.12. Time series of the amplitude envelope of the vertical displacement field, $|\xi|$, shown in a frame of reference moving with the vertical group velocity, c_{gz}, of the initial wavepacket. The time series are taken from fully nonlinear numerical simulations with $k_z/k_x =$ a) -0.4, b) -0.7 and c) -1.4. Snapshots from these simulations are shown in Fig. 4.11. The results are compared with the solution of the nonlinear Schrödinger equation (4.108) with $k_z/k_x =$ d) -0.4, e) -0.7 and f) -1.4.

Figures 4.11d, e and f, is near the critical point between modulational stability and instability. Here we see the wavepacket exhibits both broadening and sharpening during its evolution.

The results of the three simulations are summarized by the vertical time series in Figures 4.12a, b and c. When viewed in a frame of reference moving upwards with the vertical group velocity predicted by linear theory, the amplitude envelope initially stays centred about $\tilde{Z} \equiv k_x(z - c_{gz}t) = 0$. In the case $k_z = -0.4k_x$, shown in Figure 4.12a, the modulationally unstable wavepacket narrows and broadens and its vertical advance slows. This asymmetric behaviour is also observed in the marginally modulationally unstable case with $k_z = -0.7k_x$ (Figure 4.12b), but is not so pronounced for the modulationally stable wavepacket with $k_z = -1.4k_x$ (Figure 4.12c).

We can gain insight into the reason for this symmetry breaking by examining the solution of the nonlinear Schrödinger equation for these waves. The standard form of the equation is given by (4.28), in which the X-derivatives are replaced by

Z-derivatives and ω_0 is taken to be a function of the initial vertical wavenumber k_z. The coefficients are given by (4.106) and (4.105). Defining $\Theta = \tan^{-1}(k_z/k_x)$, the resulting equation for the evolution of the amplitude envelope of the vertical displacement field is

$$\imath A_t = -\frac{1}{2}\frac{N}{k_x^2}\cos^3\Theta\,(3\sin^2\Theta - 1)A_{ZZ} + \frac{1}{2}Nk_x^2\sec\Theta\,|A|^2A, \qquad (4.107)$$

in which $Z = z - c_{gz}t$.

The first term on the right-hand side is just the linear dispersion term which appears in the linear Schrödinger equation $-\omega_0''/2$, in which primes denote k_z derivatives. The second term represents Doppler-shifting by the wave-induced mean flow. This result can be derived by replacing ∂_t with $\partial_t + \langle u \rangle_L \partial_x$ in the linear Schrödinger equation, so including the advection of the waves by the wave-induced mean flow $\langle u \rangle_L$. Writing this in terms of $|A|^2$ using (3.107) and replacing $\partial_x \to \imath k_x$ gives the nonlinear term in (4.107).

Solving (4.107) gives reasonably good agreement with the results of fully non-linear simulations, at least during the early stages of the weakly nonlinear evolution of the wavepacket. However, it does not capture the symmetry breaking observed in Figure 4.12a, in which the wavepacket advances more slowly than the vertical group velocity after approximately ten buoyancy periods ($N_0 t \gtrsim 60$). Like the solution of the nonlinear Schrödinger equation shown in Figure 4.4b for weakly nonlinear interfacial waves, (4.107) is invariant upon reflection in Z and so a wavepacket initially centred about $Z = 0$ will always be centred about $Z = 0$.

The symmetry breaking occurs in the case $k_z = -0.4k_x$ because the modulationally unstable wavepacket grows to such an amplitude and narrows to such an extent that higher-order terms in the nonlinear Schrödinger equation, which up to this point have been neglected, play a competing role with the lower-order terms. For horizontally periodic internal wavepackets of finite vertical extent, the nonlinear Schrödinger equation that captures the next-order effects is

$$\imath A_t = -\frac{1}{2}\omega_0''A_{ZZ} + \imath\frac{1}{6}\omega_0'''A_{ZZZ} + \frac{1}{2}Nk_x^2\sec\Theta\,|A|^2A + \imath\frac{1}{4}Nk_x\sin\Theta\,\frac{\partial |A|^2}{\partial Z}A. \tag{4.108}$$

Solutions of (4.108) for a moderately large-amplitude Gaussian wavepacket with $k_z = -0.4$, -0.7 and -1.4 are shown in Figures 4.12d, e and f, respectively. In all three cases, the weakly nonlinear equation captures well the evolution of the wavepacket and the symmetry breaking in cases with $k_z = -0.4$ and -0.7. This symmetry breaking is a direct consequence of the second and fourth terms on the right-hand side of (4.108). Without them, the modulationally unstable case with $k_z = -0.4$ would evolve in a manner similar to that for deep water waves, illustrated in Figure 4.4.

The discrepancy at late times between fully nonlinear simulations and weakly nonlinear theory arises in part because the wavepacket is so narrow that the assumption it is quasi-monochromatic is no longer valid. Furthermore, dynamics neglected by (4.108) begin to play a role. In particular, the weakly nonlinear equation does not account for the transfer of energy from waves of wavenumber k_x to superharmonic waves with a higher horizontal wavenumber. This is the phenomenon of parametric subharmonic instability, discussed in the next section.

4.6 Breakdown of internal waves into turbulence

We have shown that nonlinear effects modulate the structure of internal wavepackets, but the amplitude of these waves was not so large that this resulted in energy dissipation and permanent momentum deposition to the background flow.

In this section we will examine different mechanisms for instability that lead to the breakdown of internal plane waves and wavepackets. Parametric subharmonic instability is a mechanism through which small-scale internal waves tap energy from a plane internal wave. This operates even for small-amplitude waves, though the time-scale for the initial growth of the disturbance is long. Once established, however, the instability provides an efficient means to transfer energy to small scales. We will also examine the conditions upon the wave amplitude that result directly or indirectly in overturning and breaking.

4.6.1 Parametric subharmonic instability

Parametric subharmonic instability can readily be demonstrated through the motion of a forced pendulum, for example, a yo-yo on a stretchable string that is periodically pulled up and down. If the natural oscillation frequency of the pendulum is half the forcing frequency, then weakly nonlinear theory predicts that the amplitude of the pendulum's motion will grow over time. The growth is not due to resonance, as would occur if the forcing and natural frequency were equal. This would lead to a linear increase in amplitude with time. Weakly nonlinear theory, which involves the approximate solution of Mathieu's equation, predicts the growth is exponential in time.

In a similar fashion, a plane internal wave of wavenumber $\vec{k}$ and frequency ω periodically forces fluid parcels to oscillate, and this forcing can excite motion at subharmonic frequencies. This has been demonstrated in laboratory experiments, as shown in Figure 4.13. The snapshots show sideviews of a tank filled with uniformly stratified fluid that oscillates vertically up and down at a fixed frequency set to equal the frequency of a mode 1–1 internal wave (e.g., see Figure 3.10a). Over time the growth in amplitude of the mode is observed through the displacement of eight horizontal dye lines that mark successive isopycnal layers between the top and

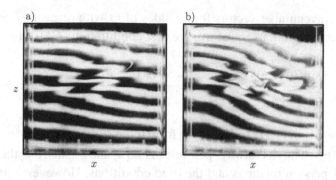

Fig. 4.13. Snapshots from experiment showing the development of small-scale disturbances from parametric subharmonic instability of a low-mode internal wave in a tank oscillating vertically at a constant frequency. [From experiments by Benielli and Sommeria, Dyn. Atmos. Ocean (1996). Reproduced, with permission of Cambridge University Press, from Fig. 6 of Bouruet-Aubertot *et al.*, J. Fluid Mech., **285**, 265–301 (1995).]

bottom of the tank. At the time shown in Figure 4.13a, a small-scale disturbance develops which oscillates with half the frequency of the 1–1 mode. This indicates that it results from parametric subharmonic instability.

Any disturbance with wavenumber $\vec{k}'$ and frequency $\omega' = \omega'(\vec{k}')$ can co-exist with a plane wave. It amplifies in time, however, if it resonantly interacts with the plane wave through the nonlinear advection terms of the equations of motion. Because these terms are quadratic in the fluctuation fields the wavenumbers and frequencies of the plane wave and disturbance must satisfy the resonant triad interaction conditions given generally by

$$\vec{k}_3 = \vec{k}_1 \pm \vec{k}_2 \text{ and } \omega_3 = \omega_1 \pm \omega_2. \tag{4.109}$$

In these expressions we can set $\vec{k}_3 = \vec{k}_{\mathrm{pw}}$ and $\omega_3 = \omega_{\mathrm{pw}}$ as the wavenumber and frequency of the plane wave so that

$$\vec{k}_{\mathrm{pw}} = \vec{k}' + \vec{k}'' \text{ and } \omega_{\mathrm{pw}} = \omega' + \omega'', \tag{4.110}$$

in which the primed quantities denote the wavenumbers and frequencies of the subharmonically excited waves.

These equations may be recast in terms of the magnitudes of the wavenumber vectors and the angles Θ formed by lines of constant phase to the vertical. Using the dispersion relation in the form $\omega = N \cos(\Theta)$, the second condition in (4.110) becomes

$$\cos(\Theta_{\mathrm{pw}}) = \cos(\Theta') + \cos(\Theta''). \tag{4.111}$$

Because the wavenumber vectors form angles of Θ with respect to the horizontal geometrical considerations transform the first condition in (4.110) to

$$\frac{\sin(\Theta' + \Theta'')}{k_{pw}} = \frac{\sin(\Theta_{pw} - \Theta')}{k'} = \frac{\sin(\Theta_{pw} + \Theta'')}{k''}. \qquad (4.112)$$

Given k_{pw} and Θ_{pw}, (4.111) and (4.112) form three equations in the four unknowns k', k'', Θ' and Θ''. Thus it is always possible to find $\vec{k}'$ and $\vec{k}''$ such that they satisfy the respective dispersion relations and the triad conditions. However, certain choices of excited resonant frequencies lead to faster growth. These form the modes of instability that will develop most rapidly from the plane wave.

Floquet analysis reveals that the fastest growing subharmonic waves have a wavenumber vector that is an integer multiple of $\vec{k}_{pw}$. Defining $K \equiv |\vec{k}_{pw}|$ and letting M be the magnitude of the disturbance wavenumber component in the along-crest direction, subharmonic disturbances have a wavenumber vector in the x–z plane given by

$$\vec{k}_n = (nK\cos\Theta - M\sin\Theta, \; nK\sin\Theta + M\cos\Theta), \qquad (4.113)$$

in which n is an integer.

Restricting each subharmonic wavenumber to be one of $\vec{k}_0, \vec{k}_1, \ldots,$ (4.113) together with (4.111) and (4.112) uniquely define the wavenumbers of the subharmonic waves.

An illustration of a resonant triad is shown in Figure 4.14. This is constructed for the particular case of an upward-propagating plane internal wave that excites two subharmonic waves, one with wavenumber $\vec{k}' \equiv \vec{k}_0$ oriented parallel to the waves and the other with wavenumber $\vec{k}'' \equiv \vec{k}_1$. From conditions (4.111) and (4.112), the angle to the horizontal of the wavenumber vector $\vec{k}_1$ is set by $\cos(\Theta'') = \sqrt{2}\cos(\Theta_{pw} + \pi/4)$ and the wavelengths are given by $\lambda'' = \lambda_{pw}\cos(\Theta'' - \Theta_{pw})$ and $\lambda' = \lambda_{pw}\cot(\Theta'' - \Theta_{pw})$. Unless other geometrical restrictions are imposed, these subharmonic waves are typically those that grow most quickly from a small-amplitude plane wave.

For example, Figure 4.15 shows the results of a fully nonlinear numerical simulation in which a plane internal wave in uniformly stratified fluid becomes unstable to subharmonic disturbances. Initially the plane wave has amplitude $A_\xi = 0.3k_x^{-1} \simeq 0.048\lambda_x$ and, in order to enhance the growth of subharmonic waves, small-amplitude noise is superimposed on the initial disturbance field. Although in theory a plane wave is assumed to extend to infinity in both the horizontal and vertical directions, this simulation restricts the plane wave to a doubly periodic domain having four wavelengths in both the horizontal and vertical.

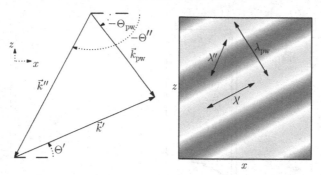

Fig. 4.14. Schematic illustrating the construction of a resonant triad formed by a plane wave of frequency ω whose wavenumber vector, $\vec{k}_{pw}$, forms an angle $\Theta_{pw} = \cos^{-1}(\omega/N)$ to the horizontal. This wave excites two subharmonic waves whose wavenumber vectors $\vec{k}'$ and $\vec{k}''$ form angles Θ' and Θ'', respectively to the horizontal. The angles satisfy the condition given by (4.111). This particular combination of wavenumbers is set so that $\vec{k}'$ lies parallel to the crests of the plane wave. The corresponding wavelengths of the plane wave and superharmonics are shown to the left, superimposed upon greyscale contours of the vertical displacement field of the plane wave.

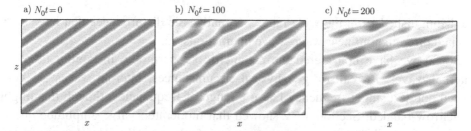

Fig. 4.15. Vertical displacement field at nondimensional times a) $N_0 t = 0$, b) $N_0 t = 100$ and c) $N_0 t = 200$ for a plane internal wave that becomes unstable to subharmonic disturbances. The frames are extracted from a fully nonlinear numerical simulation in which an initial plane wave of vertical wavenumber $k_z = -1.4 k_x$ has noise superimposed and then evolves in a two-dimensional, uniformly stratified Boussinesq fluid with buoyancy frequency N_0.

After approximately 16 buoyancy periods ($N_0 t = 100$), the development of the subharmonic disturbances is evident through the appearance of perturbations along the crests and troughs of the plane wave, as shown in Figure 4.15b. Once the subharmonic waves grow to substantial amplitude they too interact resonantly with other subharmonic waves which in turn interact with other subharmonics. Further excitation occurs rapidly as a greater range of triad interactions between waves of significant amplitude becomes possible. Figure 4.15c shows that after approximately 32 buoyancy periods the structure of the initial plane wave can no longer

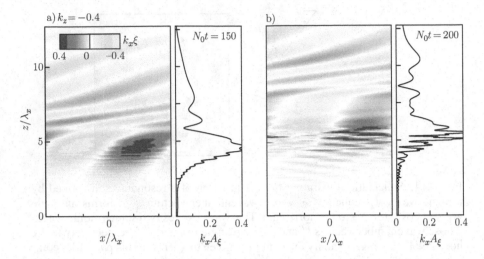

Fig. 4.16. Late-time structure of wavepacket whose initial evolution is shown in Figs. 4.11a, b and c. At the outset, $k_z = -0.4k_x$, $A_0 = 0.048\lambda_x$ ($k_x A_0 = 0.3$) and the Gaussian envelope has width $\sigma = 10/k_x$. Contours of the vertical displacement field are shown at times a) $t = 150/N_0$ and b) $t = 200/N_0$.

be distinguished from the wide range of small-scale disturbances that have been excited through nonlinear wave–wave interactions.

So far we have considered the parametric subharmonic instability of plane internal waves. In realistic geophysical circumstances, internal waves are generated not at a single frequency but at a range of frequencies, though this might be narrow-banded. Section 4.5.2 showed that moderately large-amplitude quasi-monochromatic internal wavepackets are unstable initially, not to wave–wave interactions, but to interactions between the waves and the wave-induced mean flow. However, at later times in their evolution parametric subharmonic instability can develop. This is demonstrated in Figure 4.16, which shows the structure of a moderately large-amplitude Gaussian wavepacket at late times in its evolution.

The initial disturbance has horizontal wavenumber k_x and frequency ω_0. This sets the vertical wavenumber through the dispersion relation (3.54) for two-dimensional Boussinesq waves. Explicitly

$$|k_z| = |k_x|\sqrt{\frac{N_0^2}{\omega_0^2} - 1}. \tag{4.114}$$

Through parametric subharmonic instability, self-interactions with the initial waves transfer energy most efficiently to waves with half the frequency. Assuming the horizontal wavenumber of this disturbance is the same as that of the initial wavepacket, the resulting vertical wavelength of the disturbance is given by

$$\frac{\lambda_z'}{\lambda_x} = \left[4\left(\frac{\lambda_x}{\lambda_z}\right)^2 + 3 \right]^{-1/2}. \tag{4.115}$$

Therefore, for a wavepacket composed of non-hydrostatic waves with $\omega \lesssim N_0$ and $|k_z| \ll |k_x|$, (4.115) predicts that parametric subharmonic instability will be manifest initially through excitation of waves having a vertical wavelength approximately $3^{-1/2} \simeq 0.6$ times the horizontal wavelength. For a wavepacket initially having a lower frequency with $|k_z| \gg |k_x|$, (4.115) predicts the vertical wavelength will halve through parametric subharmonic instability. In particular, for the simulation shown in Figure 4.16, we compute $\lambda_z' \simeq 0.52\lambda_x$, which is consistent with the fine-scale vertical structure that develops at late times in the simulation.

Like turbulence, wave–wave interactions extract energy from large scales and transfer it to successively smaller scales where it is dissipated. Unlike turbulence, however, the transfer process does not involve overturning and mixing by eddies over a range of scales. Eventually, the disturbances cascade energy to such small scales that they dissipate efficiently by molecular viscosity and diffusion, if not resulting in small-scale turbulent mixing.

4.6.2 Overturning instabilities

Here we consider internal waves of such large amplitude relative to their vertical and horizontal wavelength that the waves overturn and break. Ultimately wave breaking is a consequence of convective overturning, in which dense fluid overlies less dense fluid. But this circumstance can arise in a variety of ways.

Overturning occurs if the amplitude is so large that the waves carry dense fluid above light so that it becomes convectively unstable. That is, overturning occurs if

$$\partial \varrho_T/\partial z \equiv d\bar{\rho}/dz + \partial\rho/\partial z > 0 \tag{4.116}$$

somewhere in the flow field. Together with the polarization relations listed in Table 3.1, we can express this 'overturning condition' in terms of the vertical displacement amplitude, A_ξ. Explicitly, we find that internal waves are overturning if

$$A_\xi > A_{OT} \equiv \left| \frac{1}{k_z} \right| = \frac{\lambda_x}{2\pi} \cot|\Theta|, \tag{4.117}$$

in which $\lambda_x = 2\pi/k_x$ is the horizontal wavelength and $|\Theta|$ is a measure of the wave frequency through (3.57): $|\Theta| = \cos^{-1}(\omega/N_0)$.

Though conceptually straightforward, the overturning condition does not guarantee that breaking will in fact occur. For example, isopycnal surfaces are overturning at the time shown in Figure 4.13a, but convective instability takes longer to develop than the time for the wave motion to pull the isopycnals back into a statically stable state. Only at the later time shown in Figure 4.13b, has the disturbance grown to such large-amplitude that the overturned waves have time to break into small-scale convective turbulent eddies.

The maximum growth rate of disturbances arising from (Rayleigh–Taylor) convective instability associated with an inverted continuous density stratification is

$$\sigma = \sqrt{\frac{g}{\rho_0}\left(\frac{d\bar{\rho}}{dz} + \frac{\partial\rho}{\partial z}\right)}. \tag{4.118}$$

The value of σ is real where the waves drive the fluid to be overturning, as indicated by (4.116).

For convective instability to occur, we estimate $\sigma \gtrsim \omega$: the convection growth rate is faster than the wave frequency. Together with the polarization relations for internal waves, we find that the waves are not just overturning but also convectively unstable if the following convection condition is satisfied:

$$A_\xi > A_{CV} \equiv \frac{\lambda_x}{2\pi}\cot|\Theta|\left(1 + \cos^2\Theta\right). \tag{4.119}$$

In the limit $|\Theta| \to 90°$, corresponding to hydrostatic waves, $A_{CV} \simeq A_{OT}$, so the overturning condition assesses sufficiently well the stability of the waves. For nonhydrostatic (relatively fast frequency) waves, the convection condition predicts that their amplitude should be twice as large as the overturning condition for the waves to overturn and break.

Figure 4.17 illustrates isopycnal surfaces for small-amplitude waves, for waves at the point of overturning and for waves that overturn sufficiently to become convectively unstable. The marginal stability curves for overturning and convective internal waves are plotted as the solid and dashed lines, respectively, in Figure 4.18a.

4.6.3 Shear instabilities

Even if the waves are not so large that they convectively overturn, they may still turbulently break as a result of shear instability. In this case, it is not the release of potential energy that drives the instability. Rather it is the extraction of kinetic energy from the shear associated with the waves. This results directly in the mixing of density gradients when a vertical shear is sufficiently strong to overcome the buoyancy-restoring forces in a stratified fluid.

In Section 3.6.3 we found that a parallel shear flow, for which the horizontal velocity is a function of z alone, is unstable only if $\text{Ri}_g < 1/4$ somewhere in the

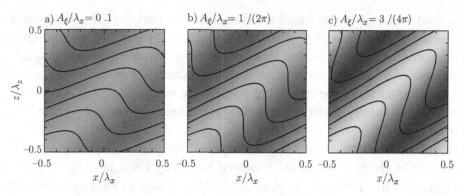

Fig. 4.17. Normalized vertical displacement field (greyscale) of a doubly periodic internal wave with superimposed contours showing displaced isopycnal surfaces. The waves are defined with wavenumbers $k_x = -k_z$ and the amplitudes, indicated, are chosen to show a) moderately large-amplitude waves, b) waves that just meet the overturning condition (4.117) and c) waves that just meet the convection condition (4.119). In all three cases, the greyscale varies from black to white as ξ/λ_x ranges from -0.3 to 0.3.

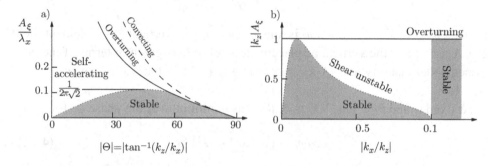

Fig. 4.18. a) Marginal stability curves associated with overturning (solid line) and convectively breaking waves (dashed line) and with wave packets that are driven to overturn through interactions with the wave-induced mean flow (self-accelerating waves) as indicated by the curve above the grey shaded region. Stability curves are given in terms of the vertical displacement amplitude relative to the horizontal wavelength and plotted as a function of $\Theta = \cos^{-1}(\omega/N_0)$. b) Marginal stability curves associated with overturning (solid line) and shear unstable inertia gravity waves (dotted line). Here the stability curves are given in terms of the product of the vertical displacement amplitude and vertical wavenumber plotted against the ratio of the horizontal to vertical wavenumber. The shear stability curve is plotted assuming $f_0/N_0 = 0.01$ and $\min(\mathrm{Ri}_g) = 1/4$.

flow field. Although the flow field associated with internal waves is neither parallel nor steady, we may nonetheless attempt to estimate the onset of shear instability by determining the amplitude of internal waves for which the minimum gradient Richardson number, Ri_g, drops below a critical value somewhere in the flow field. This assumption should be reasonable, particularly for inertia gravity waves whose flow fields are nearly horizontal and slowly evolving in time compared with the buoyancy period.

In uniformly stratified fluid with background buoyancy frequency N_0, Ri_g is defined to be the ratio of the local squared buoyancy frequency, $N_0^2 + \Delta N^2$, to the square of the shear, $\partial u / \partial z$. Here ΔN^2 is the change in buoyancy due to the stretching and compression of isopycnals by waves and the shear results from vertical gradients of the horizontal flow due to waves. For internal waves, the shear and ΔN^2 are 90° out of phase. Thus, for example, there is no shear where the vertical density gradient is weakest. So Ri_g is not necessarily a minimum where ΔN^2 is a minimum. Explicitly, for a two-dimensional monochromatic internal wave in a uniformly stratified Boussinesq fluid, the polarization relations give

$$\text{Ri}_g = \left(\frac{N_0 k_x}{\omega k_z} \right)^2 \frac{1}{\alpha^2} \frac{1 - \alpha \cos \varphi}{\sin^2 \varphi}, \tag{4.120}$$

in which $\varphi = k_x x + k_z z - \omega t$ is the phase of the vertical displacement field $\xi = A_\xi \cos \varphi$ and the vertical displacement amplitude is given in terms of the non-dimensional quantity $\alpha = |A_\xi k_z|$. According to (4.117), the wave is overturning if $\alpha > 1$.

At fixed α and $\Theta = \tan^{-1}(k_z/k_x)$, Ri_g has a minimum at the phase φ_c for which

$$\cos \varphi_c = (1 - \sqrt{1 - \alpha^2})/\alpha. \tag{4.121}$$

The minimum occurs where ΔN^2 is a minimum only if $\alpha = 1$, corresponding to internal waves at the point of overturning. For relatively small-amplitude waves, $\varphi_c \simeq \cos^{-1}(\alpha/2) \simeq 0$, in which case the minimum occurs where the shear is strongest and the background density gradient is unperturbed by waves.

Putting (4.121) into (4.120), the minimum gradient Richardson number is

$$\min(\text{Ri}_g) = \frac{1}{2} \left(\frac{N_0 k_x}{\omega k_z} \right)^2 \frac{1}{1 - \sqrt{1 - \alpha^2}}. \tag{4.122}$$

Rearranging this expression gives an explicit condition for shear instability in terms of the vertical displacement amplitude:

$$A_\xi > A_{\text{SHR}} \equiv \frac{2}{|k_z|} \left(\frac{N_0 |k_x|}{\omega |k_z|} \right) \sqrt{1 - \left(\frac{N_0 k_x}{\omega k_z} \right)^2}, \tag{4.123}$$

in which we have used (3.144) to set $\min(\mathrm{Ri}_g) = 1/4$ as the threshold for shear instability.

For internal waves that do not feel Coriolis forces, $N_0 k_x/(\omega k_z) = 1/\sin\Theta$. Hence the discriminant in (4.123) is negative, meaning that the shear instability criterion cannot be met. However, shear instability can occur for inertia gravity waves, whose dispersion relation (3.194) includes the Coriolis parameter f_0. Provided $|k_x/k_z| < \sqrt{f_0/N_0}$, $A_{\mathrm{SHR}} \leq A_{\mathrm{OT}}$, with equality occurring if $|k_x/k_z| = f_0/N_0$.

The marginal stability curves for overturning and shear unstable internal waves are plotted in Figure 4.18b for the case with $f_0/N_0 = 0.01$, characteristic of the atmosphere and ocean at mid-latitudes. For $|k_z| < 100|k_x|$ breaking results only from overturning waves. For larger $|k_z|$ the waves are shear unstable before reaching overturning amplitudes.

4.6.4 Overturning driven by self-acceleration

The overturning, convective and shear instability criteria predict that nonhydrostatic internal waves (with frequency close to N_0) are stable even at a very large vertical displacement amplitude. This is because such waves, having a relatively large vertical wavelength, can displace fluid vertically large distances without advecting dense fluid over light. Because such waves have horizontal gradients of vertical velocity they may be shear unstable but not in a way that opposes buoyancy as expressed through the Richardson number criterion (3.144).

Nonetheless, nonhydrostatic wavepackets can overturn if their amplitude is so large that their associated wave-induced mean flow eventually advects dense fluid over light. This process is known as 'self-acceleration'. Internal wavepackets are predicted to become overturning as a result of self-acceleration if the wave-induced mean flow exceeds the horizontal group velocity, as illustrated in Figure 4.19. Initially isopycnal surfaces are nearly parallel to the horizontal. At late times however, the isopycnals overturn near the top of the domain.

Using (3.62) and (3.107), self-acceleration is predicted to drive the waves to instability if

$$A_\xi > A_{\mathrm{SA}} \equiv \frac{\lambda_x}{2\pi\sqrt{2}} \sin 2\Theta. \qquad (4.124)$$

The marginal stability curve for this criterion is shown as the dotted line surrounding the shaded grey region in Figure 4.18a.

Contrary to the overturning and convection conditions, the self-acceleration condition predicts that small-amplitude wavepackets with $\Theta \simeq 0$ ($\omega \simeq N_0$) eventually overturn as a result of interactions with the wave-induced mean flow. Waves with $\omega \simeq 2^{-1/2} N_0$ are stable at the largest vertical displacement amplitudes, as great as 11% of their horizontal wavelength.

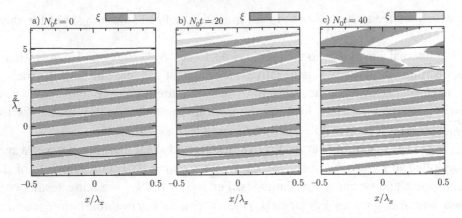

Fig. 4.19. Snapshots of a simulation at three times in which a horizontally periodic, vertically localized wavepacket breaks due to self-acceleration. Crests and troughs of the vertical displacement field ξ are illustrated with the indicated greyscale, and contours show displaced isopycnal surfaces. The wavepacket is initialized with vertical wavenumber $k_z = -k_x$ with vertical displacement amplitude $A_\xi = 0.11\lambda_x$ in uniformly stratified fluid having buoyancy frequency N_0. At time $t = 40N_0^{-1}$ isopycnals overturn near $z = 4\lambda_x$.

Exercises

4.1   Two arbitrary oscillatory functions are represented in terms of their phase φ by

$$f(\varphi) = \frac{1}{2}\mathscr{A}_f e^{\iota\varphi} + \text{cc} \quad \text{and} \quad g(\varphi) = \frac{1}{2}\mathscr{A}_g e^{\iota\varphi} + \text{cc},$$

in which cc denotes the complex conjugate and $\mathscr{A}_f$ and $\mathscr{A}_g$ are possibly complex amplitudes. Show that any quadratic combination of f and g leads to a superposition of disturbances with double and zero phase.

4.2   Assuming power series expansions of a time-periodic function $y(t) = y_0 + \epsilon y_1 + \epsilon^2 y_2 + \dots$ and the squared frequency $\omega^2 = \omega_0^2 \left(1 + \epsilon\sigma_1 + \epsilon^2\sigma_2 + \dots\right)$, obtain a perturbation approximation for the ordinary differential equation

$$y'' + y = \epsilon yy',$$

with initial conditions $y(0) = 1$ and $y'(0) = 0$. Give your result accurate to terms of order ϵ^2.

4.3   Follow the procedure described in Section 4.2.3 to derive the nonlinear Schrödinger equation (4.27).

4.4   Assuming the mean energy of one-dimensional waves can be written as $\langle E \rangle = A^2 f(k)$ for some function f, show that the amplitude equation (4.32) follows from (4.31) and (4.30).

4.5 Find the weakly nonlinear dispersion relation for deep water waves by following the steps below.

(a) Show that the surface boundary conditions can be combined to form the single condition

$$\left[\phi_{tt} + g\phi_z + 2\nabla\phi \cdot \nabla\phi_t + \frac{1}{2}\nabla\phi \cdot \nabla(|\nabla\phi|^2)\right]\bigg|_{z=\eta} = 0.$$

(b) Perform a Taylor-series expansion on your result in a) and use the recursion relation

$$\eta = -\frac{1}{g}\left[\phi_t + \frac{1}{2}|\nabla\phi|^2\right]\bigg|_{z=\eta}$$

to come up with the three leading-order terms in ϕ and its derivatives. (Note, you do not need to perform a perturbation expansion for this part of the problem.)

(c) Now let $\phi = \epsilon\phi_0 + \epsilon^2\phi_1 + \ldots$ and find the order ϵ, ϵ^2 and ϵ^3 terms. (For this problem, you should simplify the higher-order ϵ expressions using results from the lower-order ϵ terms.)

(d) Use your result in (c) and the second-order equation from (a) to formulate the secular equation and so show that $\sigma_1 = 0$ and hence $\phi_1 = 0$.

(e) Use the second-order equation relating η to ϕ to find η_1.

(f) At third order, formulate the secular equation to find σ_2, which is non-zero. Thus write the dispersion relation accurate to second order in amplitude.

4.6 Show that the cubic-amplitude correction to the surface displacement of deep water waves is $\eta_3 = 3/8 \, A^3 k^2 \cos 3(kx - \omega t)$.

4.7 Follow the steps below to show that the crests of saturated deep water waves develop into peaks that form a 120° angle at the cusps.

(a) Assuming the waves are steady, set up a co-ordinate system moving with the waves having origin at the cusp. In Cartesian co-ordinates, x is horizontal and z is upwards. In polar co-ordinates, r is the distance from the cusp and θ is measured counter-clockwise from the negative z-axis. Write x and z in terms of r and θ.

(b) Write Laplace's equation (2.2) in polar co-ordinates.

(c) Suppose near the cusp the surface displacement is $\eta = -r\cos\theta_0$, in which θ_0 is the angle we wish to determine. Write the zero surface-pressure condition (2.3) and the no-normal-flow condition (2.5) in terms of r and θ_0.

    (d) Using your result in (c), show that only one term in the general solution for ϕ is required. What is this term?

    (e) Combine your results in (c) and (d) to show that $\theta_0 = 60°$ and hence that the total angle across the cusp is $2\theta_0 = 120°$.

4.8   Using (4.61), show that Boussinesq interfacial waves in infinitely deep water are always modulationally unstable.

4.9   Write the nonlinear Schrödinger equation in the form (4.27) for interfacial waves in a two-layer fluid for which the lower layer has depth H and the upper layer is infinitely deep.

4.10 (a) Supposing u satisfies the advection equation $u_t + cu_x = 0$, show that u is constant on a characteristic curve with $x - ct$ constant.

    (b) Supposing u satisfies the nonlinear advection equation $u_t + uu_x = 0$, find the formula for the characteristic curves and determine the Riemann invariant, i.e. an expression which remains constant on the characteristic curves.

4.11 (a) Show that $\partial h/\partial u = (h/g')^{1/2}$ holds along characteristic curves for rightward-propagating nonlinear shallow water waves and determine the corresponding relationship between h and u for leftward-propagating waves.

    (b) Derive the formula for the Riemann invariants $r_{\pm}$ of the rightward- and leftward-propagating waves.

    (c) Derive the formula (4.76) describing the motion of rightward-propagating waves in the absence of leftward-propagating waves, and determine the corresponding formula for leftward-propagating waves.

4.12 (a) Derive the ordinary differential equation (4.82) from the Korteweg–de Vries equation (4.80).

    (b) Show by substitution that the solitary wave structure given by (4.83) is indeed a solution of the ordinary differential equation (4.82).

4.13 (a) Using the polarization relations for internal waves listed in Table 3.1, show that the right-hand side of the fully nonlinear equation for two-dimensional Boussinesq internal waves, given by (4.92), is identically zero.

    (b) Even with background rotation and in three dimensions plane internal waves are an exact solution of the fully nonlinear equations. Show this by Fourier transforming the mass conservation condition for incompressible fluid in the form (1.30) and using the result to show that the advection operator $\vec{u} \cdot \nabla$ is identically zero.

4.14 Derive the nonlinear Schrödinger equation for horizontally periodic internal wavepackets of finite vertical extent by following the argument below (4.107) in which the linear Schrödinger equation is modified to account for the Doppler-shifting of the waves by the wave-induced mean flow.

4.15 A plane internal wave propagates right and upwards with its constant-phase lines forming a 45° angle to the vertical. Find the wavenumber vectors of the resonant subharmonic pair of waves that have half the frequency of the plane wave. Give your answer relative to the magnitude of the wavenumber vector of the plane wave.

4.16 From (4.114), and under the assumption that parametric instability transfers energy to a disturbance with half the frequency but the same horizontal wavenumber, derive the formula (4.115) for the vertical wavelength of the disturbance.

4.17 Use the polarization relations for Boussinesq internal waves to show that the gradient Richardson number associated with plane waves is given by (4.120).

4.18 Assuming that inertia gravity waves are shear unstable if the minimum gradient Richardson number is less than $1/4$ somewhere in the flow field, show that this occurs if $|k_z| > \sqrt{f_0/N_0}|k_x|$.

5

Generation mechanisms

5.1 Introduction

In this chapter we examine the various mechanisms through which interfacial waves and internal waves can be created. Broadly speaking, the waves can be generated either by solid bodies or by disturbances within the fluid such as convection, imbalance of large-scale flows, gravity currents and turbulence. Here we focus upon generation by solid bodies, the theory for which is better established.

In a uniformly stratified fluid, we have already seen that an oscillating body creates a cross-shaped pattern of waves. Here, we study this problem in more detail by examining the structure of the wave beams that emanate from a vertically oscillating cylinder and sphere. Although this mechanism has little bearing on geophysical flows, it is amenable to study in laboratory experiments which can test the limitations of theory applied to this seemingly simple circumstance.

More realistic is the study of uniform and tidal flow over topography. This subject has been well studied in theory and by way of experiments, and the results have been compared with ample observational data. Because much of this work is discussed elsewhere, only the salient points will be presented here.

5.2 Oscillating bodies

In Section 3.3 we saw that any body oscillating with frequency ω in uniformly stratified fluid with buoyancy frequency N_0 creates a cross-shaped pattern of internal waves if $\omega < N_0$. Here we examine how the structure of the wave beams is affected by the shape of the body in two and three dimensions. In all cases we assume that the oscillations of the body are small so that internal waves are generated with small amplitudes. We will see that it is necessary to include viscous effects in order to eliminate singularities in the displacement fields that are predicted by the theory of waves in inviscid fluid.

5.2.1 Oscillating cylinder in inviscid fluid

Because oscillating cylinder experiments are performed typically in salt-stratified fluid, we assume the fluid is Boussinesq and uniformly stratified with (constant) buoyancy frequency N_0, given by (3.2). We suppose the cylinder has a circular cross-section with radius a and that it oscillates vertically up and down at frequency ω. The amplitude of its oscillation (half the peak-to-peak vertical displacement) is denoted by A; the maximum vertical velocity is $W = A\omega$.

We will not look at the transient start-up problem here. Instead, we will assume the cylinder has been oscillating for such a long time that the fluctuation fields themselves oscillate at frequency ω everywhere in space. We do not expect the disturbances to have any structure in the y-direction, which is parallel to the cylinder axis. Thus we can represent the fluid motion in terms of the streamfunction by

$$\psi(x,z,t) = \Psi(x,z)e^{-\iota\omega t}. \tag{5.1}$$

Here $\Psi(x,z)$ is the possibly complex-valued 'streamfunction amplitude' and it is understood that the actual streamfunction is the real part of (5.1). The magnitude of the streamfunction amplitude $|\Psi|$ represents the streamfunction amplitude envelope. This is the maximum value of the streamfunction field at any point in space for all time.

Once Ψ is determined, the velocity field is given by (the real parts of)

$$u = -\frac{\partial\Psi}{\partial z}e^{-\iota\omega t} \quad \text{and} \quad w = \frac{\partial\Psi}{\partial x}e^{-\iota\omega t}. \tag{5.2}$$

We begin by neglecting the effects of viscosity, later including them to model beam attenuation. In inviscid fluid, we found that ψ satisfies the differential equation (3.40). Using (5.1), this gives an equation for the streamfunction amplitude:

$$N_0{}^2\frac{\partial^2\Psi}{\partial x^2} - \omega^2\nabla^2\Psi = 0. \tag{5.3}$$

The boundary conditions require that the fluid on the boundary of the cylinder moves vertically with the speed of the cylinder. Taking the maximum vertical speed to be $W > 0$, the streamfunction on the cylinder is

$$\Psi|_C = Wx|_C = -Wa\sin(\theta). \tag{5.4}$$

Here the cylinder boundary $C = \left\{(x,z) \mid x^2 + z^2 = a^2\right\}$ is taken to be stationary, consistent with the assumption of small-amplitude oscillations, $A \ll a$. It follows from (5.2) that the actual vertical velocity on C at some time t is $\Re\{W\exp(-\iota\omega t)\} = W\cos(\omega t)$. So, with $W > 0$, $t = 0$ corresponds to the time at which the cylinder moves upwards through its equilibrium position.

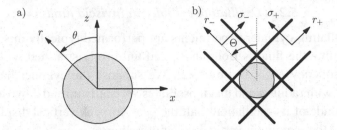

Fig. 5.1. a) Co-ordinate system used to model potential flow around a cylinder and b) the co-ordinate system used to model the across-beam ($\sigma_\pm$) and along-beam ($r_\pm$) axes.

The second expression in (5.4) has been recast in polar co-ordinates with θ being taken counter-clockwise from the z-axis, as shown in Figure 5.1a. We also impose the condition that the waves either vanish far from the cylinder if $\omega > N_0$, or that energy is transported away from the cylinder if $\omega < N_0$.

Before solving the boundary value problem explicitly, we examine the solution in the case $N_0 = 0$, corresponding to the disturbance field that arises in uniform-density fluid. In this case, (5.3) reduces to the potential equation

$$\nabla^2 \Psi = 0. \tag{5.5}$$

The general solution may be found through the method of separation of variables invoked in cylindrical co-ordinates. Requiring vanishing solutions for large r, we have

$$\Psi(r,\theta) = \sum_{n=1}^{\infty} r^{-n} [A_n \cos(n\theta) + B_n \sin(n\theta)] = \sum_{n=1}^{\infty} \mathcal{A}_n \mathcal{Z}^{-n} + \text{cc}, \tag{5.6}$$

in which A_n and B_n are real constants. In the second expression we have recast the result in terms of the complex variable $\mathcal{Z} = \imath r e^{\imath\theta}$ using the polar co-ordinate system shown in Figure 5.2a. $\mathcal{A}_n$ are generally complex constants and cc denotes the complex conjugate of the sum.

Applying the boundary condition (5.4) written in polar co-ordinates, we see immediately that Ψ must satisfy

$$\Psi(r,\theta) = -W \frac{a^2}{r} \sin\theta = \frac{1}{2} W \frac{a^2}{\mathcal{Z}} + \text{cc}. \tag{5.7}$$

This result is what we would get from solving the problem of irrotational flow around a cylinder translating at speed W. For the problem of the oscillating cylinder, the flow alternately moves upwards and downwards around the cylinder with frequency ω, as prescribed by the real part of the right-hand side of (5.1).

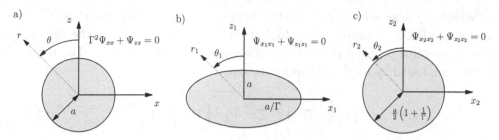

Fig. 5.2. a) Schematic illustrating the boundary value problem in x–z space, b) transformation to the potential equation and corresponding boundary in x_1–z_1 space and c) the corresponding boundary problem for flow around a cylinder in x_2–z_2 space determined under the Joukowski transformation.

Our plan first is to extend this result to the case $\omega > N_0$. This will likewise give a potential flow equation that can be solved by co-ordinate transformations involving the method of conformal mapping. Thereafter, by the method of analytic continuation, we will derive the expression for the streamfunction amplitude in the case $\omega < N_0$.

If $\omega > N_0$, (5.3) can be written

$$\Gamma^2 \frac{\partial^2 \Psi}{\partial x^2} + \frac{\partial^2 \Psi}{\partial z^2} = 0, \tag{5.8}$$

in which $\Gamma^2 \equiv 1 - N_0^2/\omega^2 > 0$. This is recast into the form of the potential equation (5.5) through the co-ordinate transformation $(x, z) = (\Gamma x_1, z_1)$. The corresponding boundary condition becomes

$$\Psi|_{C_1} = W\Gamma x_1 \,|_{C_1} = W\Gamma \frac{a}{2\iota \exp(\iota \theta_1)} + \text{cc.} \tag{5.9}$$

In the last expression, the transformed complex co-ordinate $\mathcal{Z}_1 \equiv x_1 + \iota z_1 = \iota r_1 e^{\iota \theta_1}$ has been evaluated on the elliptical boundary $C_1 = \left\{ (x_1, z_1) \mid (\Gamma x_1)^2 + z_1^2 = a^2 \right\}$. So the problem reduces to that of finding the potential flow around an ellipse centred at the origin with major axis a/Γ lying on the x-axis and minor axis a lying on the z-axis. The resulting boundary value problem is illustrated schematically in Figure 5.2b.

By the method of conformal mapping, we can transform the problem of potential flow around a cylinder of radius $a_2 = a(1 + 1/\Gamma)/2$ to that of flow around the ellipse C_1. This is done through the Joukowski transformation:

$$\mathcal{Z}_1 = \mathcal{Z}_2 + \frac{a^2}{4}\left(\frac{1}{\Gamma^2} - 1\right)\frac{1}{\mathcal{Z}_2}, \tag{5.10}$$

in which $\mathcal{Z}_2 \equiv x_2 + \imath z_2 = r_2 e^{\imath \theta_2}$ is the complex co-ordinate system in which the circular cylinder of radius a_2 centred at the origin is embedded. The resulting boundary value problem is illustrated schematically in Figure 5.2c. The inverse transform, found by solving the quadratic equation, is

$$\mathcal{Z}_2 = \frac{1}{2}\left[\mathcal{Z}_1 + \sqrt{\mathcal{Z}_1^2 + a^2\left(1 - \frac{1}{\Gamma^2}\right)}\right]. \qquad (5.11)$$

It is straightforward to check that substituting the parametric representation of the ellipse $C_1 = \{\mathcal{Z}_1 = -(a/\Gamma)\sin\theta + \imath a\cos\theta \mid 0 \le \theta < 2\pi\}$ into (5.11) gives the circle $C_2 = \{\mathcal{Z}_2 = \imath a(1 + 1/\Gamma)/2 \exp(\imath\theta_2) \mid 0 \le \theta < 2\pi\}$.

From (5.10), we can write x_1 in terms of x_2 and so the boundary condition (5.9) on C_2 becomes

$$\Psi|_{C_2} = W\frac{2\Gamma}{\Gamma+1}x_2\bigg|_{C_2} = W\frac{\Gamma}{\Gamma+1}\frac{a_2}{\imath\exp(\imath\theta_2)} + \text{cc}. \qquad (5.12)$$

At this point we can use our solution for potential flow around a cylinder (5.7) and so find the streamfunction amplitude in the complex x_2–z_2 plane:

$$\Psi(\mathcal{Z}_2) = W\frac{\Gamma}{\Gamma+1}\frac{a_2^2}{\mathcal{Z}_2} + \text{cc} = \frac{1}{4}W\left(1+\frac{1}{\Gamma}\right)\frac{a^2}{\mathcal{Z}_2} + \text{cc} \qquad (5.13)$$

for all $\mathcal{Z}_2$.

Using (5.11) to recast this result in terms of $\mathcal{Z}_1$ and then in terms of $\mathcal{Z} = (x/\Gamma) + \imath z$, we have

$$\Psi(x,z) = \frac{1}{2}Wa^2\left(1+\frac{1}{\Gamma}\right)\left[\mathcal{Z} + \sqrt{\mathcal{Z}^2 + a^2(1 - 1/\Gamma^2)}\right]^{-1} + \text{cc}. \qquad (5.14)$$

Though algebraically complicated if written explicitly in terms of x and z, (5.14) nonetheless gives an analytic expression for the streamfunction amplitude that describes the flow surrounding a cylinder that oscillates vertically with frequency larger than the buoyancy frequency. In doing so, care must be taken in evaluating the square root of the complex argument in the denominator of (5.14) so as to ensure bounded solutions.

We now wish to find Ψ if the cylinder oscillates at frequency $\omega < N_0$. In this case the elliptic partial differential equation (5.8) becomes the hyperbolic differential equation

$$-\gamma^2\frac{\partial^2\Psi}{\partial x^2} + \frac{\partial^2\Psi}{\partial z^2} = 0, \qquad (5.15)$$

in which $\gamma^2 = N_0^2/\omega^2 - 1 > 0$. We recognize that this is the same equation as (5.8) but with Γ replaced by $\imath\gamma$.

Through the method of analytic continuation, we can immediately solve (5.15) with boundary condition (5.9). Because we have assumed that Ψ is an analytic function, the solution (5.14) should hold whether or not Γ is real or imaginary. Thus the streamfunction amplitude is given by (5.14) in which $\Gamma = \imath\gamma$ and $Z = (-\imath x/\gamma) + \imath z$. This must be done in a way that ensures solutions are bounded and that disturbance energy propagates away from the cylinder.

Nice things happen. The argument to the square root in (5.14) becomes a real number and so the representation of Ψ explicitly in terms of real variables becomes less complicated. For this purpose, it is convenient to recast the result in terms of across-beam co-ordinates σ_+ and σ_- instead of x and z. Using the definition $|\Theta| = \cos(\omega/N_0)$, which as usual represents the magnitude of the angle formed between the vertical and lines of constant phase for waves with frequency ω, we set $\gamma = -\tan\Theta$. Here it is assumed that $\Theta < 0$ so that γ gives the (positive) slope of the wave beams in the upper right-hand and lower left-hand quadrant. (In Section 3.3.3 we saw that these waves have negative vertical and positive horizontal wavenumber if they move away from the origin, and so Θ as given by (3.56) is negative.)

Lines with slope γ are given by constant values of

$$\sigma_+ = -x\cos\Theta - z\sin\Theta. \tag{5.16}$$

Perpendicular to these are lines with constant values of

$$r_+ = -x\sin\Theta + z\cos\Theta. \tag{5.17}$$

With $\Theta < 0$, the pair (σ_+, r_+) give the across-beam and along-beam co-ordinates for waves in the first and third quadrants, as shown in Figure 5.1b.

Similarly, we define the across-beam and along-beam co-ordinates in the second and fourth quadrants by

$$\sigma_- = x\cos\Theta - z\sin\Theta \tag{5.18}$$

and

$$r_- = x\sin\Theta + z\cos\Theta, \tag{5.19}$$

respectively.

In terms of Θ, the complex co-ordinate Z is

$$Z = -\imath\sigma_+/\sin\Theta.$$

Substituting this into (5.14) and noting that $a^2(1 - 1/\Gamma^2) = a^2/\sin^2\Theta$, the first term simplifies to $\Psi_+ = (1/2)Wa^2 e^{-\imath\Theta}[-\sigma_+ + (\sigma_+^2 - a^2)^{1/2}]^{-1}$. To ensure bounded and outward-propagating solutions, the branch-cut to the square root must be taken so that, after simplifying, we have

$$
\Psi_+(x,z) = \begin{cases} \dfrac{1}{2}Wae^{-\iota\Theta}\left[-\dfrac{\sigma_+}{a} + \mathrm{sign}\left(\dfrac{\sigma_+}{a}\right)\sqrt{\left(\dfrac{\sigma_+}{a}\right)^2 - 1}\right] & |\sigma_+/a| > 1 \\[4mm] \dfrac{1}{2}Wae^{-\iota\Theta}\left[-\dfrac{\sigma_+}{a} - \iota\sqrt{1 - \left(\dfrac{\sigma_+}{a}\right)^2}\right] & |\sigma_+/a| < 1. \end{cases}
$$

$$(5.20)$$

This is the streamfunction amplitude for waves propagating upwards and rightwards with $r_+ > 0$ and it is assumed that $-\pi/2 < \Theta < 0$. The real and imaginary parts of Ψ_+ in the first quadrant are shown in Figure 5.3a.

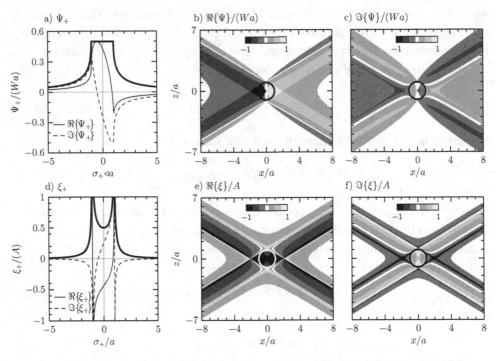

Fig. 5.3. a) Real (thin solid line), imaginary (dashed line) and magnitude (thick solid line) of the streamfunction amplitude envelope in the first quadrant, $\Psi_+(\sigma_+)$. b) Real and c) imaginary parts of the total streamfunction amplitude envelope $\Psi_+ + \Psi_-$. d) Corresponding vertical displacement field in the first quadrant, $\xi_+(\sigma_+)$ and the e) real and f) imaginary parts of the total displacement field $\xi_+ + \xi_-$. The thick circles in b), c), e) and f) indicate the location and radius of the cylinder. The real part of the field gives the flow pattern when the cylinder moves upwards through its equilibrium position (e.g. at time $t = 0$) and the imaginary part gives the flow pattern when the cylinder reaches its maximum vertical displacement (e.g. at time $t = (\pi/2)/\omega$).

There are two alternate ways of writing (5.20). Putting the σ-dependent terms in a complex exponential, we have

$$\Psi_+ = \frac{1}{2}Wae^{-\imath\Theta}\exp[-\imath\cos^{-1}(-\sigma_+/a)]. \tag{5.21}$$

The argument to the arccosine has been written as negative-valued to emphasize the necessary branch-cut to recover the second expression in (5.20). This expression makes it obvious that the magnitude of Ψ_+ is constant where $|\sigma_+| \leq a$. It also reveals that the phase of the waves decreases as σ_+ increases. This is consistent with the expected downward-propagating phase (and upward-propagating energy) for waves in the first quadrant.

Using the integral transform

$$\int_0^\infty \frac{1}{k}J_\nu(k)e^{\imath kx}\,dk = \frac{1}{\nu}\left[\sqrt{1-x^2}+\imath x\right]^\nu, \tag{5.22}$$

(5.20) is equivalently given by

$$\Psi_+(\sigma_+) = -\imath\frac{1}{2}Wae^{-\imath\Theta}\int_0^\infty \frac{J_1(k_\sigma a)}{k_\sigma}e^{-\imath k_\sigma\sigma_+}\,dk_\sigma. \tag{5.23}$$

Here J_1 is the first-order Bessel function of the first kind. Note that the integral only sums over contributions from positive across-beam wavenumbers k_σ. The negative complex exponent in the integrand is consistent with waves in the first quadrant transporting energy upwards and so having negative wavenumber. Although the structure of the waves represented by (5.23) is less immediately apparent, this expression will prove useful later in our consideration of beam attenuation due to viscous effects.

The second term (the complex conjugate) in (5.14) is found by substituting $\Gamma = -\imath\gamma$ and simplifying as above. This gives the streamfunction amplitude Ψ_- of waves that propagate upwards and leftwards, determined so that

$$\Psi_-(\sigma_-, r_-) = -\Psi_+(\sigma_+, r_+), \tag{5.24}$$

for $r_- > 0$. By symmetry, the wave beam in the third quadrant is given by $\Psi_+(-\sigma_+, -r_+) = -\Psi_+(\sigma_+, r_+)$, and in the fourth quadrant $\Psi_-(-\sigma_-, -r_-) = -\Psi_-(\sigma_-, r_-) = \Psi_+(\sigma_+, r_+)$. Thus the streamfunction amplitude is seen to be composed of a superposition of outward-propagating beams represented by $\Psi = \Psi_+ + \Psi_-$. It is straightforward to show that Ψ satisfies the boundary condition (5.4).

Through analytic continuation, the resulting solution for Ψ is complex, the real and imaginary parts of which are plotted in Figures 5.3b and c, respectively. It is understood that the actual structure of the waves is given by the real part of (5.1). Thus Figure 5.3b represents the streamfunction as the centre of the cylinder moves upwards through the origin and Figure 5.3c represents the streamfunction when the cylinder moves to its maximum vertical displacement a quarter-period later.

Although the streamfunction is continuous, derivatives with respect to $\sigma_\pm$ are not. In particular, (5.20) predicts that the along-beam velocity is infinite where $\sigma_\pm = a$. Likewise, the result predicts that the vertical displacement amplitude, given by $\xi = (\iota/\omega)\Psi_x$, is infinite. This field is plotted in Figure 5.3d, e and f. The approach to infinity along the tangents to the cylinder is so rapid that the integrated instantaneous mass flux is finite. Nonetheless, this counterintuitive and unphysical result reflects a deficiency in the approximations used to derive the solution. This is addressed in the next section.

5.2.2 Oscillating cylinder in viscous fluid

Primarily the singularities in the displacement and velocity field arise because we have neglected viscosity. The boundary condition (5.4) requires no-normal flow across the cylinder but does permit free-slip flow along the cylinder. Furthermore, the large across-beam variations of ψ near tangents to the cylinder imply that viscosity will act to smooth the gradients and keep the along-beam velocities and displacements finite. Thus viscosity plays a non-negligible role even for flows with very large Reynolds numbers.

The boundary value problem that includes no-slip conditions on the cylinder is difficult to solve analytically. However, the spreading of wave beams due to viscosity, an effect known as 'viscous attenuation', can be accounted for through a technique known as boundary layer theory.

In a Newtonian fluid, viscous effects are accounted for by adding the diffusion term $\mu\nabla^2\vec{u}$ to the right-hand side of the momentum equations. This has been done for a liquid in (3.37). Explicitly, in the Boussinesq approximation the momentum equations are

$$\frac{D\vec{u}}{Dt} = -\frac{1}{\varrho_0}\nabla p - \frac{1}{\varrho_0}g\rho\hat{z} + \nu\nabla^2\vec{u}. \tag{5.25}$$

The constant μ is the 'molecular viscosity', which effectively plays the role of friction acting within the fluid. When divided by the density ϱ_0 we have the 'kinematic viscosity' $\nu \equiv \mu/\varrho_0$. This is a measure of the diffusivity of velocity in a spatially varying shear flow. Its characteristic value for water is $\nu \simeq 10^{-6}\ \mathrm{m^2/s}$ and for air is $\nu \simeq 10^{-5}\ \mathrm{m^2/s}$.

For small amplitude waves, we may combine the linearized form of the momentum equations (5.25) together with the continuity equation for an incompressible fluid (1.30) to give a single, fourth-order partial differential equation in Ψ:

$$N_0^2\frac{\partial^2\Psi}{\partial x^2} - \omega^2\nabla^2\Psi + \iota\omega\nu\nabla^4\Psi = 0. \tag{5.26}$$

In deriving this equation, we have included the diffusivity of momentum but we have neglected the diffusivity of salinity. Particularly in laboratory experiments performed using salt-stratified solutions, this assumption is reasonable considering that salt-diffusivity is three orders of magnitude smaller than ν.

Focusing upon the upward-propagating beam alone, we use (5.16) and (5.17) to transform (5.26) into across- and along-beam co-ordinates. We further assume that across-beam variations are more significant than along-beam variations. Thus we neglect r_+ derivatives of order two or greater in the resulting partial differential equations. This assumption is analogous to making the boundary-layer approximation for viscous flow over a rigid surface. Dropping the subscripts from σ_+, r_+ and Ψ_+, the resulting equation for the streamfunction in the first quadrant is

$$\frac{\partial^2 \Psi}{\partial \sigma \partial r} + \frac{\iota \nu}{2N_0 \sin \Theta} \frac{\partial^4 \Psi}{\partial \sigma^4} = 0. \tag{5.27}$$

We anticipate that viscous effects will become increasingly important in the far field, $r \gg a$, but that the inviscid solution (5.20) adequately represents the structure near the cylinder itself. Thus an approximate solution of (5.27) can be found by Fourier transforming in σ, solving the resulting ordinary differential equation in r, inverse transforming, and matching the solution at $r = 0$ with the integral representation of the inviscid waves, (5.23). Thus we find

$$\Psi_{(\sigma)} = -\iota \frac{1}{2} Wae^{-\iota\Theta} \int_0^\infty \frac{J_1(K)}{K} \exp\left(-K^3 \tilde{\nu} \frac{r}{a} - \iota K \frac{\sigma}{a}\right) dK, \tag{5.28}$$

in which $K = k_\sigma a$ is a nondimensional measure of the across-beam wavenumber k_σ, and the nondimensional quantity

$$\tilde{\nu} = \frac{\nu}{2a^2 N_0 \sin |\Theta|} \tag{5.29}$$

is a measure of the rate of viscous attenuation with along-beam distance away from the cylinder.

From (5.28), we may go on to find the velocity and displacement fields. In particular, the vertical displacement field $\xi(x,z)$ is given by

$$\xi = \frac{1}{-\iota\omega} \frac{\partial \Psi(x,z)}{\partial x} \simeq \frac{-\iota}{N_0} \frac{\partial \Psi(r,\sigma)}{\partial \sigma}. \tag{5.30}$$

In this last expression, we have used the dispersion relation $\omega = N_0 \cos(\Theta)$ and the boundary-layer approximation has been invoked, resulting in neglect of the $\partial_r \Psi$ term.

Figure 5.4 shows the predicted structure of upward- and rightward-propagating waves generated by an oscillating cylinder. The calculated vertical displacement

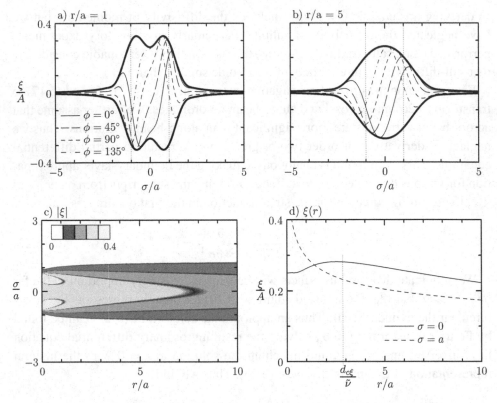

Fig. 5.4. Values of the vertical displacement field, ξ, computed in the first quadrant for a circular cylinder of radius a oscillating with relative frequency $\omega/N = 0.5$ and maximum vertical displacement $A = W/\omega$. The kinematic viscosity is set to be $v = 0.01a^2N$ so the effects of viscous attenuation are governed by the parameter $\tilde{v} \simeq 0.0058$. Instantaneous values of ξ are evaluated at along-beam distances a) $r/a = 1$ and b) $r/a = 5$. The plots show both the envelope (thick solid lines) and the instantaneous amplitude of the waves at phase $\phi = 0$ (thin solid line), $\phi = \pi/4$ (long-dashed line), $\phi = \pi/2$ (short-dashed line), $\phi = 3\pi/4$ (dashed-dotted line). The dotted vertical lines in both plots illustrate the values $\sigma = \pm a$. c) Contours of the amplitude envelope $|\xi(r,\sigma)|$ show the transition from bimodal to unimodal beams. d) Values of $|\xi|$ as a function of along-beam distance evaluated at $\sigma = 0$ (solid line) and $\sigma = a$ (dashed line). The positions $d_{\star\xi}$, where the centreline reaches maximum amplitude, and $d_{c\xi}$, where the across-beam curvature is zero, are indicated.

field is normalized by the vertical displacement A of the cylinder. The plots of the across-beam amplitude at different phases ϕ of oscillation are shown in Figures 5.4a and b. As expected for waves with upward group velocity, the phase lines move downwards (to decreasing σ) as ϕ increases. The amplitude envelope at $r = a$ exhibits two peaks near $\sigma = \pm a$. Further from the cylinder, these peaks merge as the wave beam spreads due to viscosity. At $r = 5a$, the peak amplitude occurs along the centreline of the beam where $\sigma = 0$.

The change in structure from bimodal (having peak amplitudes on either side of the centreline of the beam) to unimodal (having peak amplitude along the centreline) is evident in the structure of the amplitude envelope of ξ, shown as a function of r and σ in Figure 5.4c. After the beam adopts a unimodal structure the centreline amplitude gradually decreases and the beam widens as r increases. This broadening and weakening of the beam due to viscosity is what is referred to as viscous attenuation.

One way to characterize the bimodal-to-unimodal transition is through the across-beam curvature of the amplitude envelope measured at the centreline. This is positive (concave upward) for bimodal beams and negative (concave downward) for unimodal beams. We define r_c to be the along-beam distance where the centreline curvature of the streamfunction amplitude envelope is zero. That is,

$$\frac{\partial^2 |\Psi|}{\partial \sigma^2}\bigg|_{(\sigma,r)=(0,r_c)} = 0. \tag{5.31}$$

In order to evaluate r_c, we define

$$I_n(d;p) = \int_0^\infty K^{p-1} J_n(K) e^{-K^3 d}\, dK, \tag{5.32}$$

in which $d = \tilde{v} r/a$ with $\tilde{v}$ given by (5.29). Substituting (5.28) into (5.31), the condition for r_c is given through the implicit algebraic relationship

$$I_1^2(d_c;1) = I_1(d_c;0) I_1(d_c;2), \tag{5.33}$$

in which $d_c = \tilde{v} r_c/a$. The integrals in (5.32) may be evaluated numerically and from these it is found that $d_c \simeq 0.00605$.

Where the streamfunction changes from bimodal to unimodal behaviour does not necessarily occur at the same location for other fields. For example, the transition of the vertical displacement field occurs at $r_{c\xi}$ where $\partial_{\sigma\sigma} |\xi| = 0$. Using this condition together with (5.30), the transition distance is given by

$$I_1^2(d_{c\xi};2) = I_1(d_{c\xi};1) I_1(d_{c\xi};3). \tag{5.34}$$

Numerically, one finds that $d_{c\xi} \simeq 0.01650$, as indicated in Figure 5.4d. This is the same value computed for the transition distance of the velocity fields.

Asymptotic approximations to the integrals (5.32) can be used to determine the rate of decay of the amplitude envelopes far from the cylinder. For example, the

Table 5.1. *Theoretically predicted values of the non-dimensional along-beam distance $d_c = \tilde{v}r/a$ at which the across-beam curvature is zero along the centreline. Also listed are power law decay rates ($\propto r^{-p_v}$) of the centreline beam amplitudes as $r \to \infty$.*

field	cylinder	sphere
streamfunction: Ψ	$d_c = 0.00605$	$d_c = 0.00507$
	$p_v = 1/3$	$p_v = 1$
velocity: $\vec{u}$	$d_c = 0.01650$	$d_c = 0.01086$
displacement: $\vec{\xi}$	$p_v = 2/3$	$p_v = 4/3$

streamfunction amplitude envelope along the centreline is given by evaluating (5.28) at $\sigma = 0$:

$$|\Psi|_{\sigma=0} = \frac{1}{2}Wa \int_0^\infty \frac{J_1(K)}{K} e^{-K^3 d}\, dK, \tag{5.35}$$

in which $d = \tilde{v}r/a$, as before.

If d is large, the integral is dominated by the behaviour of the integrand for small K. This inspires us to perform a Taylor-series expansion of the first-order Bessel function about $K = 0$: $J_1 \simeq K/2$. The integration variable K in the resulting approximation can be replaced by $\tilde{K} = dK^3$ and so the definite integral is converted into the form of a Gamma function. Explicitly we find

$$|\Psi|_{\sigma=0} = \frac{1}{6}\Gamma(1/3)Wa\, d^{-1/3} \propto r^{-1/3}. \tag{5.36}$$

And so we have found that the amplitude decreases as the negative one-third power of the distance far from the cylinder along the centreline of the beam.

Likewise we can compute the power law rate of decay, r^{-p_v}, for other fields of interest. The power law exponents, p_v, are listed in Table 5.1. In particular, for the displacement and velocity fields, we find $p_v = 2/3$, meaning that the amplitudes of these fields decay with distance more rapidly than the streamfunction.

5.2.3 Oscillating sphere

The analysis of internal waves generated by an oscillating sphere is similar to that for an oscillating cylinder. However, there are significant qualitative differences in the resulting predictions that the circumstance is worth considering here in some detail.

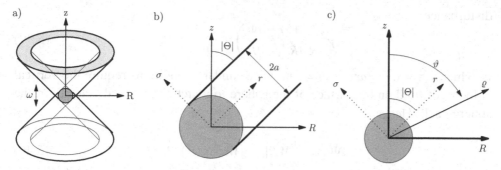

Fig. 5.5. a) Geometry of the wave-cones generated by an oscillating sphere in a uniformly stratified medium. b) Along-beam and across-beam co-ordinate system, (r, σ), used to model the upward-propagating cone. c) Radial and azimuthal spherical co-ordinates, (ϱ, ϑ), also used to model the upward-propagating wave cone.

As before, we suppose the sphere oscillates vertically with small amplitude at a constant frequency, ω, which is less than the ambient buoyancy frequency N_0. Thus the sphere generates propagating internal waves that transport energy away from the source along beams forming a constant angle $|\Theta|$ with the vertical. Unlike the case of an oscillating cylinder, waves emanating from a sphere are not confined to the x–z plane. Rather they spread radially outwards in the horizontal direction as they propagate vertically upwards and downwards away from the sphere. The beams are thus manifest in the form of cones with vertices near the origin and exhibiting rotational symmetry about the z-axis, as shown in Figure 5.5a.

Because the horizontal cross-sectional area of a cone increases linearly with distance from its apex we expect, even for inviscid waves, that the amplitude will decrease with distance, r, from the source. Indeed, for the momentum flux (which is proportional to amplitude-squared) to remain constant at each vertical level, the amplitude should decrease as $r^{-1/2}$.

Assuming the wave field is axisymmetric, it can be characterized in R–z co-ordinates, in which $R = (x^2 + y^2)^{1/2}$ is the radial distance from the z-axis in the horizontal plane. Using these co-ordinates, we define the axisymmetric streamfunction amplitude $\Psi(R, z)$ so that the radial, u_R, and vertical, u_z, components of the velocity field are given by

$$u_R = -\frac{\partial \Psi}{\partial z} \tag{5.37}$$

and

$$u_z = \frac{1}{R} \frac{\partial (R\Psi)}{\partial R}. \tag{5.38}$$

The inviscid equations of motion may recast in terms of the streamfunction amplitude as in (5.15) but written in cylindrical co-ordinates for axisymmetric disturbances:

$$-\gamma^2 \frac{1}{R}\frac{\partial}{\partial R}\left(R\frac{\partial \Psi}{\partial R}\right) + \frac{\partial^2 \Psi}{\partial z^2} = 0, \tag{5.39}$$

in which $\gamma^2 = (N_0^2/\omega^2 - 1) = \tan^2\Theta$. Boundary conditions require the vertical velocity of fluid on the surface of the sphere to match the vertical velocity of the sphere itself. Hence

$$\Psi|_C = \frac{1}{2}WR|_C = Wa\sin(\vartheta)/2, \tag{5.40}$$

in which C represents the co-ordinates on the surface of the sphere and ϑ is the azimuthal angle taken to increase from 0 to π going from the positive to negative z-axis.

As with the oscillating cylinder, we can alternately express the wave field in terms of along-beam (r) and across-beam (σ) co-ordinates. This is illustrated in Figure 5.5b for the upward-propagating wave-cone. By analogy with (5.16) and (5.17), these are written in terms of the cylindrical co-ordinates by

$$\sigma = -R\cos|\Theta| + z\sin|\Theta|, \qquad r = R\sin|\Theta| + z\cos|\Theta|. \tag{5.41}$$

The potential theory approach used for the oscillating cylinder facilitated by the use of Green's functions may be used to solve the boundary value problem (5.39) together with (5.40) and the condition that the waves are bounded and propagate outwards from the sphere. After some algebra the streamfunction for the upward propagating conical wavefronts is found to be

$$\Psi = \frac{1}{4}Wae^{-\imath(\Theta+3\pi/4)}\sqrt{\frac{a}{\mathcal{Z}_\rho}}\begin{cases} \imath\left[\frac{\sigma}{a} - \sqrt{\left(\frac{\sigma}{a}\right)^2 - 1}\right]^{3/2} & \sigma > a. \\ \left[-\frac{\sigma}{a} - \imath\sqrt{1 - \left(\frac{\sigma}{a}\right)^2}\right]^{3/2} & |\sigma| < a \\ -\left[-\frac{\sigma}{a} - \sqrt{\left(\frac{\sigma}{a}\right)^2 - 1}\right]^{3/2} & \sigma < -a. \end{cases} \tag{5.42}$$

Here W is the maximum vertical velocity of the sphere and a is its radius. For convenience we have introduced the complex radial co-ordinate $\mathcal{Z}_\rho \equiv r + \imath\sigma$. The magnitude of $\mathcal{Z}_\rho$ is the radial distance from the origin to the point (r, σ). Explicitly in terms of spherical co-ordinates,

$$\mathcal{Z}_\rho = \varrho\exp[\imath(\Theta - \vartheta)], \tag{5.43}$$

in which $\varrho = |\mathcal{Z}_\rho|$ is the radial co-ordinate and ϑ is the azimuthal angle, as shown in Figure 5.5c.

Using (5.43), the streamfunction given by (5.42) may alternately be written

$$\Psi = \frac{1}{4} W a e^{-\iota(3\Theta/2 \,-\vartheta/2+\, 3\pi/4)} \sqrt{\frac{a}{\varrho}} \exp\left[-\iota\frac{3}{2}\cos^{-1}(-\sigma/a)\right], \tag{5.44}$$

which may be compared with the streamfunction for the oscillating cylinder (5.21). Here, for mathematical convenience, we have used a hybrid of the along-beam and across-beam co-ordinate system (r,σ) together with the spherical co-ordinate system (ϱ,ϑ), as illustrated in Figure 5.5c.

Sufficiently far from the sphere and within the beam where $|\sigma| < a$, the radial distance is approximately equal to the along-beam distance $\varrho \simeq r$. Thus the magnitude of the streamfunction is constant across the beam and the phase decreases as σ increases. Unlike waves generated by an oscillating cylinder, here the amplitude decreases as the inverse square root of the along-beam distance. Such an amplitude decay is necessary for the total energy and momentum transport to remain constant as the cross-sectional area of the wave-cone increases linearly with distance from the sphere.

The streamfunction for the waves generated by an oscillating sphere is plotted in Figure 5.6. This includes both the upward- and downward-propagating wave-cones. The structure of the latter is deduced by reflection symmetry about the horizontal plane. One can confirm from (5.44) that the superposition of the streamfunctions evaluated on the surface of the sphere satisfy the boundary condition (5.40).

The vertical displacement field, ξ, is given explicitly by

$$\xi = \iota\frac{1}{\omega}\frac{1}{R}\frac{\partial}{\partial R}(R\Psi) \simeq -\iota\frac{1}{\omega}\cos(\Theta)\frac{\partial\Psi}{\partial\sigma}. \tag{5.45}$$

The last expression, plotted in Figure 5.6, is a good approximation within the wave-cone far from the sphere.

The formula for the far-field streamfunction of the upward-propagating wave-cone, given by (5.44) with $\varrho \simeq r$ and $\vartheta \simeq \Theta$, may equivalently be written as a Fourier transform in semi-infinite wavenumber space through use of the identity (5.22). Explicitly, we find

$$\Psi \simeq \frac{3}{8}\iota\, W a e^{-\iota\Theta}\sqrt{\frac{a}{r}}\int_0^\infty \frac{1}{K}J_{3/2}(K)e^{-\iota K\sigma/a}dK. \tag{5.46}$$

This should be compared with the corresponding integral equation for the stream-function of internal waves generated by an oscillating cylinder (5.23). In (5.46), the expression involves the along-beam distance, r, but this appears outside the integral and the integrand itself now involves the spherical Bessel function $J_{3/2}$.

The σ-dependent integral is exploited to develop a theory for viscously attenuated wave-beams. Assuming the amplitude of the internal waves is small, by linearizing

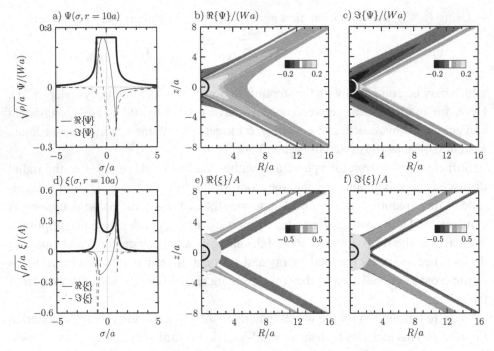

Fig. 5.6. a) Real (thin solid line) and imaginary (dashed line) parts of the stream-function amplitude $\Psi(\sigma, r)$ evaluated at $r = 10a$. b) Real and c) imaginary parts of the total streamfunction amplitude evaluated for the upward- and downward-propagating waves. d) Corresponding vertical displacement field $\xi(\sigma), r)$ and the e) real and f) imaginary parts of the total displacement field for the upward- and downward-propagating waves. The thick circles in b), c), e) and f) indicate the location and radius of the cylinder. A light grey circle is drawn about the origin in e) and f) as a reminder that the plotted solutions are inaccurate near the sphere itself.

the Navier–Stokes and continuity equations for incompressible fluid in cylindrical co-ordinates, it can be shown that $\Psi(R, z)$ satisfies

$$N_0^2 D_R^2 \Psi - \omega^2 \left[D_R^2 + \partial_{zz} \right] \Psi + \imath \omega \nu \nabla^2 \left[D_R^2 + \partial_{zz} \right] \Psi = 0, \qquad (5.47)$$

in which ν is the kinematic viscosity,

$$D_R^2 \Psi \equiv \frac{\partial}{\partial R} \left(\frac{1}{R} \frac{\partial R \Psi}{\partial R} \right),$$

and the axisymmetric form of the Laplacian in cylindrical co-ordinates acting on the azimuthal component, f, of a vector function is

$$\nabla^2 f = \frac{1}{R} \frac{\partial}{\partial R} \left(R \frac{\partial f}{\partial R} \right) + \frac{\partial^2 f}{\partial z^2} = \left[D_R^2 - \frac{1}{R^2} \right] f + \frac{\partial^2 f}{\partial z^2}.$$

Analysis of (5.47) is facilitated by using (5.41) to transform from cylindrical to along- and across-beam co-ordinates. We then implicitly define ψ by

$$\Psi(r,\sigma) = \sqrt{\frac{a}{r}}\psi(r,\sigma).$$

This is substituted into (5.47) after applying the coordinate transformation (5.41). Using the boundary layer approximation, we assume along-beam changes are negligibly small compared with across-beam changes. For $r \gg \sigma$ and $r \gg a$, the leading-order terms of the resulting equation give

$$\frac{\partial^2 \psi}{\partial r \partial \sigma} - \frac{\iota \nu}{2N \sin \Theta} \frac{\partial^4 \psi}{\partial \sigma^4} = 0. \tag{5.48}$$

The solution of (5.48) is found by applying a Fourier transform in σ:

$$\psi(r,\sigma) \sim \int_0^\infty F(K) \exp\left(-\tilde{\nu}K^3\frac{r}{a} - \iota K\frac{\sigma}{a}\right) dK, \tag{5.49}$$

in which $K = k_\sigma a$ is a nondimensional measure of the across-beam wavenumber and the kinematic viscosity is represented nondimensionally by

$$\tilde{\nu} = \frac{\nu}{2a^2 N \sin \Theta}. \tag{5.50}$$

This number must be small for the boundary-layer approximation to be valid. In (5.49) we have retained only those Fourier components with $K \geq 0$ because only these correspond to upward wave propagation.

By analogy with the case of waves generated by an oscillating cylinder, we evaluate the result (5.49) in the limit $\nu \to 0^+$ and compare this result with the inviscid solution (5.46). This gives the Fourier transform coefficient, $F(K)$, which acts approximately to require no-normal-flow boundary conditions on the surface of the sphere. Thus for $r \gg a$, the streamfunction is given by

$$\Psi \simeq \frac{3}{8}\iota Wae^{-\iota\Theta}\sqrt{\frac{a}{r}}\int_0^\infty \frac{1}{K}J_{3/2}(K)\exp\left(-\tilde{\nu}K^3\frac{r}{a} - \iota K\frac{\sigma}{a}\right) dK. \tag{5.51}$$

As in the case of the oscillating cylinder, the solution of (5.51) shows that wave-beams generated by an oscillating sphere exhibit a transition from bimodal to unimodal structure. This is illustrated, for example, in Figure 5.7 which shows the vertical displacement field $\xi \simeq -(\iota/\omega)\partial_\sigma \Psi$. Consistent with the expected behaviour for upward-propagating waves, the crests move to decreasing σ with increasing phase.

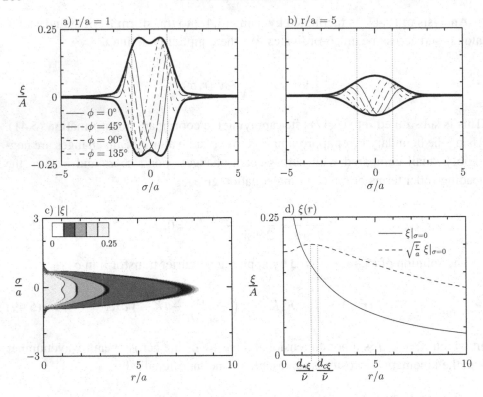

Fig. 5.7. The vertical displacement field, ξ, for the wave-cone generated by an oscillating sphere computed using (5.51). The sphere has radius a, relative frequency $\omega/N = 0.5$ and maximum vertical displacement $A = W/\omega$. The kinematic viscosity is set to be $\nu = 0.01a^2N$, so the effects of viscous attenuation are governed by the parameter $\tilde{\nu} \simeq 0.0058$. For comparison with Figure 5.4, values of ξ are evaluated at a) $r/a = 1$ and b) $r/a = 5$, though it should be noted that the approximations leading to (5.51) require $r \gg a$. The plots show both the envelope (thick solid lines) and the instantaneous amplitude of the waves at phase $\phi = 0$ (thin solid line), $\phi = \pi/4$ (long-dashed line), $\phi = \pi/2$ (short-dashed line), and $\phi = 3\pi/4$ (dashed-dotted line). The dotted vertical lines in both plots illustrate the values $\sigma = \pm a$. c) Contours of the amplitude envelope $|\xi(r,\sigma)|$ show the transition from bimodal to unimodal beams. d) Values of $|\xi|$ (solid line) and $\sqrt{r/a}|\xi|$ (dashed line) as functions of along-beam distance evaluated at $\sigma = 0$. The positions $d_{\star\xi}$, where the centreline of $\sqrt{r/a}|\xi|$ reaches maximum amplitude, and $d_{c\xi}$, where the across-beam curvature is zero, are indicated.

As with beams from an oscillating cylinder, we define $d_c = \tilde{\nu}r_c/a$ as the non-dimensional distance where the beam changes from bimodal to unimodal structure through the change in curvature at the centreline of the cone. For the vertical displacement field, this is given by

$$I_{3/2}{}^2(d_{c\xi};2) = I_{3/2}(d_{c\xi};1)I_{3/2}(d_{c\xi};3), \qquad (5.52)$$

in which I_n is defined in (5.32). The values of d_c for the vertical displacement and other fields are listed in Table 5.1.

The peaks on either flank of the centreline of the bimodal wave-beam broaden due to viscous attenuation and this acts both to decrease the amplitude of the peaks and to increase the amplitude of the beam along the centreline. Thus, near the sphere the amplitude of the waves should decrease less quickly than $r^{-1/2}$. After the peaks have merged, the centreline amplitude should decrease faster than $r^{-1/2}$ due to the combined effects of the expanding wave-cone and viscous attenuation.

As was done in (5.36), we may form an asymptotic approximation for large d to the integral (5.51) along the centreline of the beams. In particular, we find that the leading order far-field streamfunction decreases as $r^{-1/2}d^{-1/2} \propto r^{-1}$. That is, the effects of viscous attenuation are as important as the expansion of the wave-cone in determining the resulting decrease in amplitude.

5.3 One- and two-layer flow over topography

A great deal has been written on this subject not only for its geophysical importance but also because many aspects are amenable to analytic theory and may be examined in laboratory experiments. The intent here is to develop mathematical techniques used to arrive at the essential results. This section focuses upon the generation of interfacial waves in one- and two-layer fluids with the assumption that the topography underlies the fluid. Disturbances at the interface take place when the fluid moves over the hills.

We consider two different flow configurations. In one, the upstream flow velocity is constant with depth throughout the upper and lower layers of the fluid. We treat this problem first by considering a one-layer fluid. At the start of Section 5.3.4 we explain how these results are trivially extended to a two-layer fluid provided that the upstream velocity is the same in both layers.

In the second configuration, the focus of Section 5.3.4, we consider a one-and-a-half-layer fluid in which the upstream speed is constant over the depth of the lower layer, but is zero in the upper layer.

5.3.1 Equations of motion for a one-layer fluid

We begin by examining the flow of a finite-depth layer of inviscid, uniform density, incompressible fluid moving over a two-dimensional obstacle, as illustrated in Figure 5.8.

Neglecting the effects of rotation and viscosity in (1.35), and taking the density $\varrho = \varrho_0$ to be constant, the fully nonlinear momentum and continuity equations for

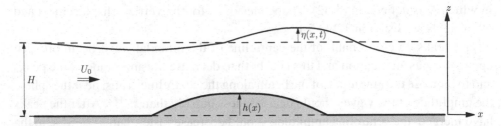

Fig. 5.8. Variables and co-ordinate system used to model the flow of a one-layer fluid over an obstacle.

incompressible fluid in two dimensions are

$$\varrho_0 \frac{Du}{Dt} = -\frac{\partial P}{\partial x} \tag{5.53}$$

$$\varrho_0 \frac{Dw}{Dt} = -\frac{\partial P}{\partial z} - \varrho_0 g \tag{5.54}$$

$$\frac{\partial u}{\partial x} + \frac{\partial w}{\partial z} = 0, \tag{5.55}$$

in which P is the total pressure and the material derivative $D/Dt = \partial_t + (U_0 + u) \partial_x + w \partial_z$ includes the background horizontal flow U_0 in the advective terms.

The boundary conditions for the free surface are

$$w = \frac{D\eta}{Dt} \quad \text{at} \quad z = H + \eta, \tag{5.56}$$

$$P = 0 \quad \text{at} \quad z = H + \eta. \tag{5.57}$$

The latter condition effectively requires that the pressure is everywhere constant on the surface and we arbitrarily set this constant to zero. The bottom boundary condition is

$$u \cdot \hat{n} = 0 \quad \text{at} \quad z = h(x). \tag{5.58}$$

The allows for free-slip flow along but not through the rigid bottom boundary.

5.3.2 One-layer small-amplitude hydrostatic flow

Analytic solutions can be found if we assume surface disturbances are small ($||\eta(x,t)|| \ll H$) and also that the obstacle is sufficiently small ($||h(x,t)|| \ll H$).

For now we further assume the flow is hydrostatic, so that the vertical pressure gradient exactly balances the buoyancy term and does not give rise to significant vertical accelerations. Equivalently, this means that vertical accelerations are small compared with gravity. That is

$$\left\|\frac{\partial w}{\partial t}\right\| \lesssim H\omega^2 \lesssim gH^2k^2 \ll g,$$

in which we have used the fact that the fastest small-amplitude waves move with the shallow water wave speed $(gH)^{1/2}$. Thus the hydrostatic approximation requires

$$(kH)^2 \ll 1. \tag{5.59}$$

So the flow is hydrostatic if the horizontal scale of the obstacle and the waves it generates are much longer than the depth.

This is shallow water theory. As we saw in Section 2.2.10, the horizontal velocity field is effectively z-independent ($u(x,z,t) \simeq u(x,t)$) and the vertical velocity varies linearly with z.

With these assumptions, the equations of motion simplify as follows. The vertical momentum equation (5.54) together with the surface pressure condition (5.57) give

$$p = \varrho_0 g(H + \eta - z). \tag{5.60}$$

For small-amplitude perturbations, $\|u\| \ll U_0$. So, using (5.60), the linearized horizontal momentum equation becomes

$$\frac{\partial u}{\partial t} + U_0\frac{\partial u}{\partial x} = -g\frac{\partial \eta}{\partial x}. \tag{5.61}$$

Through some trigonometry, the condition for no-normal flow through the lower boundary (5.58) requires the vertical velocity to equal the product of the horizontal velocity and the slope. Integrating the continuity equation for incompressible fluid (5.55) therefore gives

$$w \simeq U_0\frac{dh}{dx} - \left(\frac{\partial u}{\partial x}\right)(z - h),$$

in which we have used $\|u\| \ll U_0$ in the first term on the right-hand side.

We use this result in the surface boundary condition (5.56) and keep only those terms that are linear in small quantities, u, η and h. Thus we find

$$\frac{\partial \eta}{\partial t} + U_0\frac{\partial \eta}{\partial x} + H\frac{\partial u}{\partial x} = U_0\frac{dh}{dx}. \tag{5.62}$$

Finally, we combine (5.61) and (5.62) to eliminate u, giving an equation in η alone:

$$\left(\frac{\partial}{\partial t} + U_0\frac{\partial}{\partial x}\right)^2\eta - c^2\frac{\partial^2}{\partial x^2}\eta = U_0^2\frac{d^2h}{dx^2}. \tag{5.63}$$

Here we have defined $c \equiv \sqrt{gH}$, which is the shallow water wave speed. If h and U_0 are both zero, (5.63) reduces to the shallow water wave equation (2.70). If h is zero but U_0 non-zero, then (5.63) reduces to (2.70) through the transformation $x \to x - U_0 t$, which corresponds to a co-ordinate system moving at the speed of the background flow. Without these approximations, (5.63) is the forced wave equation in a frame of reference that is stationary with respect to an obstacle as the flow moves over it.

Fourier transform methods are used to solve the initial value problem. Before time $t = 0$ we suppose the flow moves at constant speed U_0 along a flat bottom and that at $t = 0$ a hill is instantaneously uplifted from below, uniformly lifting the column of water above it. Explicitly, the initial conditions on the surface displacement are

$$\eta(x,0) = h(x) \text{ and } \eta_t(x,0) = 0. \tag{5.64}$$

The solution of (5.63) with this boundary condition is

$$\eta(x,t) = \frac{1}{2}\left[\frac{1}{1+\mathrm{Fr}}h(x - c_+ t) + \frac{1}{1-\mathrm{Fr}}h(x - c_- t)\right] + \frac{\mathrm{Fr}^2}{\mathrm{Fr}^2 - 1}h(x), \tag{5.65}$$

in which the 'Froude number', given by

$$\mathrm{Fr} \equiv \frac{U_0}{\sqrt{gH}} \tag{5.66}$$

represents the ratio of the flow speed to the shallow water wave speed, and $c_\pm = U_0 \pm \sqrt{gH}$ represents the speed of the leftward- and rightward-propagating waves superimposed on the background flow.

The solution reveals two classes of flow behaviour, which are called 'subcritical' or 'supercritical', depending on whether $\mathrm{Fr} < 1$ or $\mathrm{Fr} > 1$, respectively. In this sense, the Froude number is conceptually similar to a Mach number, which represents a ratio of flow speed to sound speed. Flow is subsonic or supersonic, depending on whether the Mach number is respectively less than or greater than one.

Just as sound waves propagate in all directions from a source moving at subsonic speed, so do shallow water waves propagate upstream and downstream away from an initial disturbance if the flow is subcritical, as shown in Figure 5.9a. After a sufficiently long time has passed the flow directly over the obstacle becomes steady. But the surface is not flat. Instead it deflects downwards over the obstacle by an amount $[\mathrm{Fr}^2/(1 - \mathrm{Fr}^2)]h(x)$. In contrast, waves in supercritical flow are swept downstream and the steady disturbance that remains over the obstacle is an upward deflection of the surface, as shown in Figure 5.9b.

We can appreciate the transition in behaviour by considering the energetics of the system. In slow moving, subcritical flow the fluid does not have enough kinetic

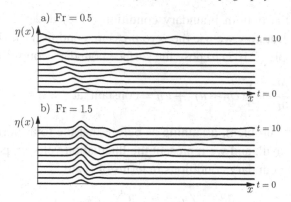

Fig. 5.9. At time $t = 0$, a topographical disturbance appears underneath a flow, travelling at a speed U_0. This causes the surface to be instantaneously displaced by the same scale and shape of the disturbance (in this case a Gaussian). In a) the flow speed is subcritical and the waves travel off in different directions. In b) the flow is supercritical causing the travelling waves to both propagate in the same direction at different speeds. Each vertically offset displacement plot in both figures corresponds to a time step of 1.

energy on its own to move the whole fluid column up and over the obstacle. The downward deflection of the surface provides the mechanism through which potential energy is transferred to kinetic energy, giving the fluid the boost it needs to traverse the obstacle. On the other hand, fast moving, supercritical flow has enough kinetic energy that it can sacrifice some to potential energy in the course of lifting the whole fluid column up and over the obstacle.

5.3.3 One-layer small-amplitude nonhydrostatic flow

In the above discussion, the hydrostatic approximation allowed us to reduce the governing equations to the forced shallow water wave equations. Here we relax this assumption and instead use Bernoulli's principle for irrotational flows in order to study flows over obstacles having height comparable to breadth. Such motion can be nonhydrostatic. However, we still assume the obstacle height is small compared to the fluid depth so that we can take advantage of linear theory to arrive at analytic solutions.

Because the fluid is irrotational, its motion may be represented by gradients of the velocity potential Φ. Together with the mass conservation law for an incompressible fluid (1.30), we therefore require

$$\nabla^2 \Phi = 0, \tag{5.67}$$

just as we found in (2.2).

We still have the bottom boundary condition (5.58) and the upper boundary condition (5.56). But now we use the zero pressure condition (5.57) together with Bernoulli's principle (1.102) to pose the surface boundary condition on ϕ:

$$\frac{\partial \Phi}{\partial t} + \frac{1}{2}(U_0 + u)^2 + g\eta = \text{constant} \quad \text{at } z = \eta + H. \tag{5.68}$$

We linearize the problem, assuming as before that the total horizontal velocity is $U_0 + u$ with u small, and we represent the fluctuation velocity potential by ϕ so that $\vec{u} = \nabla \phi$. The boundary conditions in terms of ϕ are

$$\frac{\partial \phi}{\partial z} = \left(\frac{\partial}{\partial t} + U_0 \frac{\partial}{\partial x}\right)\eta \quad \text{at } z = H \tag{5.69}$$

$$\left(\frac{\partial}{\partial t} + U_0 \frac{\partial}{\partial x}\right)\phi + g\eta = 0 \quad \text{at } z = H \tag{5.70}$$

$$\frac{\partial \phi}{\partial z} = U_0 \frac{dh}{dx} \quad \text{at } z = 0. \tag{5.71}$$

We seek only steady solutions here so that the time derivatives in the boundary conditions vanish. Proceeding as before, we Fourier transform the equations in x and solve the resulting ordinary differential equation in z. Fourier transforming back to real space gives

$$\phi(x,z) = \iota U_0 \int_{-\infty}^{\infty} \left[\sinh kz + \cosh kz \left(\frac{\text{Fr}^2 kH \tanh kH - 1}{\tanh kH - \text{Fr}^2 kH}\right)\right] \hat{h}(k)e^{\iota kx}\, dk, \tag{5.72}$$

in which Fr is the Froude number, defined by (5.66).

The solutions once again fall into two classes depending on whether the flow is subcritical (Fr < 1) or supercritical (Fr > 1). In subcritical flow, the integral is dominated by values of k near the singularity where $\tanh kH = \text{Fr}^2 kH$. Using standard integral approximation techniques together with (5.70) to relate ϕ to η, we find

$$\eta(x) \simeq \frac{\pi \text{Fr}^2 k_0 (\text{Fr}^4 (k_0 H)^2 - 1)\cosh k_0 H}{\text{Fr}^2 - 1 + \text{Fr}^4 (k_0 H)^2} \left[\iota \hat{h}(k_0)e^{\iota k_0 x} + \text{cc}\right] \quad \text{for } x > 0, \tag{5.73}$$

in which k_0 is the wavenumber that satisfies $\tanh k_0 H = \text{Fr}^2 k_0 H$ and cc denotes the complex conjugate. Thus subcritical nonhydrostatic flow gives rise to undular waves downstream of the obstacle. Such steady downstream waves do not occur in subcritical hydrostatic flow.

Steady downstream waves do not occur in supercritical nonhydrostatic flow because the integrand of (5.72) is not singular for any value of k if Fr > 1. In this case the steady solution corresponds to an upward surface displacement over the obstacle, similar to the solution we found for hydrostatic waves.

5.3.4 Two-layer flow

The analysis above, restricted to low hills that create small-amplitude waves, is straightforwardly extended to interfacial waves in a two-layer fluid in which the upstream horizontal flow speed is U_0 in both the upper and lower layers. This is done by redefining the Froude number in (5.66) so that it is based upon the shallow water speed in a two-layer fluid. Explicitly, using (2.98), we define

$$\text{Fr} \equiv \frac{U_0}{\sqrt{g'\bar{H}}}, \tag{5.74}$$

in which g' is the reduced gravity given by (2.78) and $\bar{H}$ is the harmonic mean of the upper and lower layer fluids given by (2.99). To distinguish the Froude number associated with surface waves from that associated with interfacial waves, Fr given by (5.66) is sometimes called the 'external Froude number' and Fr given by (5.74) is called the 'internal Froude number'.

In a one-and-a-half-layer fluid the depth of the upper-layer fluid is infinite. This is a good approximation if the horizontal scale of the interfacial disturbances is much smaller than the actual upper-layer depth. In this case $\bar{H}$ is simply the depth of the lower-layer fluid but the effects of gravity are still expressed through g'.

In what follows we consider interfacial disturbances in a one-and-a-half-layer fluid in which the lower layer is assumed to move over an obstacle with a constant upstream horizontal speed U_0, but the upper layer velocity is assumed to be zero. The lower layer, of upstream depth H, is assumed to have density ρ_2 and the upper layer has density ρ_1, moderately smaller than ρ_2. This is illustrated schematically in Figure 5.10.

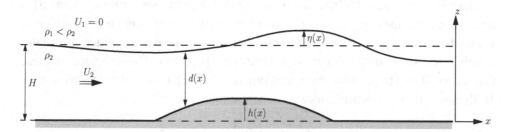

Fig. 5.10. Variables and co-ordinate system used to model the flow of a one-and-a-half-layer fluid over an obstacle. The lower layer moves with uniform upstream speed U_0 and the upper layer is stationary.

As well as illustrating the dynamics introduced by having a shear layer at the interface, the approach below serves to demonstrate an alternative mathematical approach which allows us to gain insight into how the solutions of the equations of motion behave without explicitly solving the equations.

Because the upper-layer density is non-negligible, we cannot assume the pressure p_0 at the interface is constant. Instead, we must apply Bernoulli's principle to both the upper- and lower-layer fluids. Assuming the flow is steady, we have on each side of the interface

$$\frac{p_0}{\rho_1} + g(H + \eta) = \text{constant}$$

$$\frac{p_0}{\rho_2} + \frac{1}{2}U^2 + g(H + \eta) = \text{constant}.$$

Here $U(x)$ is the total speed of the lower-layer fluid, which increases or decreases as the depth of the lower layer decreases or increases, respectively. The depth, $d(x)$, of the lower layer is given explicitly in terms of the interface displacement $\eta(x)$ and the height $h(x)$ of topography by

$$d \equiv H + \eta - h. \tag{5.75}$$

Thus, by continuity for an incompressible fluid, the total volume flux $Q_0 \equiv Ud$ is constant.

The equations are simplified by eliminating p_0 and by replacing η and U with d and Q_0. Thus we arrive at the single equation

$$\frac{1}{2}\frac{Q_0^2}{d^2} + g'd = E_0 - g'h \equiv E(x), \tag{5.76}$$

in which the reduced gravity is $g' \equiv g(\rho_2 - \rho_1)/\rho_2$.

In (5.76) we have also defined the constant E_0 to be the energy per unit mass of the system far upstream where $h = \eta = 0$. Knowing $h(x)$ and the constants Q_0 and E_0, we can use (5.76) to find $d(x)$ and hence the interfacial displacement $\eta(x)$.

The left-hand side of (5.76) is a function $E(x)$ that decreases with increasing $d(x)$ up to a critical value $d_c \equiv (Q_0^2/g')^{1/3}$, above which it increases again. The minimum value of E is $(3/2)g'd_c$. Whether E is an increasing or decreasing function dictates whether the flow is subcritical or supercritical, respectively. The structures of these flows correspond to the subcritical and supercritical regimes in the one-layer case. To illustrate this, we define a local Froude number by

$$\text{Fr}_d \equiv \frac{U}{\sqrt{g'd}} = \sqrt{\frac{Q_0^2}{g'd^3}} = \left(\frac{d_c}{d}\right)^{3/2}. \tag{5.77}$$

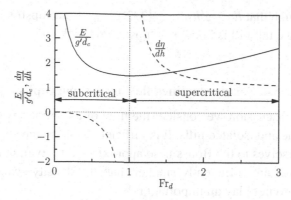

Fig. 5.11. Subcritical and supercritical flow regimes characterized by local Froude numbers over which the d-dependent energy terms in (5.76) are, respectively, decreasing and increasing (solid line), and rate of change of interface displacement with obstacle height (dashed line).

The subscript is meant to emphasize that Fr_d is not a constant but varies with x as d varies with x. As such, it represents a local Froude number based upon the instantaneous depth $d(x)$ and total speed $U(x)$, rather than the characteristic depth H and speed U_0. Figure 5.11 plots E as a function of Fr_d. This shows that E is a minimum at the critical Froude number where $Fr_d = 1$.

Now suppose far upstream of an obstacle the flow is supercritical. As the flow moves over the obstacle, h increases and so, according to (5.76), E decreases. Because we are in the supercritical regime Fr_d consequently decreases and hence, according to (5.77), d must increase. Finally, because $\eta = d + h - H$, we see that the interface height increases. Consistent with the shallow water prediction (5.65) but with much less mathematical labour, we have found that the interface deflects upwards as supercritical flow moves over an obstacle.

If instead the upstream flow is subcritical, then increasing h gives decreasing E. This in turn gives increasing Fr_d and therefore decreasing d. One can show, in general, that d changes with h as $1/(Fr_d^2 - 1)$ and so η changes with h as $Fr_d^2/(Fr_d^2 - 1)$. Therefore, if $Fr_d < 1$, then both d and η decrease with increasing h. Again this result is consistent with the shallow water results.

It may be possible that the obstacle is so tall that $E = E_0 - g'h$ is smaller than the minimum energy $(3/2)g'h$. In this case, no steady solutions exist: because the fluid does not have enough total energy to pass over the obstacle it instead piles up on its upstream side. The incident kinetic energy is thus stored as potential energy on the upstream side. Eventually, enough potential energy will be built up that the flow can surmount the obstacle. Effectively, E_0 based on the new upstream depth will be large enough that $E_0 - g'h_{\max} \gtrsim (3/2)g'h_{\max}$. The upstream adjustment to flow approaching a tall hill is known as 'upstream blocking'.

5.4 Steady stratified flow over topography

In the following discussion we consider the flow of a continuously stratified fluid over both periodic and isolated hills. It is sufficient to study the generation process by restricting ourselves to the Boussinesq approximation. Even in the atmosphere, typical hill heights are sufficiently smaller than the density scale height so that anelastic effects do not play an important role.

For the most part, the theory presented below is restricted to the study of uniformly stratified fluid with buoyancy frequency N_0 moving at constant speed U_0 over the hills. As in our study of one- and two-layer fluids, we find that the response of the flow can be classified as subcritical or supercritical depending upon a non-dimensional number. But specifically how this number is defined depends upon the circumstance being considered. This is discussed below, followed by special consideration of flow over periodic and localized hills of small and large amplitude.

5.4.1 Froude and long numbers

If we were to remain consistent with the definition of a Froude number in layered fluids, we would define its counterpart in a stratified fluid to be the ratio of flow speed to the fastest wave speed. Thus in a uniformly stratified fluid of total depth H containing internal waves with a horizontal scale much longer than H, it may seem natural to extend the definition of Froude number to

$$\mathrm{Fr}_H = \frac{U_0}{\sqrt{g'H}} = \frac{U_0}{N_0 H}, \qquad (5.78)$$

in which g' is the reduced gravity, now based upon the density difference between the bottom and top of the domain and $N_0{}^2 \equiv -(g/\varrho_0)\,d\bar{\rho}/dz = g'/H$.

In most atmospheric and oceanographic circumstances, the definition (5.78) is not very useful because the domain height is irrelevant to the dynamics. The relevant length scales are instead set by the height, h_0, and breadth, l_0, of the hills. The former determines the amplitude of the waves and the latter determines the characteristic horizontal wavelength.

By analogy with (5.78), we define the (horizontal) Froude number to be

$$\mathrm{Fr} \equiv \frac{U_0}{N_0 l_0}. \qquad (5.79)$$

This definition of a Froude number is not exactly like a Mach number or a Froude number defined for one-layer shallow water flow in that the value of Fr separating subcritical and supercritical flow regimes is not necessarily unity. This is because $N_0 l_0$ is not necessarily the maximum speed of waves in the medium. This speed would depend upon the horizontal wavelength, which is related to l_0, and the wave frequency which depends upon l_0 and U_0. However, Fr in (5.79) is analogous to the Mach and Froude numbers in that the flow behaviour can be classified as subcritical or supercritical depending upon whether Fr is small or large, respectively.

Also by analogy with (5.78) we may define a (vertical) Froude number based upon the hill height as

$$\mathrm{Fr}_v \equiv \frac{U_0}{N_0 h_0}.$$

This provides a measure of nonlinearity: taller hills can generate larger amplitude waves and can lead to effects not captured by linear theory, such as flow blocking and boundary-layer separation. However, the definition of Fr_v poses some conceptual difficulties. First, it is not a Froude number in the classical sense because it is not a ratio of flow to wave speeds and the parameter is irrelevant in the linear theory regime. Also, opposite to an intuitive definition of a nonlinearity parameter, Fr_v becomes smaller as nonlinear effects become more important.

For convenience, we take the liberty of introducing a non-dimensional parameter called the 'Long number' after Robert Long, who performed many pioneering experimental and theoretical investigations into stratified flow over topography. The Long number is defined by

$$\mathrm{Lo} \equiv \frac{N_0 h_0}{U_0}, \tag{5.80}$$

and is equivalent to Fr_v^{-1}.

The parameters Fr and Lo can be used to classify a broad range of phenomena pertaining to inviscid, non-rotating stratified flow over topography.

5.4.2 Flow over small-amplitude periodic hills

We first consider the simple case, illustrated in Figure 5.12a, of uniformly stratified Boussinesq fluid with buoyancy frequency N_0 moving with constant background horizontal flow speed U_0 over periodic sinusoidal hills with height given by

$$h(x) = h_0 \cos(k_0 x). \tag{5.81}$$

The hill crests are separated by $\lambda = 2\pi/k_0 > 0$ and the half hill-to-valley distance is $h_0 > 0$. The condition that the hills are small is expressed by $h_0 k_0 \ll 1$.

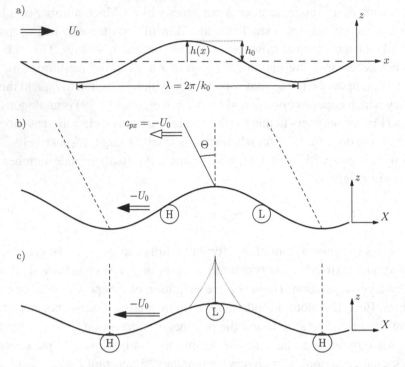

Fig. 5.12. a) Variables and co-ordinate system used to model the uniform flow of uniformly stratified fluid over small-amplitude topography. The hills have wavelength $\lambda = 2\pi/k_0$. b) Tilted phase lines in frame of reference moving with the background wind ($X = x + U_0 t$) in the circumstance where $U_0 < N/k_0$. c) Vertical phase lines associated with waves with exponentially decreasing amplitude with height in the circumstance where $U_0 > N/k_0$.

For mathematical convenience, $h(x)$ may be represented in complex form by

$$h(x) = h_0 e^{\imath k_x x}, \tag{5.82}$$

in which $|k_x| = k_0$. Whether or not k_x is positive, the real part of (5.82) gives (5.81). Below we will see that the sign of k_x depends upon the sign of U_0.

In this circumstance, it is useful to define the characteristic horizontal length scale by $l_0 \equiv k_0^{-1}$ so that the suitable Froude number for the flow is

$$\mathrm{Fr} = \frac{|U_0| k_0}{N_0}. \tag{5.83}$$

Allowing U_0 to be positive for rightward flow and negative for leftward flow, the absolute value on U_0 has been introduced to ensure $\mathrm{Fr} \geq 0$. The Long number is $\mathrm{Lo} = h_0 N_0 / |U_0|$, which is small for flow over sufficiently small amplitude hills.

To solve this problem, we first transform into a frame of reference moving with the flow and then use the properties of waves in uniformly stratified, stationary fluid to find solutions that match the transformed surface boundary conditions. Finally, we switch back to the frame of reference in which the hills are stationary.

In a frame of reference that moves with the background flow, the horizontal co-ordinate is $X = x + U_0 t$. Thus we seek the structure of waves in a stationary fluid forced by hills that move at speed $-U_0$, as illustrated in Figure 5.12b.

From the discussion in Section 3.3.1, we know that stationary fluid supports waves satisfying the differential equation (3.42). For flow over two-dimensional topography, we expect the waves themselves to be two-dimensional, having no structure in the y-direction. Generally, solutions are a superposition of periodic waves of the form

$$b(X, z, t) = A_b \exp[\imath (k_x X + k_z z - \Omega t)], \tag{5.84}$$

in which b represents a fluctuation quantity such as the vertical velocity or pressure field. Here we have introduced Ω, which represents the extrinsic wave frequency as recorded by an observer moving with the background flow.

The free-slip condition for no normal flow through the bottom boundary is given generally by (5.58). Equivalently, we may require that the displacement of fluid at the surface is given by the displacement of the hills $h(x)$. For small-amplitude hills moving at speed $-U_0$ with respect to the perceived stationary fluid above, this condition is

$$\xi(X, 0, t) \simeq h(x + U_0 t) = h_0 e^{\imath k_x (x + U_0 t)}, \tag{5.85}$$

in which we have used the complex representation of the hills (5.82).

Using (5.84) to extend this result to positive z, we find

$$\xi = \mathcal{A}_\xi \exp[\imath (k_x x + k_z z - \Omega t)]. \tag{5.86}$$

To satisfy the boundary condition (5.85), the amplitude of the vertical displacement field must be the same as the hill height:

$$\mathcal{A}_\xi = h_0. \tag{5.87}$$

Also, the relative frequency must be $-k_x U_0$. Insisting that this is positive, we must have

$$k_x = -\text{sign}(U_0) k_0. \tag{5.88}$$

Hence

$$\Omega = -k_x U_0 = k_0 |U_0|. \tag{5.89}$$

In particular, if the flow moves left-to-right (as shown, for example, in Figure 5.12), then U_0 is positive and therefore k_x is negative. This means simply that in a frame

Fig. 5.13. A rare double lenticular cloud observed over Mount Shasta, California. [Reproduced with permission, ©2007, Cindy Diaz, ShastaPhotografix.com.]

of reference moving with the flow, the hills and the waves they generate move right-to-left.

In the frame of reference of the hills the waves are observed to have intrinsic frequency ω which is related to the extrinsic frequency Ω through $\Omega = \omega - k_x U_0$. By comparison with (5.89), we see that $\omega = 0$: the waves are stationary with respect to the hills. This explains why a lenticular cloud, such as that shown in Figure 5.13, appears to be stationary over mountain-tops while winds rush through them.

The solution of (3.42) with (5.84) gives the dispersion relation

$$\Omega^2 = N_0^2 \frac{k_x^2}{k_x^2 + k_z^2}. \tag{5.90}$$

This allows us to compute explicitly the vertical structure for ξ in (5.86) by solving (5.90) for k_z. Using (5.89) and the definition of the Froude number (5.83) we can rearrange (5.90) to find

$$k_z^2 = \left(\frac{N_0}{U_0}\right)^2 - k_x^2 = \left(\mathrm{Fr}^{-2} - 1\right) k_0^2. \tag{5.91}$$

This formula reveals a qualitative change in the vertical structure, depending upon whether Fr is smaller or greater than unity. We classify the flow as being subcritical if Fr < 1 and supercritical if Fr > 1.

In the subcritical case the hills push upon the stratified fluid with a forcing frequency $\Omega = |U_0| k_0 < N_0$. Consequently, the right-hand side of (5.91) is positive and so k_z is real-valued. This means that k_z can indeed be interpreted as a wavenumber and the z-dependent term, $\exp(\imath k_z z)$, in (5.86) corresponds to a vertically periodic disturbance with wavelength $2\pi/|k_z|$. Because the sign of k_z is opposite to that of the vertical group velocity, upon taking the square root of (5.91) the appropriate solution is that with the negative sign: energy must be transported upwards and away from the hills, not towards them.

In the supercritical case the stratified fluid is forced at a frequency larger than the buoyancy frequency and so the hills do not generate vertically propagating waves. Explicitly, because the right-hand side of (5.91) is negative if $\Omega = |U_0|k_0 > N_0$, k_z is a pure imaginary number. Writing $k_z = \pm\imath\gamma$ with γ a positive real number, the z-dependent term in (5.86) becomes the exponential function $\exp(\mp\gamma z)$. The increasing exponential with height is not a physical solution in a domain with no upper boundary, so we must have $k_z = +\imath\gamma$.

In summary, we have seen that solutions in the upper-half plane are given by (5.86) in which

$$k_z = \begin{cases} -\sqrt{\left(\frac{N_0}{U_0}\right)^2 - k_0{}^2} = -k_0\sqrt{Fr^{-2} - 1} & Fr < 1 \\ \imath\gamma = \imath\sqrt{k_0{}^2 - \left(\frac{N_0}{U_0}\right)^2} = \imath k_0\sqrt{1 - Fr^{-2}} & Fr > 1 \end{cases} \tag{5.92}$$

Switching back to the frame of reference of the hills, the vertical displacement field is given by (5.86) in which $X \to x$ and $\Omega \to 0$. Explicitly taking the real part of the expressions gives

$$\xi = h_0 e^{\imath(k_x x + k_z z)} \to \begin{cases} h_0\cos(k_x x + k_z z) & Fr < 1 \\ h_0\cos(k_x x)e^{-\gamma z} & Fr > 1, \end{cases} \tag{5.93}$$

in which k_z and γ are given by (5.92).

In subcritical flow, lines of constant phase are given by values x and z satisfying $k_x x + k_z z = $ constant. These have slope $-k_x/k_z$. Recall from (5.89) that $\Omega = -U_0 k_x$ is positive so that $k_x < 0$ if the background flow moves from left-to-right, and $k_x > 0$ if the background flow moves right-to-left. Therefore propagating waves have phase lines that tilt into the ambient flow. For example, Figure 5.12b shows that the phase lines of upward-propagating waves have negative slope when $U_0 > 0$, in which case $k_x < 0$.

Supercritical flow over hills is said to create 'evanescent disturbances', as discussed in Section 3.3.7. Lines of constant phase are vertical and, moving upwards along one of these lines, the amplitude of the disturbance decreases exponentially, as illustrated in Figure 5.12c. The rate of decay in the vertical is given by the

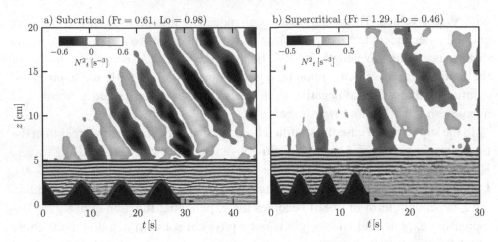

Fig. 5.14. A composite of time-series images from two laboratory experiments showing internal wave generation by a) subcritical and b) supercritical flow over and in the lee of topography. The model sinusoidal hills are towed at a fixed speed U_0 starting at time $t = 0$. This generates disturbances above the hills shortly thereafter. A camera looks through the tank at an image of horizontal black and white lines, the raw image of which is shown up to a short distance above the hill crests. The distortions of the lines are used to compute the rate of change of the vertical density gradient, which is cast in terms of the rate of change of the total squared buoyancy frequency $N^2{}_t \equiv -(g/\varrho_0)\partial_{zt}\rho$. This field is illustrated with the greyscale colour map as indicated. [Adapted, by permission of Elsevier Ltd, from Figure 6 of Aguilar *et al.*, Deep-Sea Res., **53**, 96–115 (2006).]

e-folding depth $1/\gamma$. For very fast or weakly stratified flows, the e-folding depth is approximately $k_0{}^{-1}$. This is the same vertical decay rate from horizontally periodic disturbances in uniform-density, irrotational flow, as in the case of the flow beneath deep surface waves.

The patterns of disturbance by subcritical and supercritical flow over topography has been observed in laboratory experiments, as illustrated in Figure 5.14. Here a train of sinusoidal hills is towed at constant speed U_0 through a uniformly stratified fluid. In both cases the towing starts at time $t = 0$ and disturbances are visualized using a Schlieren method which directly measures changes to vertical density gradients in the fluid.

Consistent with theory, vertically propagating waves appear above the hills if the hills are towed sufficiently slowly (Fr $= 0.61$). This is evident through the tilt in the phase lines of the waves; that the vertical phase speed c_{Pz} is downwards is consistent with upward-propagating waves having positive vertical group speed. Indeed, the front of the waves over the hills moves upwards at speed c_{gz} based upon the wavenumber k_0 associated with the hills and the relative frequency $|\Omega| = U_0 k_0$. If the hills are towed quickly (Fr $= 1.29$), the flow is supercritical and evanescent disturbances appear over the hills. Beyond the theory presented so far, this experiment also shows the launching of vertically propagating waves in the lee of the hills, which occurs in part because the Long number is not small.

So far we have found the solution for the vertical displacement field. Using the polarization relations we may derive solutions for any of the other basic state fields. This may be done for subcritical flow using the formulae in the third column of Table 3.1, in which $\Theta = -\text{sign}(k_x)\cos^{-1}(\Omega/N_0)$ and the phase of the waves, $\varphi = k_x x + k_z z$, is independent of time.

In particular, the vertical velocity field is found from (5.93) by multiplying the complex representation of ξ by $-\iota\Omega$. That is

$$w = -\iota\Omega h_0 e^{\iota(k_x x + k_z z)} \Rightarrow \begin{cases} \Omega h_0 \sin(k_x x + k_z z) & \text{Fr} < 1 \\ \Omega h_0 \sin(k_x x)e^{-\gamma z} & \text{Fr} > 1 \end{cases}. \qquad (5.94)$$

The horizontal velocity field is

$$u = \iota\frac{k_z}{k_x}\Omega h_0 e^{\iota(k_x x + k_z z)} \Rightarrow \begin{cases} -N_0 h_0\sqrt{1 - \text{Fr}^2}\sin(k_x x + k_z z) & \text{Fr} < 1 \\ N_0 h_0\sqrt{\text{Fr}^2 - 1}\cos(k_x x)e^{-\gamma z} & \text{Fr} > 1 \end{cases}. \qquad (5.95)$$

Using (5.88) and (5.89), the vertical velocity field at the bottom boundary is $w|_{z=0} = -U_0 k_0 h_0 \sin(k_0 x)$. It can easily be checked that this satisfies the free-slip boundary condition (5.58) for small-amplitude waves. This is a maximum at the midpoint of the hill facing the flow and is a minimum at its midpoint in the lee. Because $h_0 k_0 \ll 1$ for small-amplitude waves, (5.94) shows that the maximum vertical velocity is much smaller than the background flow speed.

The fluctuation horizontal velocity field at the bottom boundary changes depending upon whether the flow is subcritical or supercritical. If $\text{Fr} < 1$, $u|_{z=0} = \text{sign}(U_0)N_0 h_0(1 - \text{Fr}^2)^{1/2}\sin(k_0 x)$ indicating that u opposes the ambient wind at the midpoint of the hill facing the flow. If $\text{Fr} > 1$, $u|_{z=0} = N_0 h_0(\text{Fr}^2 - 1)^{1/2}\cos(k_0 x)$, indicating that the total wind is fastest over the hill crests.

Whether subcritical or supercritical, the pressure field is related to the horizontal velocity field by

$$p = \varrho_0(\Omega/k_x)u = -\varrho_0 U_0 u, \qquad (5.96)$$

in which ϱ_0 is the characteristic density of the fluid. So for subcritical flows the pressure is highest at the midpoint of the hill facing the ambient wind and in supercritical flows the pressure is highest in the valley.

From these polarization relations we can evaluate momentum and energy transport by computing appropriate correlations between basic state fields. In the Boussinesq approximation, the momentum flux is $\varrho_0 \langle uw \rangle$. For subcritical flow, this becomes

$$\varrho_0 \langle uw \rangle = -\frac{1}{2}\frac{k_z}{k_x}\Omega^2 h_0^2 = -\frac{1}{2}\text{sign}(U_0)\,(N_0 h_0)^2 \text{Fr}\sqrt{1 - \text{Fr}^2}, \quad \text{Fr} < 1. \quad (5.97)$$

If the flow moves from left-to-right over the hills ($U_0 > 0$), the momentum flux is negative meaning that horizontal momentum (which is negative because $k_x < 0$) is transported upwards by propagating waves. Where these waves break at upper levels they decelerate the background winds.

The fact that internal waves transport momentum away from the hills means that they exert a drag on the hills themselves. The average horizontal force per unit area acting on the hills is $-\varrho_0 \langle uw \rangle$. For example, a stratified wind with $N_0 = 0.01\,\text{s}^{-1}$ that blows with a speed of $U_0 = 10\,\text{m/s}$ over sinusoidal hills having wavelength $\lambda = 30\,\text{km}$ and peak-to-valley distance $2h_0 = 1\,\text{km}$ will exert a mean stress of $2.5\,\text{Pa}$ on the ground. Over one wavelength the force per unit along-crest distance is $75\,\text{N/m}$. This force acts to accelerate the Earth itself in the direction of the winds.

The magnitude of the drag on the ground may be represented in terms of a nondimensional drag coefficient,

$$C_D \equiv |\langle uw \rangle|/U_0^2 = \frac{1}{2}\text{Lo}^2 \text{Fr}\sqrt{1 - \text{Fr}^2}, \quad \text{Fr} < 1. \quad (5.98)$$

At fixed Long number $\text{Lo} = N_0 h_0/|U_0|$, the momentum flux approaches zero as Fr approaches zero or unity and reaches a maximum if $\text{Fr} = 1/\sqrt{2}$, as shown in Figure 5.15. This corresponds to waves that propagate with lines of constant phase forming an angle of $|\Theta| = \cos(\Omega/N_0) = 45°$ to the vertical. The hills generate waves with the largest vertical flux of horizontal momentum in this case. If $\text{Fr} > 1$, the flow is supercritical and u and w are $90°$ out of phase. As a consequence, the drag is identically zero, meaning that no momentum is transported vertically by evanescent disturbances.

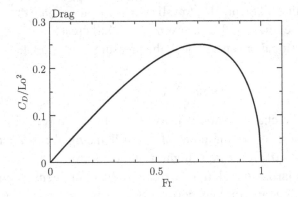

Fig. 5.15. Normalized drag exerted on sinusoidal hills by waves generated through subcritical flow of uniformly stratified fluid over the hills.

5.4.3 Internal waves over small-amplitude localized hills

Using the results above for periodic hills, we can use standard Fourier transform methods to predict the structure of waves generated by small-amplitude hills of arbitrary shape. Here we consider the one-dimensional circumstance in which the hill height is prescribed by $h(x)$. The two-dimensional case is considered in Section 5.4.5.

The Fourier transform of $h(x)$ is

$$h(x) = \int_{-\infty}^{\infty} \hat{h}(k_x) e^{\imath k_x x} \, dk_x, \tag{5.99}$$

in which

$$\hat{h}(k_x) = \frac{1}{2\pi} \int_{-\infty}^{\infty} h(x) e^{-\imath k_x x} \, dx. \tag{5.100}$$

Because $h(x)$ is a real function, it follows from (5.100) that $\hat{h}(-k_x) = \hat{h}^{\star}(k_x)$, in which the star denotes the complex conjugate.

Thus the disturbance generated by flow over the hills is a superposition of periodic waves generated by flow over topography having wavenumber k_x and corresponding amplitude $\hat{h}(k_x)$. In the case of the sinusoidal hills considered in the previous section, we set $h(x) = h_0 \exp[\imath k_0 x]$. Substitution into (5.100) gives $\hat{h}(k_x) = h_0 \delta(k_x - k_0)$, in which δ is the Dirac delta function.

The corresponding vertical displacement field of waves generated by the hills is

$$\xi(x,z) = \int_{-\infty}^{\infty} \hat{h} e^{\imath (k_x x + k_z z)} \, dk_x, \tag{5.101}$$

in which $\hat{h}$ is given by (5.100). In the exponential of (5.101) k_z is a function of k_x through (5.92). Information about the flow speed U_0 and stratification N_0 is buried in the formulae for k_z.

Although explicit analytic solutions of (5.99) generally cannot be found, the results can be efficiently determined numerically using fast-Fourier transform methods. At each level z, one simply computes $\xi(x,z)$ from the inverse transform of the Fourier amplitudes $\hat{h}(k_x) \exp[\imath k_z(k_x) z]$. The vertical velocity field is determined by inverse transforming $-\imath U_0 k_x \hat{h} \exp(\imath k_z z)$ and, likewise, one can determine the other fields.

For example, consider the excitation of waves by uniform flow over an isolated hill having the 'Witch of Agnesi' shape given by

$$h(x) = A \left[1 + \left(\frac{x}{L} \right)^2 \right]^{-1}. \tag{5.102}$$

An analytic expression for its Fourier transform can be found explicitly:

$$\hat{h}(k_x) = \frac{1}{2}Ae^{-L|k_x|}. \tag{5.103}$$

In particular, the amplitude of the waves generated by the hills with horizontal wavenumber k_x is negligibly small if $|k_x| \gg 1/L$. Although the integral (5.101) involves a superposition of subcritical and supercritical waves, the dominant contribution comes from waves with horizontal wavenumber approximately less than $1/L$. This inspires us to define the Froude number for this flow as $\mathrm{Fr} = U_0/(N_0L)$. If $\mathrm{Fr} \ll 1$, (5.101) involves a significant contribution from subcritical (vertically propagating) waves with wavelengths of the scale of the hill. If $\mathrm{Fr} \gg 1$, only waves that are very long compared to the extent of the hills are propagating. The disturbance directly above the hill is evanescent.

This is shown in Figure 5.16, which plots contours of isopycnals at evenly spaced intervals above the hills. Vertically propagating internal waves are launched over the hill in the subcritical case with $\mathrm{Fr} = 0.2$ and evanescent disturbances occur directly over the hill in the supercritical case with $\mathrm{Fr} = 5.0$.

In order to visualize the isopycnal displacements for this plot, an amplitude of $A = 0.5$ was chosen for this calculation. However, such a large amplitude in comparison with the hill width $L = 1$ is unrealistic for the linear theory approximations used to derive (5.101). In the next section a different approach is taken to find the disturbance field resulting from internal waves launched by large-amplitude hills.

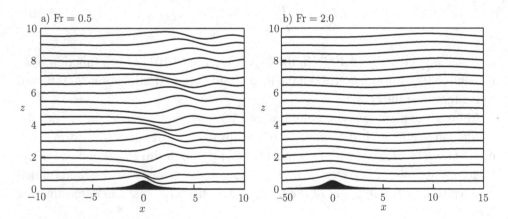

Fig. 5.16. Sketch of waves generated by flow over small-amplitude topography with a) $\mathrm{Fr} = 0.5$ and b) $\mathrm{Fr} = 2.0$. The displacement of isopycnals, solid lines, is computed using (5.101). The hill height is given by the 'Witch of Agnesi' profile with amplitude $A = 0.5$ and width $L = 1$, shown as the solid object at the bottom of both plots. The Froude numbers are defined by $\mathrm{Fr} = U_0/(N_0L)$.

5.4.4 Internal waves over large-amplitude localized hills

Again we consider steady flow over two-dimensional hills but now remove the small-amplitude restriction. We nonetheless restrict our examination to hills that are sufficiently smooth to allow free-slip flow over the bottom boundary. Boundary-layer separation does not occur. The approach taken is referred to as 'Long's model'.

Two-dimensional, Boussinesq (hence incompressible) flow evolves according to the coupled fully nonlinear vorticity and internal energy equations, (1.86) and (1.19), respectively. Our objective is to combine these under the assumption of steady-state flow in order to derive a single equation in one variable.

In steady state, (1.19) can be written

$$J(\Psi, \varrho) = 0. \tag{5.104}$$

Here $\varrho \equiv \bar{\rho}(z) + \rho$ is the total density field and Ψ is the total streamfunction, which satisfies

$$-\Psi_z = \bar{U} + u \text{ and } \Psi_x = w. \tag{5.105}$$

We have also introduced the Jacobian determinant, J. Generally, if a and b are functions of x and z, then $J(a,b) = a_x b_z - b_x a_z$, which is the determinant of the matrix formed by first-order derivatives of a and b. Using (5.105), the Jacobian of Ψ and ϱ is $(\bar{U} + u)\varrho_x + w\varrho_z$, the advective terms in the material derivative acting on ϱ.

The point of introducing the Jacobian is to exploit the useful property that $J(a,b) = 0$, which means a is linearly dependent upon b for all x and z. Thus (5.104) demonstrates that the total density is constant along streamlines. Hence $\varrho(\Psi)$ can be determined from the upstream flow conditions where Ψ and ϱ are prescribed functions of z.

In steady state the vorticity equation (1.86) becomes $J(\Psi, \zeta) = \frac{g}{\varrho_0}\varrho_x$, in which $\zeta \equiv (\nabla \times \vec{u}) \cdot \hat{y}$ is the y-component of the vorticity vector. Using (5.105), we can write ζ in terms of Ψ:

$$\zeta = -\nabla^2 \Psi. \tag{5.106}$$

The prescribed relationship $\varrho(\Psi)$ also lets us eliminate ϱ on the right-hand side of the vorticity equation, thereby giving

$$J(\Psi, -\nabla^2 \Psi) = \frac{N^2}{\bar{U}}\Psi_x. \tag{5.107}$$

Here we have represented the unperturbed upstream conditions by the horizontal velocity profile $\bar{U}(z) = -\Psi_z$ and the total squared buoyancy frequency $N^2(z) = -(g/\varrho_0)\partial\varrho/\partial z$, which is written in terms of Ψ using $d\varrho/d\Psi = (\partial\varrho/\partial z)/(\partial\Psi/\partial z)$. Thus we have derived a nonlinear partial differential equation in Ψ alone which completely characterizes steady flow over hills with arbitrarily prescribed profiles of upstream velocity and buoyancy frequency.

If the background flow speed and stratification are uniform, (5.107) can be manipulated further to give a linear partial differential equation. We decompose Ψ into the sum of the background and fluctuation components $-U_0 z + \psi$ and we note that $J(\Psi, \psi) = U_0 \psi_x$. Therefore, we can combine both sides of (5.107) into a single Jacobian determinant:

$$J\left(\Psi, \nabla^2 \psi + \left(\frac{N_0}{U_0}\right)^2 \psi\right) = 0. \tag{5.108}$$

So the second term in J is linearly dependent upon Ψ. However, in the upstream unperturbed region the second term is identically zero. Therefore, we have found that everywhere in the domain, ψ satisfies the Helmholtz equation

$$\nabla^2 \psi + \left(\frac{N_0}{U_0}\right)^2 \psi = 0. \tag{5.109}$$

Although (5.109) is a linear partial differential equation, it nonetheless describes nonlinear phenomena through specification of the free-slip condition for no-normal flow through the bottom boundary. This is equivalent to requiring that a streamline traces out the hills:

$$\psi(x, h(x)) = U_0 h(x). \tag{5.110}$$

Taking x-derivatives on both sides gives $w = (U_0 + u)h'$ at $z = h(x)$.

Next suppose the hills are periodic and of small amplitude, being prescribed with complex variables so that $h(x) = h_0 \exp(\imath kx)$. The linearized bottom boundary condition becomes

$$\psi(x, 0) = U_0 h(x), \tag{5.111}$$

which is likewise periodic in x. We may therefore assume $\psi(x, z) = \hat{\psi}(z) \exp(\imath k_x x)$ throughout the domain, with $\hat{\psi}(0) = U_0 h_0$. Substitution in (5.109) gives

$$\hat{\psi}'' + k_z^2 \psi = 0. \tag{5.112}$$

As in (5.91), $k_z^2 = (N_0/U_0)^2 - k_x^2$ defines the vertical wavenumber k_z as a function of the horizontal wavenumber k_x.

Next we wish to consider nonlinear effects by allowing $h(x)$ to have non-negligible height. That is, we suppose that the characteristic hill height h_0 is sufficiently large that $\text{Lo} \equiv N_0 h_0 / U_0$ is no longer negligibly small. Solutions can be determined either by Fourier techniques, similar to those above, or by the method of Green's functions. Which mathematical approach is best is established generally by the Froude number $\text{Fr} = U_0 / (N_0 l_0)$ in which l_0 is a characteristic width of the hill. Hydrostatic flows (with $\text{Fr} \ll 1$) are best modelled by Fourier transforms but if Lo is sufficiently large, it turns out that nonhydrostatic flows (with $\text{Fr} \gg 1$) are best modelled by Green's functions.

First we seek a Fourier representation in the form

$$\psi(x, z) = \int_{-\infty}^{\infty} U_0 \hat{\eta}(k_x) e^{\iota (k_x x + k_z(k_x)z)} \, dk_x. \tag{5.113}$$

To satisfy (5.109) with the requirement for upward radiation and boundedness, $k_z(k_x)$ is again given by (5.92). The value of the unknown Fourier amplitude $\hat{\eta}$ is determined implicitly from the exact boundary condition (5.110), which gives the requirement

$$h(x) = \int_{-\infty}^{\infty} \hat{\eta} e^{\iota (k_x x + k_z h(x))} \, dk_x. \tag{5.114}$$

In particular, we see that $\hat{\eta} = \hat{h}$ in the small-amplitude limit, $\|k_z h\| \ll 1$.

Analytic solutions for $\hat{\eta}$ can be found in special cases. Otherwise, one can attempt to solve (5.114) numerically, for example, through discretization and matrix inversion.

However, if the flow is nonhydrostatic, the matrix is nearly singular (or, more precisely, ill-conditioned) and so cannot be inverted with acceptable floating-point accuracy. Effectively, this is because the integrand becomes increasingly dominated by exponential rather than oscillatory behaviour in k_x when the peak contribution to the integral occurs for values of $k_x \simeq l_0$ for which k_z is imaginary.

We make this more explicit by converting x and h to nondimensional variables through the assignments $x = l_0 \tilde{x}$ and $h = h_0 \tilde{h}$. The complex exponent thus becomes $\iota k_x l_0 (\tilde{x} + (h_0/l_0)(k_z/k_x)\tilde{h})$. If $\text{Fr} \ll 1$, then from (5.92) $|k_z| \simeq |k|/\text{Fr}$. Hence the coefficient of $\tilde{h}$ is of order $\text{Lo} = N_0 h_0 / U_0$. If $\text{Fr} \gg 1$, it is of order Fr Lo. Thus the condition for the problem to be solvable numerically is $\text{Fr Lo} = h_0/l_0 \ll 1$: the aspect-ratio of the hills must be small.

Conversely, the method of Green's functions permits well-conditioned numerical solutions for high aspect-ratio hills. Here the strategy is first to solve (5.109) in which the right-hand side is replaced by a delta-function. The resulting solutions effectively give the streamfunction response to a delta-function hill. Next, an arbitrary hill is represented by a continuous set of delta-function hills that span its surface. The resulting streamfunction response is the integral of the responses from each delta-function hill.

The response to a single delta-function is a Bessel series with terms chosen to ensure upward and downstream wave propagation. In the case $\mathrm{Fr} = 1$ the Green's function in polar co-ordinates is

$$G(\tilde{r}, \theta) = \frac{1}{2} Y_1(\tilde{r}) \sin \theta + \frac{1}{\pi} \sum_{n=1}^{\infty} \frac{4n}{4n^2 - 1} J_{2n}(\tilde{r}) \sin 2n\theta, \qquad (5.115)$$

in which $\tilde{r} = r/l_0$.

An example of flow over an isolated hill is shown in Figure 5.17. In this case the hill is two-dimensional with semi-circular cross-section having radius, R_0. In this particular geometry, the height and length scales are equal so that

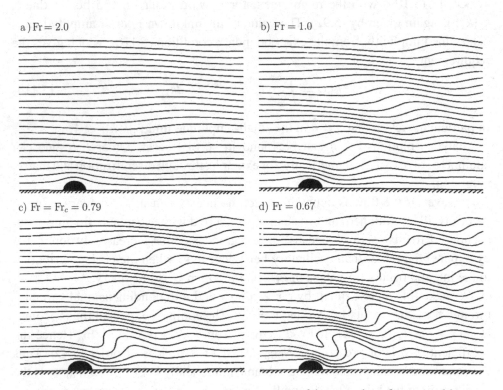

a) $\mathrm{Fr} = 2.0$

b) $\mathrm{Fr} = 1.0$

c) $\mathrm{Fr} = \mathrm{Fr}_c = 0.79$

d) $\mathrm{Fr} = 0.67$

Fig. 5.17. Contours showing the displacement of isopycnal surfaces resulting from uniformly stratified flow upstream of speed U_0 moving over a semi-circular hill of radius R_0. Vertically propagating internal waves are launched more efficiently as the Froude number $\mathrm{Fr} = U_0/(N_0 R_0)$ decreases. At the critical Froude number, $\mathrm{Fr}_c = 0.79$, isopycnals are on the point of overturning above the hill. [Reproduced, by permission of Cambridge University Press, from Figures A.1–4 of Miles & Huppert, J. Fluid Mech., **33**, 803–814 (1968).]

$Fr = Lo^{-1} = U_0/(N_0 R_0)$, in which U_0 is the upstream flow speed and N_0 is the buoyancy frequency of the uniformly stratified ambient.

At fast flow speeds, for which $Fr = 2.0$ (Figure 5.17a), the hill primarily excites evanescent disturbances over the hill with a long-wavelength, small-amplitude wave in its lee. At smaller Froude numbers the vertical displacement of isopycnals becomes more pronounced. For the semi-circle, the waves are at the point of over-turning when the Froude number is at the critical value $Fr_c = 0.79$ (Figure 5.17c). At still smaller Froude numbers one does not expect the predicted flow field to occur in reality because the internal waves break over the hill, creating unsteady motion. Indeed, this phenomenon is responsible for the occurrence of downslope windstorms, as discussed in Section 5.4.6.

5.4.5 Internal waves over localized two-dimensional hills

So far we have focused upon internal waves in the x–z plane generated by flow over one-dimensional hills. Here we extend these considerations to examine the fully three-dimensional evolution of waves generated by flow over a small-amplitude hill that varies horizontally in both the x and y directions.

If the height of the hill is small compared with its horizontal extent, the approach taken in Section 5.4.3 is straightforwardly extended to a hill whose height is given by $h(x, y)$. The double Fourier transform of h is $\hat{h}(k_x, k_y)$, which satisfies

$$h(x, y) = \int \int \hat{h}(k_x, k_y) e^{i(k_x x + k_y y)} \, dk_y \, dk_x. \qquad (5.116)$$

The corresponding vertical displacement field above the hills is

$$\xi(x, y, z) = \int \int \hat{h} e^{i k_z z} e^{i(k_x x + k_y y)} \, dk_y \, dk_x. \qquad (5.117)$$

The Fourier amplitude is a function of k_x and k_y both through the value of $\hat{h}$ and through the dependence of k_z upon the horizontal wavenumber components. Assuming the ambient flow U_0 is oriented in the x-direction, by analogy with (5.91) k_z is

$$k_z^2 = \left(k_x^2 + k_y^2\right) \frac{1 - Fr_k^2}{Fr_k^2}, \qquad (5.118)$$

in which $Fr_k \equiv |U_0 k_x|/N_0$, as in (5.83). For upward-propagating waves, if $Fr_k < 1$ then k_z is given explicitly by taking the negative square root of the right-hand side of (5.118) and multiplying this by the sign of k_x. If $Fr_k > 1$, the branch cut is taken in the upper half-plane corresponding to exponentially decreasing amplitude with z.

For example, the vertical displacement field associated with flow over an axisymmetric 'Witch of Agnesi' hill is plotted in Figure 5.18. The hill is

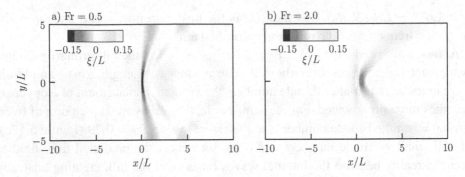

Fig. 5.18. Vertical displacement field, relative to the hill width L, for waves generated by an axisymmetric 'Witch of Agnesi' hill centred at $x = y = 0$ for flows with Froude number, Fr $= U_0/(N_0L)$, equal to a) 0.5 and b) 2.0. The hill amplitude is $A = 0.5L$ and the displacement field is shown in a horizontal plane a distance $z = L$ above the hill. For comparison, see the isopycnal displacements for flow over the two-dimensional 'Witch of Agnesi' hill plotted in Figure 5.16.

prescribed by

$$h(x,y) = A\left[1 + \frac{x^2 + y^2}{L^2}\right]^{-1},$$

(5.119)

with $A = 0.5$ and $L = 1$. The grey scale shows values of vertical displacement a distance $z = 1$ above the hill for a subcritical case with Fr $\equiv U_0/(N_0L) = 0.5$ and an approximately supercritical case with Fr $= 2$.

As in the two-dimensional counterpart, shown in Figure 5.16, vertically propagating waves emanate above the hill in the subcritical case, but the disturbance field is more localized directly over the hill as the Froude number increases. The waves do not simply propagate downstream but also outwards in the spanwise direction forming what appears as an elongated wake structure, similar in appearance to surface waves produced by the bow of a moving ship. Because the waves spread horizontally as they propagate upwards their amplitude decreases with height faster than in the two-dimensional case.

5.4.6 Downslope windstorms

Under certain atmospheric conditions the generation of internal waves originating from flow over mountains results in storms upstream that both dry and heat the air that is carried over the hills. In their lee, the resulting flow is known in North America as a Chinook wind or in Europe as a Foehn. In its mild manifestation,

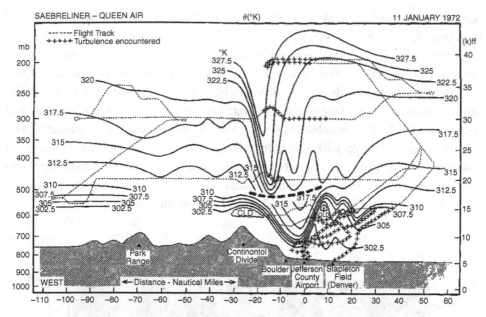

Fig. 5.19. Potential temperature contours (solid lines) from aircraft flight data taken over the Rocky Mountains during a severe downslope windstorm event in Boulder, Colorado on January 11, 1972. The dashed lines represent the path of the aircraft with regions of turbulence shown by crosses. [Reproduced, with permission ©American Meteorological Society, from Figure 7 of Lilly, J. Atmos. Sci., **35**, 59–77 (1978).]

these winds can provide welcome relief from winter's chill. However, if the waves are generated at such large amplitude that they break aloft, they block the flow at mid-levels of the troposphere and force the flow to rush even more quickly down the lee slopes of the mountains. The resulting ground level gusts, known as 'downslope windstorms', can reach hurricane force.

Figure 5.19 shows contours of potential temperature measured by aircraft in the troposphere along a west-to-east cross-section above the Colorado Rocky Mountains. At the time of the event a particularly intense downslope windstorm occurred in Boulder, Colorado. The cause of the intense surface winds is evident from the steeply descending isentropes above the mountains, a result of breaking mountain-generated waves.

The contours may be compared with the theoretical displacement of isopycnal surfaces in subcritical flow over a semi-circular hill, as shown in Figure 5.17d. The observed isentropic displacements exhibit a more complicated pattern in part because of the complex topography, the non-uniformity of the background wind and stratification and also because the breaking waves result in non-steady flow.

Nonetheless, the location of the breaking aloft of the continental divide is consistent with theory.

Simulations have used the upstream profiles of potential temperature and wind measured during the storm depicted in Figure 5.19. The numerical model represents the hill by a simple two-dimensional 'Witch of Agnesi' structure, as in (5.102). The initial condition was set from the steady-state prediction of Long's model, discussed in Section 5.4.4. The unstable flow was then allowed to evolve nonlinearly. The results shown in Figure 5.20 demonstrate the generation of a blocking region in the mid-troposphere which drives turbulent bursts of air towards the ground. Remarkably, this simplified model was able to reproduce the approximate period of the intensely pulsating winds that were observed during the windstorm.

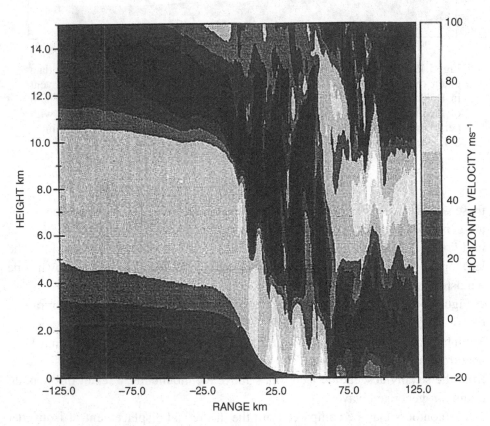

Fig. 5.20. Snapshot from a simulation of a downslope windstorm over a two-dimensional hill with upstream wind and potential temperature conditions set by observations of a downslope windstorm event near Boulder, Colorado on January 11, 1972. [Reproduced, with permission ©American Meteorological Society, from Figure 18 of Scinocca & Peltier, J. Atmos. Sci., **46**, 2885–2914 (1989).]

5.5 Tidal flow over topography

In the previous section, we focused upon a flow with uniform speed moving over topography. In the 1990s another important mechanism for internal wave generation in the ocean was recognized, that due to the oscillatory motion of tides moving over underwater hills called 'sills'. This wave generation mechanism is sometimes referred to as 'baroclinic conversion' because it provides a way partially to transform the energy of barotropic tides (in which the entire ocean column moves vertically up and down) to baroclinic internal waves. Through other processes, the internal waves then convert this energy to successively smaller scales, ultimately dissipating their energy in part through mixing.

Here, after an introduction to the barotropic tide, we present the basic theory for wave generation by oscillatory flow over one-dimensional hills. The period of these waves is established by the tidal period and in this sense one can draw an analogy between these waves and those generated by oscillating bodies, as considered in Section 5.2. For tidally generated waves, however, it is crucial to include the effects of Coriolis forces.

5.5.1 Barotropic tides

Tides result from the centrifugal and gravitational forces of the moon, sun and less significantly the other planets. During the Earth's rotation and its orbit around the Sun, the forces act periodically with a range of frequencies. The most significant of these is the principal lunar tide, otherwise known as the 'M2 tide'. The tide results from the Moon pulling on the oceans on the near-moon side and the oceans accelerating away from the Moon on the opposite side of the Earth due to centrifugal forces. The latter occurs because the centre of mass of the rotating Earth–Moon system lies between the centres of the Earth and Moon. The period of the M2 tide is somewhat less than a half-day and so is sometimes called the 'semi-diurnal lunar tide'.

The diurnal 'M1' tide has double the period, nearly that of one day. This has smaller amplitude than the M2 tide and, when superimposed with it, leads to 'higher high tides' or 'lower high tides' depending upon the coupling of phases between the tides. The influence of the Sun and other factors further modulate this signal. Because the M2 tide provides the strongest influence upon the oceans, in the discussion here we will focus upon forcing at semi-diurnal lunar period alone.

The tide effectively is a shallow inertial wave, as described in Section 2.7.2. Although the ocean is stratified, the motion behaves as though the ocean had uniform density and, in this sense, the wave is barotropic. In this discussion, however, the wave is forced and the bottom topography does not have uniform depth.

Near sills, island chains and the continental margins, the barotropic motion of the tide can force the flow to move from deeper to shallower water. Depending upon the geometry of the bottom topography and the characteristics of the tide, such motion can excite internal, baroclinic motions. We have already seen how this occurs for steady flow over hills in Section 5.3.

5.5.2 *Inertia gravity wave beams*

Here we suppose the horizontal extent of the bottom bathymetry is long compared to the tidal excursion, which is the maximum horizontal distance traversed by fluid as the tide moves it back and forth. For simplicity, we consider a uniformly stratified fluid that moves back and forth at speed $U = U_0 \cos(\omega_0 t)$ over one-dimensional sinusoidal hills with wavenumber k_0 in the x-direction.

This circumstance is different from that examined in Section 5.4.2 for which the flow was unidirectional and the hills were small-amplitude. In that case, the extrinsic frequency Ω was set by the flow speed U_0 and the horizontal wavenumber of the hills k_0. In this case, the tide introduces a new time-scale through the tidal frequency ω_0. The resulting wave-generation process is similar to that for internal waves generated by oscillating bodies in that their spatial structure is set by the oscillation frequency, not the amplitude, provided that this is sufficiently small.

Particularly at mid-latitudes, the tidal frequency is comparable to the Coriolis frequency f_0. So the motion excites inertia gravity waves. Restricting ourselves to their structure in the (x, z) plane, the evolution equation (3.196) given in terms of the streamfunction, ψ, is

$$\nabla^2 \psi_{tt} + N_0{}^2 \psi_{xx} + f_0{}^2 \psi_{zz} = 0. \tag{5.120}$$

Here $\nabla^2 = \partial_{xx} + \partial_{zz}$ and subscripts have been used to denote derivatives in time and space.

The condition for free-slip flow over the moving hills is given by

$$\psi(x, h, t) = U_0 \cos(\omega_0 t) \, h(x). \tag{5.121}$$

Note that ψ is evaluated at $z = h$, not $z = 0$ as would be done for small-amplitude hills. The hills themselves are prescribed by

$$h = h_0 \cos(k_0 x), \tag{5.122}$$

which has maximum slope $m_h \equiv h_0 k_0$.

Because the waves are generated at a monochromatic frequency the waves are formed so that lines of constant phase have a fixed angle Θ to the vertical. The slope m_w of these lines is given by (3.197). It turns out that the important parameter in this study is the ratio

$$\epsilon \equiv \frac{m_h}{m_w} = h_0 k_0 \sqrt{\frac{N_0{}^2 - \omega^2}{\omega^2 - f_0{}^2}}. \tag{5.123}$$

In Section 3.5.3 we classified the topographic slope as subcritical if $\epsilon < 1$ and supercritical if $\epsilon > 1$. Incident downward-propagating waves reflect upwards from a subcritical slope and reflect downwards from a supercritical slope. Importantly, the phase lines of the reflected waves become closely spaced in the critical case $\epsilon = 1$.

In the case of tidal generation, there are no incident waves. But the condition $\epsilon < 1$ guarantees that waves generated by the hills propagate upwards. We will

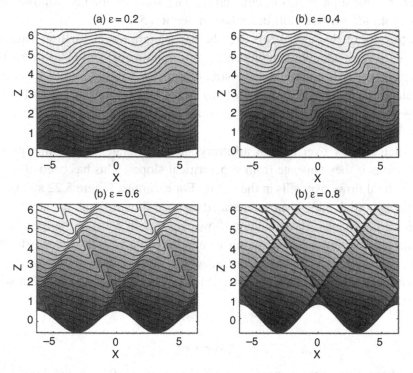

Fig. 5.21. Generation of internal tides by oscillatory flow over sinusoidal topography with relative hill-to-wave slope given by a) $\epsilon = 0.2$, b) $\epsilon = 0.4$, c) $\epsilon = 0.6$ and d) $\epsilon = 0.8$. Contours denote displaced isopycnal surfaces. In all four cases the nondimensional amplitude of the forcing is $\epsilon U_0 k_0 / \omega_0 = 0.5$. [Reproduced, with permission ©American Meteorological Society, from Figure 3 of Balmforth *et al.*, J. Phys. Oceanogr., **32**, 2900–2914 (2002).]

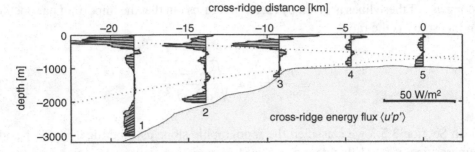

Fig. 5.22. Measured energy flux as a function of depth taken from observation sites above a sill between Kauai and Oahu in the Hawaiian Islands. Dotted lines indicated the two predicted paths of wave-beams whose slopes are tangent to the sill at the critical slope where $\epsilon = 1$. [Reproduced, with permission ©American Meteorological Society, from Figure 5a of Nash *et al.*, J. Phys. Oceanogr., **36** 1123–1135 (2006).]

restrict our attention to this circumstance. The solution method follows that for large-amplitude periodic hills described in Section 5.4.4.

The results of this calculation are shown in Figure 5.21 for four values of ϵ and fixed forcing amplitude represented as a nondimensional measure of U_0. As ϵ increases towards unity, the wave structure more closely resembles beams that emanate from the steepest slopes of the hills. The behaviour is similar to that of internal wave-beams emanating from an oscillating cylinder, as illustrated in Figure 5.1.

Oscillatory flow over hills of arbitrary shape generates waves that propagate along beams if they emanate from supercritical slopes. This has been observed to occur in tidal flows over sills in the ocean. For example, Figure 5.22 shows observations of inertial wave-beams generated by tidal flow through the strait between two Hawaiian islands. The oscillatory flow over the sill between the islands creates upward- and downward-propagating beams that follow paths tangent to the sill at the critical angle set by the tidal frequency. The presence of waves is evident through measurements of energy flux which is large at depths coincident with the path of the waves.

Exercises

5.1   Show that the ellipse $C_1 = \{\mathcal{Z}_1 = (a/\Gamma)\cos\theta + \imath a\sin\theta \mid 0 \le \theta < 2\pi\}$ transforms into the circle $C_2 = \{\mathcal{Z}_2 = a(1 + 1/\Gamma)/2\exp(\imath\theta) \mid 0 \le \theta < 2\pi\}$, through the inverse Joukowski transformation, (5.11).

5.2   Show that the streamfunction amplitude envelope for upward- and rightward-propagating waves generated by an oscillating cylinder in inviscid fluid, given by (5.20), satisfies the partial differential equation (5.15) with $\gamma^2 = (N^2/\omega^2) - 1 > 0$, and show that the superposition of the beams in the four quadrants satisfies the boundary condition (5.4).

5.3 Find the vertical displacement field, ξ, associated with the upward- and rightward-propagating wavebeam that emanates from an oscillating cylinder in inviscid fluid and show that ξ is infinite along lines of the beam tangent to the cylinder.

5.4 (a) Derive the formula for internal waves in viscous fluid, given by (5.26).
   (b) Transform to (r,σ) co-ordinates using (5.16) and (5.17), and invoke the boundary layer approximation to derive the partial differential equation for viscous upward- and rightward-propagating wavebeams, (5.27).

5.5 Fourier transform (5.27) with respect to σ and so derive an ordinary differential equation for the transformed amplitude $\hat{\Psi}(r; k_\sigma)$. Solve this equation and use the integral representation for $\Psi(0,\sigma)$ given by (5.23) to derive the integral formula, (5.28), for a viscously attenuating beam generated by an oscillating cylinder.

5.6 Using the bimodal to unimodal transition condition (5.31) together with the integral definition of the streamfunction amplitude (5.28), show that the transition distance is given by (5.33).

5.7 A cylinder of radius 2 cm oscillates with frequency $\omega = 0.5\,\mathrm{s}^{-1}$ in uniformly salt-stratified water with buoyancy frequency $N_0 = 1\,\mathrm{s}^{-1}$. At which radial distance from the centre of the cylinder does the vertical displacement field transition from bimodal to unimodal?

5.8 (a) Use (5.30) with (5.28) to find the integral representation for the vertical displacement field, ξ, and its magnitude $|\xi|$ along the centreline of the beam where $\sigma = 0$.
   (b) Find the leading-order asymptotic approximation to the integral for $|\xi|$ at $\sigma = 0$ for large r. In particular, show that the beam decays as $r^{-2/3}$.

5.9 (a) Invoking reflectional symmetry about the horizontal plane, show that the streamfunction for the downward-propagating wave-cone emanating from a vertically oscillating sphere is given by

$$\Psi_- = \frac{1}{4}Wae^{-\imath(3\Theta/2 + \vartheta/2 - \pi/4)}\sqrt{\frac{a}{\varrho}}\exp\left[-\imath\frac{3}{2}\cos^{-1}(-\sigma_-/a)\right],$$

(E5.1)

in which $\sigma_- = -\varrho\sin(\Theta + \vartheta)$ and $r_- = -\varrho\cos(\Theta + \vartheta)$ so that $\varrho = (r_-^2 + \sigma_-^2)^{1/2}$.

b) Superimpose this result with (5.44) and evaluate the resulting stream-function on the surface of the sphere. Thus show that the boundary condition (5.40) is satisfied.

5.10 Use the general form of the integral transform (5.22) to show that the algebraic representation, (5.44), of the streamfunction for the upward-propagating wave-cone from an oscillating sphere can be written in terms of the Fourier integral (5.46).

5.11 Use Fourier transforms to solve (5.63) subject to the initial conditions (5.64) and assuming a) $Fr \neq 1$ and b) $Fr = 1$.

5.12 Write a formula for U_0/c_{gx} in terms of the Froude number given by (5.79).

5.13 Show that waves represented by negative frequencies in (5.84) are equivalent to those with positive frequencies under a change of frame of reference $x \to -x$, so that $U_0 \to -U_0$ and $k_x \to -k_x$.

5.14 Show that uniform density, irrotational flow over small-amplitude sinusoidal topography with wavelength $\lambda = 2\pi/k$ has a vertical displacement field given by the second expression in (5.93) with $\gamma = k$.

5.15 Show that downward-propagating waves generated by topography towed along the surface of a fluid are represented by (5.92) with the branch-cut taken in the second quadrant of the complex plane when determining the square root defining k_z.

5.16 What is the correct sign before the following expression for u in subcritical flow: $\pm\sqrt{[N/(U_0 k)]^2 - 1}\, U_0 k h_0 \sin(k_x x + k_z z)$?

5.17 Estimate the mean width and height of the Rocky Mountain range in western North America. At mid-latitudes the Jet Stream induces an average eastward flow of 30 m/s. Taking a characteristic buoyancy frequency of $N_0 = 10^{-2}\,\text{s}^{-1}$, estimate the extrinsic frequency and amplitude of the waves. What is the drag exerted on the ground over an approximately 2000 km latitudinal extent of the winds.

5.18 The M2-tide moves over a one-dimensional Gaussian hill with height $h(x) = h_0 \exp(-x^2/2\sigma^2)$. For what width, σ, and height, h_0, is the flow critical?

6

Wave propagation and spectra

6.1 Introduction

So far we have focused mainly upon the dynamics of waves in uniform and stationary ambient fluids. In the case of interfacial waves, we assumed that horizontal boundaries, if present, were flat so that the depth of the ambient was constant. Since interfacial wave speeds are a function of depth the question arises as to how the waves propagate in a medium that, for example, gets shallower approaching a beach. Likewise, for internal waves we have usually assumed that the background stratification is uniform. But we have seen that the stratification varies vertically in both the atmosphere and ocean. Except in the consideration of unstable shear flows, for the most part we have also neglected the presence of background winds and currents. In this chapter we examine how waves propagate in media that are non-uniform in the sense that the depth or stratification and the background flow varies.

We begin with the general treatment of small-amplitude wave propagation in non-uniform, but slowly varying media. This is known as ray theory. A simplified set of equations arising from ray theory assumes the motion is two-dimensional and steady, and that the background varies in only one spatial dimension. The theory is applied to study surface and interfacial waves approaching a beach and to examine internal waves approaching critical and reflection levels. In a somewhat different mathematical approach, the study of tunnelling examines the partial transmission and reflection of small-amplitude internal waves through weakly stratified regions.

Finally, we briefly discuss the observed spectra of waves in the atmosphere and ocean. These result from a combination of wave–wave interactions and instability processes whose details are not yet well established. However, empirical models of their structure have been useful in interpreting observed data and in developing internal wave drag parameterization schemes in general circulation models.

315

6.2 Extrinsic and intrinsic frequencies

In Chapters 2 and 3 we found plane wave solutions for small-amplitude interfacial and internal waves whose variation in space and time were represented in terms of complex exponentials. In particular, the vertical displacement of one-dimensional surface waves in a stationary ambient is

$$\eta = A \exp[\iota (kx - \omega t)], \tag{6.1}$$

in which ω is related to k through the dispersion relation. In a uniformly moving current of speed U_0 this field is slightly modified. In a frame of reference moving with the flow we write

$$\eta = A \exp[\iota (kx - \Omega t)]. \tag{6.2}$$

In this frame the ambient is perceived to be stationary. Therefore it is Ω that satisfies the dispersion relation for waves in a stationary ambient. A stationary observer on the shore observes the propagation of both the waves and the flow so that crests are seen to advance at speed $(\Omega/k) + U_0$. Their frequency is recorded to be ω, which is related to Ω by

$$\Omega = \omega - U_0 k_x. \tag{6.3}$$

The horizontal speed of crests is $\Omega/k_x = c_{Px} - U_0$: an observer moving forward at speed U_0 sees the waves recede at this speed.

When describing waves in an ambient flow it is important to distinguish between the definitions of the frequencies given by ω and Ω. Generally it is assumed that the observer is stationary and records the frequency ω as the waves are advected by the background flow. This is called the 'intrinsic frequency', 'absolute frequency', or 'ground-based frequency'.

Conversely an observer moving with the background flow records waves with frequency Ω, called the 'extrinsic frequency', 'relative frequency', or 'Doppler-shifted frequency'.

In general, for an ambient flow moving at velocity $\vec{U}_0$, the extrinsic frequency is given in terms of the intrinsic frequency by

$$\Omega = \omega - \vec{U}_0 \cdot \vec{k}, \tag{6.4}$$

in which $\Omega = \Omega(\vec{k})$ is the dispersion relation for waves in a stationary ambient.

6.3 Ray theory

The dispersion relation for interfacial waves in a two-layer, Boussinesq fluid with no ambient flow is given by the generalization of (2.94) to the x–y plane:

$$\omega^2 = g' |\vec{k}_h| \frac{1}{\coth(|\vec{k}_h|H_1) + \coth(|\vec{k}_h|H_2)}. \tag{6.5}$$

Here $g' = g(\rho_2 - \rho_1)/\rho_2$ is the reduced gravity based upon the density difference between the lower and upper fluid. The formula neglects background flows and assumes that the depths H_1 and H_2 of the upper and lower layer fluids are constant.

The dispersion relation for Boussinesq internal waves assumes the fluid is stationary and uniformly stratified. The generalization of (3.54) to three dimensions is

$$\omega^2 = N_0^2 \frac{|\vec{k}_h|^2}{|\vec{k}|^2}, \tag{6.6}$$

in which $|\vec{k}|^2 = |\vec{k}_h|^2 + k_z^2$.

Here we consider how the propagation and structure of plane waves change if the flow speed, fluid depths and stratification are not constant but vary in space and time. Provided the variations are small, at leading order we expect the local frequency and wavenumber of the waves should vary little from their values in a uniform ambient. Perturbation theory can determine the changes to the waves resulting from the ambient variations. The particular mathematical method is known as 'ray theory', 'ray tracing', or 'WKB theory'. The last of these designations refers to Wentzel, Kramers and Brillouin who developed the method in the study of problems in quantum mechanics. Some acknowledge the additional contribution of Jeffries by referring to WKBJ theory. However, the mathematics predates all four scientists, the methods having been developed by Liouville and Green almost a century earlier. Here we use the more descriptive terminology of ray theory because its results predict the path followed by a wavepacket as it is refracted by the non-uniform medium through which it propagates. That is, the wavepackets are treated as if they were particles moving along rays.

The treatment here is less general than WKB theory in that we will consider only the leading-order terms in the formal perturbation expansion of the wave equations that include the fast-time evolution of the waves and the slow-time evolution of their amplitude envelope.

6.3.1 General theory

The essence of ray theory is to suppose that the wave structure can be represented by the product of a slowly varying amplitude function multiplied by a rapidly varying oscillatory function. Thus, for example, the displacement field is written

$$\xi(\vec{x}, t) = A_\xi(\vec{x}, t) e^{i\varphi(\vec{x}, t)}. \tag{6.7}$$

Here the function A_ξ varies slowly in space and time compared with the wavelength and period, respectively, of the waves. Generally being complex-valued, A_ξ determines the relative phase, amplitude and spatial extent of the wavepacket as it evolves over time.

The fast variations associated with the undulations of the waves within the amplitude envelope are captured by the phase factor φ. At leading order, φ is linear in $\vec{x}$ and t so that

$$\nabla \varphi = \vec{k} \quad \text{and} \quad \frac{\partial \varphi}{\partial t} = -\omega. \tag{6.8}$$

Note that the second equation is given in terms of the intrinsic (ground-based) frequency.

In a spatially inhomogeneous and time-varying medium, ω is a function generally of position, wavenumber and time, so that

$$\omega \equiv \omega(\vec{k}, \vec{x}, t), \tag{6.9}$$

in which the wavenumber vector varies implicitly in space and time as the wavepacket moves through the flow.

Eliminating φ from (6.8) gives a formula for the change of wavenumber in time:

$$\frac{\partial \vec{k}}{\partial t} + \left(\nabla_{\vec{k}} \omega \right) \cdot \left(\nabla \vec{k} \right) = -\nabla \omega. \tag{6.10}$$

Here the symbol $\nabla_{\vec{k}}$ denotes derivatives taken with respect to the components of $\vec{k}$, whereas ∇ is the usual gradient operator that takes derivatives with respect to the components of $\vec{x}$. Using (6.4), the second term on the left-hand side of (6.10) can be written in terms of the group velocity:

$$\nabla_{\vec{k}} \omega = \vec{c}_g + \vec{U}. \tag{6.11}$$

Here $\vec{c}_g \equiv \nabla_{\vec{k}} \Omega$ is defined as $\vec{k}$ derivatives of the extrinsic frequency, which is that measured with respect to a (locally) stationary ambient. The advection of the waves by the background flow at velocity $\vec{U}$ gives the second term on the right-hand side of (6.11).

Finally we move to a frame of reference in which we continually track the wavepacket that is advected by the background flow while moving at its group velocity. With respect to a stationary observer, the position $\vec{x}$ changes in time according to $d\vec{x}/dt = \vec{c}_g + \vec{U}$, which we recognize as the right-hand side of (6.11).

Therefore the i-component of the vector $\vec{x}$ changes according to

$$\frac{dx_i}{dt} = \frac{\partial \omega}{\partial k_i}. \tag{6.12}$$

In the frame moving with the wavepacket the left-hand side of (6.10) is replaced with the ordinary derivative, in the same way that advection terms are absorbed into the partial time derivative to form the material derivative, which describes the evolution of a fluid parcel in the Lagrangian frame of reference. Therefore the i-component of the wavenumber vector $\vec{k}$ changes in time according to

$$\frac{dk_i}{dt} = -\frac{\partial \omega}{\partial x_i}. \tag{6.13}$$

The coupled system (6.12) and (6.13) forms six equations in the six unknowns $\vec{x}(t)$ and $\vec{k}(t)$. These are similar in structure to Hamilton's equations from classical mechanics:

$$\frac{dq_i}{dt} = \frac{\partial H}{\partial p_i} \quad \text{and} \quad \frac{dp_i}{dt} = -\frac{\partial H}{\partial q_i},$$

in which the Hamiltonian $H(\vec{q}, \vec{p})$ represents, for example, the energy of a particle at position $\vec{q}$ with momentum $\vec{p}$ and the subscript i denotes the components of the vector. Indeed, the analogy between Hamilton's equations and the wavepacket evolution equations is more explicit when compared with the Hamiltonian representation of quantum mechanics. The energy and momentum of a free 'wave-particle' are $H = \hbar \omega$ and $\vec{p} = \hbar \vec{k}$, respectively. Substituting these in Hamilton's equations, cancelling Planck's constant, $\hbar$, and letting $\vec{q} = \vec{x}$, we recover (6.12) and (6.13).

As expected, (6.12) and (6.13) share the symmetry properties of Hamiltonian equations. Two examples are particularly relevant here. First, it is a consequence of the time invariance of a physical system that energy, H, is conserved. Likewise, in steady background flow the intrinsic frequency of waves is constant during their evolution. This follows immediately by taking the ordinary time derivative of ω in (6.9), in which time is no longer an explicit variable. Using (6.12) and (6.13) gives

$$\frac{d\omega}{dt} = \sum_i \left(\frac{\partial \omega}{\partial k_i} \frac{dk_i}{dt} \right) + \left(\frac{\partial \omega}{\partial x_i} \frac{dx_i}{dt} \right) = 0. \tag{6.14}$$

Second, if a Hamiltonian system has no explicit spatial dependence in the x-direction, say, then the x-component of momentum is conserved. Likewise, if waves propagate in a fluid that has no explicit variation in the x-direction, then the horizontal wavenumber k_x does not change following the motion of the wavepacket. This follows directly from (6.13).

Another convenient feature of (6.12) and (6.13) is that the structure of the amplitude envelope of the waves does not appear explicitly in these equations. Rather, the path followed by the waves depends only upon the dispersion relation and how this and the intrinsic frequency vary with changes to the background flow and other parameters that determine the stratification and/or geometry of the system.

How the amplitude, A, varies following the motion of a wavepacket in stationary fluid was considered in Section 4.2.4. Energy, which is proportional to $|A|^2$, is transported at the group velocity. So, if the group velocity of the wavepacket decreases, the energy associated with it piles up. Depending upon how energy relates to wavenumber this can result in an increase in the wave amplitude. As discussed in Section 3.4.3, it is not energy E, but wave action $\mathcal{A} = E/\Omega$ that is conserved in a non-stationary ambient. So the amplitude changes explicitly in response to changes in both the group velocity and extrinsic frequency.

6.3.2 Ray theory for waves in two dimensions

For an interfacial wavepacket confined to the x–y plane, the equations for its position are given by the horizontal components of (6.12):

$$\frac{dx}{dt} = \frac{\partial \omega}{\partial k_x} \quad \text{and} \quad \frac{dy}{dt} = \frac{\partial \omega}{\partial k_y}. \tag{6.15}$$

While traversing this path, the wavenumber components change according to the horizontal wavenumber components of (6.13):

$$\frac{dk_x}{dt} = -\frac{\partial \omega}{\partial x} \quad \text{and} \quad \frac{dk_y}{dt} = -\frac{\partial \omega}{\partial y}. \tag{6.16}$$

In the special case of waves in the (x, y) plane for which the dispersion relation is independent both of time and of one of the two spatial variables, the mathematics simplifies substantially. For example, suppose that $\omega = \omega(k_x, x)$ is independent of y and t. This would be the case for interfacial waves in a two-layer fluid whose upper-layer depth H_1 is constant and whose lower-layer depth varies in the x-direction alone. Because the flow is time independent, it follows from (6.14) that $\omega = \omega_0$ is constant. Likewise from (6.16), it follows that $k_y = k_{y0}$ is constant. Thus k_x is implicitly given as a function of x through the relation

$$\omega(k_x, x) = \omega_0. \tag{6.17}$$

If one cares only about the path followed by a wavepacket and not the time taken to traverse it, t can be eliminated from (6.15) to give an explicit differential equation for y in terms of x:

$$\frac{dy}{dx} = \frac{\partial \omega / \partial k_y}{\partial \omega / \partial k_x}. \tag{6.18}$$

On the right-hand side, the ratio is a function of x given explicitly through the dependence of ω on x and implicitly through the dependence of k_x on x. Since the right-hand side does not depend upon y, an explicit integral solution can be found:

$$y(x) = y_0 + \int_{x_0}^{x} \frac{\partial \omega / \partial k_y}{\partial \omega / \partial k_x} dx. \tag{6.19}$$

Except in special circumstances, the integrand is often too algebraically complicated to make analytic solutions insightful. However the form of (6.19) is amenable to asymptotic or numerical solutions. Specific examples are discussed in Sections 6.4 and 6.5.

6.4 Ray theory for interfacial waves

Here we examine the propagation of surface and interfacial waves in a fluid that has spatially varying bottom depth. First this will be done for the special case of a surface wave approaching a beach. The results will then be generalized to interfacial waves in a two-layer fluid.

6.4.1 Surface waves approaching a beach

Consider a wavepacket in a one-layer fluid that is incident upon a beach. For simplicity, we assume the slope of the beach is constant and varies in the x-direction alone, as illustrated in Figure 6.1. Explicitly, the depth of the fluid is given by $h(x) = s_0 x$ for $x < 0$, in which $s_0 > 0$ is the slope. We further assume that the background flow is stationary and so, for example, we neglect the presence of along-shore currents.

Suppose at time $t = 0$ a quasi-monochromatic wavepacket is situated a distance $x_0 = -L$ from the beach and has dominant wavenumber (k_{x0}, k_{y0}) with both wavenumber components being positive. We wish to determine the path followed by the waves and to find how the wavenumber vector changes as the waves follow this path.

In general, the position in time is found by solving the coupled set of differential equations (6.15) and (6.16). Because the background flow is stationary the implicit frequency equals the explicit frequency. Thus, from the generalization of (2.17) to two horizontal dimensions, the dispersion relation is

$$\omega = \sqrt{gk_h \tanh[k_h h(x)]}, \tag{6.20}$$

in which $k_h = (k_x{}^2 + k_y{}^2)^{1/2}$.

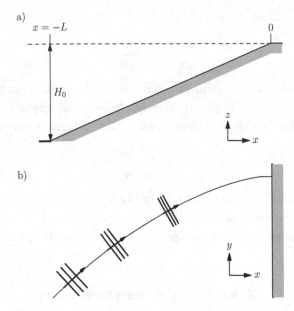

Fig. 6.1. a) Schematic of a one-layer fluid, bounded by a sloping bottom floor increasing in the *x*-direction. b) Top view showing the path (thin line) taken by a wavepacket approaching the beach and the corresponding change of the wave-number vector (thick parallel lines). Note that the crests become more parallel to the shore as the wavepacket approaches the beach.

Because $\omega = \omega_0$ is independent of time, the frequency does not change as it approaches the beach. That is,

$$\omega_0 \equiv \sqrt{gk_{h0}\tanh[k_{h0}(H_0)]}, \tag{6.21}$$

in which $k_{h0} = (k_{x0}^2 + k_{y0}^2)^{1/2}$ is the initial magnitude of the wavenumber vector and we have evaluated $H_0 \equiv |h(x_0)| = s_0 L$.

Because ω is independent of y, the second equation of (6.16) predicts that $k_y = k_{y0}$ is a constant following the motion of the wavepacket. On the other hand, k_x varies with the distance $|x|$ from the beach implicitly according to

$$\omega_0 = \sqrt{g\sqrt{k_x^2 + k_{y0}^2}\tanh\left[\sqrt{k_x^2 + k_{y0}^2}s_0|x|\right]}. \tag{6.22}$$

Qualitatively, we may interpret the behaviour of the wavepacket as follows. Consider the wavepacket approaching a beach, as shown in Figure 6.1b. The crests on the shoreward side of the path are in more shallow water and so move more slowly than the waves further out from the shore. Because the dispersion relation (6.20) is a monotonically increasing function of k_h, k_x must increase as $|x|$ decreases

Fig. 6.2. Aerial photograph of waves approaching a shoreline. [Taken near Kiberg, Norway on 12 June 1976 by Fjellanger Widerøe.]

in order for ω_0 in (6.22) to remain constant. This means that the wavenumber vector rotates to point towards the beach: independent of the orientation of the waves far out at sea, the wave crests become aligned with the shoreline during their approach. This behaviour is commonly observed, as shown for example in Figure 6.2.

The wavepacket's position is determined solely by the group velocity through (6.15):

$$\frac{dx}{dt} = c_{gx} \quad \text{and} \quad \frac{dy}{dt} = c_{gy}, \tag{6.23}$$

in which c_{gx} and c_{gy} are given by the k_x and k_y derivatives, respectively, of (6.20). Because the fluid depth varies with x, the components of the group velocity are themselves functions of x.

Let us now assume further that the waves are long compared with the depth so that the dispersion relation simplifies to that for shallow water waves. Thus k_x is represented implicitly in terms of x through the dispersion relation

$$\omega_0 = [(k_x^2 + k_{y0}^2)gs_0|x|]^{1/2} = [(k_{x0}^2 + k_{y0}^2)gH_0]^{1/2}. \tag{6.24}$$

Using (6.19) and evaluating the components of the group velocity for shallow water waves, the path followed by a wavepacket initially at $(x_0, y_0) = (-L, 0)$ is given by

$$y(x) = y_0 + \int_{x_0}^{x} \frac{k_{y0}}{k_x} d\tilde{x} = \int_{-L}^{x} \left[\frac{\omega_0^2}{gs_0|\tilde{x}|k_{y0}^2} - 1 \right]^{-1/2} d\tilde{x}. \tag{6.25}$$

The solution of this integral is shown as the path followed by the wavepacket in Figure 6.1b.

When interpreting these results, the approximations of ray theory must always be kept in mind. We have also assumed that the slope does not rise so quickly that the depth varies by a substantial fraction of itself over a distance k_x^{-1}. The predictions of ray theory become less valid as the change in depth over the extent of the waves, measured by $s_0|k_x|^{-1}$, becomes comparable to the local depth, $s_0|x|$, at a distance $|x|$ from the shore.

Furthermore, ray theory assumes the waves are small amplitude, so the behaviour of growing and breaking waves is not accounted for. The amplitude of the waves is set by the condition that the energy flux does not diverge. Using (2.38) and (2.41) and assuming the incident waves do not spread substantially in the y-direction as they approach the shoreline, their amplitude A is given by the condition

$$c_{gx}A^2 = \text{constant.} \tag{6.26}$$

Because the horizontal group velocity decreases as the waves move into shallower water, the amplitude must increase. Meanwhile, their horizontal wavelength decreases and so ray theory can be applied provided $k_xA \ll 1$. Where this condition breaks down can be determined from the initial conditions.

When the waves break they deposit momentum, some of which is absorbed by the beach but some of which can result in cross-shore currents and rip-tides. In the analyses above, we have neglected the presence of such currents which would additionally modify the propagation and structure of the waves.

6.4.2 *Interfacial waves approaching a slope*

Next we consider the propagation of interfacial waves in a two-layer fluid. As we did for surface waves, we suppose the waves move over a constant slope approaching a beach as illustrated in Figure 6.3a. The upper-layer fluid has density ρ_1 and a constant depth H_1 far from shore. The lower-layer fluid has density ρ_2 and the depth decreases as $h_2(x) = s_0|x| - H_1$, in which $s_0 > 0$ is the bottom slope. The interface intersects the slope where $x = -H_1/s_0$.

We assume an interfacial wave is centred initially about a position $(x_0, y_0) = (-L, 0)$ where the lower-layer fluid depth is $H_2 \equiv s_0 L$. Here its horizontal wavenumber is $\vec{k}_h = (k_{x0}, k_{y0})$. Because the ambient fluid is stationary and the topography is independent of y it follows that $\omega = \omega_0$ and $k_y = k_{y0}$ are constant following the motion of the waves. However, k_x is not constant. The implicit relationship between k_x and x is found from the dispersion relation (6.5) with H_2

replaced by $h_2(x)$:

$$\omega^2 = \frac{g'k_h}{\coth k_h H_1 + \coth k_h h_2(x)},\tag{6.27}$$

in which $k_h = (k_x^2 + k_{y0}^2)^{1/2}$. Because g' and H_1 are constant, k_x must change as h_2 varies with x in order to ensure that the frequency $\omega = \omega_0$ on the right-hand side of (6.27) remains constant.

The ambient is stationary, and so the trajectory of the waves (6.15) is given in terms of the group velocity, $dx/dt = c_{gx}$ and $dy/dt = c_{gy}$, where c_{gx} and c_{gy} are the k_x and k_y derivatives, respectively, of ω given by (6.27). Forming the ratio of these equations as in (6.18) gives the equation for the path of the waves independent of their position along the path in time.

For mathematical simplicity, suppose the waves are long ($k_h H_1 \ll 1$ and $k_h h_2 \ll 1$) so that the shallow water approximation applies. The dispersion relation is therefore given by

$$\omega = \sqrt{g'\bar{h}}\, k_h \quad \text{with} \quad \bar{h}(x) = \frac{H_1 h_2}{H_1 + h_2}.\tag{6.28}$$

The path of the interfacial wavepacket, computed as in (6.25) but with $gh = gs|x|$ replaced by $g'\bar{h}$, is illustrated in Figure 6.3b. Far offshore where $h_2(x) \gg H_1, \bar{h} \simeq H_1$,

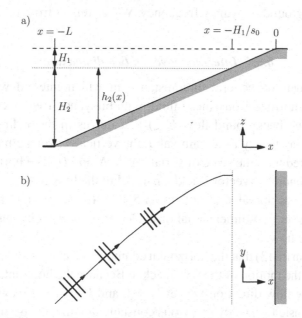

Fig. 6.3. a) Schematic of a two-layer fluid, bounded by a sloping bottom floor increasing in the x-direction. b) The path taken by interfacial waves approaching the beach.

the waves do not veer significantly towards the shore. Only as the lower-layer depth becomes comparable to H_1 do the waves feel the influence of the bottom slope. At this point, just as in the case for surface waves, the crests of the interfacial waves rotate so as to become more parallel to the beach. Of course, the waves never reach the shore. In reality they would grow in amplitude and develop into nonlinear waves. These would eventually break, partially reflect or would otherwise be affected by near-slope currents where the interface intersects the bottom slope.

6.5 Ray theory for internal waves

Here we consider the vertical and horizontal propagation of internal waves in non-uniformly stratified fluid in which we also include the effects of vertically varying background winds. For mathematical simplicity, we focus upon Boussinesq internal waves and we ignore the effects of background rotation.

We begin by looking at internal waves restricted to the x–z plane, in which case the ray theory equations can be reduced to a simple integral expression analogous to (6.19). Two cases arise that are of particular interest. In one, internal waves asymptotically approach a vertical level where their extrinsic frequency is zero or, equivalently, where the horizontal crest speed of the waves matches the speed of the background flow. This is known as a critical level. In the other case, the background wind and stratification are prescribed so that at some height the extrinsic frequency equals the background buoyancy frequency. Waves reflect from this level.

6.5.1 Internal waves in two dimensions

Here we consider the general circumstance in which internal waves move in non-uniformly stratified Boussinesq fluid, with buoyancy frequency $N(z)$, and in vertically varying background flow $\bar{U}(z)$ that moves in the x-direction. For the assumptions of ray theory to remain valid, the vertical wavelength of waves must be short compared with the scale of variations of N and $\bar{U}$. The horizontal component of the wavenumber vector is taken parallel to the flow in the x-direction. This circumstance is not unrealistic: in Section 5.4 we found that internal waves generated by flow over two-dimensional obstacles have their horizontal wavenumber aligned with the flow.

It follows from (6.13) that the horizontal components of the wavenumber are constant following the motion of the wavepacket. Because the horizontal wavenumber is parallel to the flow direction we set $k_y = 0$ and $k_x = k_{x0}$. Likewise, for steady motion the intrinsic frequency $\omega = \omega_0$ is constant although the extrinsic frequency $\Omega = \omega_0 - k_{x0}\bar{U}$ can change following the wave motion because the background flow Doppler-shifts the waves.

From (6.12), the x- and z-positions of the wavepacket vary in time according to

$$\frac{dx}{dt} = c_{gx} + \bar{U} \tag{6.29}$$

and

$$\frac{dz}{dt} = c_{gz}. \tag{6.30}$$

Combining these equations, the path of the waves is given by solving the differential equation

$$\frac{dz}{dx} = \frac{c_{gz}}{c_{gx} + \bar{U}}, \tag{6.31}$$

in which the components of the group velocity are given by (3.62). Explicitly,

$$\frac{dz}{dx} = \frac{-N \sin \Theta \cos^2 \Theta}{N \sin^2 \Theta \cos \Theta + k_{x0} \bar{U}}, \tag{6.32}$$

in which, from (3.56), $\Theta(z) = \tan^{-1}(k_z/k_{x0})$ represents the angle formed between lines of constant phase and the vertical.

The solution to (6.32) is nontrivial in that N and $\bar{U}$ are functions of z, and Θ itself is a function of z through its dependence upon the vertically varying wavenumber k_z. Rearranging the dispersion relation (3.54), k_z is given explicitly in terms of N and $\bar{U}$ by

$$k_z(z) = -k_{x0}\sqrt{\frac{N^2}{(\omega_0 - k_{x0}\bar{U})^2} - 1}. \tag{6.33}$$

Here we have chosen the sign of the square root to correspond to upward- and rightward-propagating waves with $\Omega = \omega_0 - k_{x0}\bar{U} < N$. Thus $|k_z|$ decreases and the vertical wavelength increases as the waves move to heights where either the buoyancy frequency decreases or the extrinsic frequency Ω increases through Doppler-shifting by the background winds.

If $\bar{U} = 0$, the path of internal waves is given simply by

$$\frac{dz}{dx} = -\cot \Theta = -\frac{k_{x0}}{k_z}. \tag{6.34}$$

This predicts that the slope of the path increases to infinity as the extrinsic frequency of upward-propagating waves increases to N, in which case $|\Theta| \to 0$ and $|k_z| \to 0$.

In Section 6.4.1 we found that conservation of energy requires the amplitude of surface waves to increase as their horizontal group velocity decreases. A similar principle holds for internal waves. For waves propagating in a background shear flow, wave action, not energy, is conserved. Wave action, $\mathcal{A}$, is the ratio of energy to the extrinsic frequency, as defined by (3.94). For internal waves with vertical displacement amplitude A_ξ, (3.87) predicts the average energy per unit mass is $\langle E \rangle = (N A_\xi)^2 / 2$, which is independent of wavenumber and frequency. Assuming that the waves do not spread substantially in the horizontal as they move vertically (as is the case for horizontally periodic waves), the requirement that the wave action flux is non-divergent means that

$$c_{gz} \frac{N^2 A_\xi^2}{\Omega} = c_{gz} \frac{N^2 A_\xi^2}{\omega_0 - k_{x0}\bar{U}} = \text{constant}. \tag{6.35}$$

So the amplitude increases as N or c_{gz} decreases or as the extrinsic frequency Ω increases. Whether nonlinear effects become important so that ray theory predictions become unreliable is assessed by the breaking conditions described in Section 4.6.

Because the vertical group velocity goes to zero as $|\Theta|$ approaches either 0 or $\pi/2$ the question arises as to where the energy ends up. Next we examine these two circumstances in detail beginning with a study of waves approaching a critical level, where $|\Theta| \to \pi/2$, followed by a study of waves approaching a reflection level, where $|\Theta| \to 0$.

6.5.2 Critical levels

In continuously stratified fluid a critical level is the height at which the extrinsic frequency, Ω, of internal waves is zero. Equivalently, it is where the horizontal speed ω/k_x of wave crests measured by a stationary observer matches the ambient flow speed, $\bar{U}$.

There is a subtle difference between this definition of critical levels and that which arises in the study of shear flow instability. In the latter case, horizontally periodic perturbations are determined as stable or unstable modes having the same phase speed as a point in the background flow. Where this matching occurs is called a critical level. In stability theory, the waves originate about the critical level itself. Conversely, here we are concerned with internal waves whose properties are set independently of the background flow and which move towards, rather than being situated at, a critical level.

In a well-studied case, one assumes that $N = N_0$ is constant and the ambient flow $\bar{U}$ increases linearly with height as $\bar{U}(z) = s_0 z$, in which the constant shear s_0 is positive. The two-dimensional internal waves are situated initially at the origin with horizontal wavenumber $k_{x0} > 0$ (which does not change in time) and initial vertical wavenumber $k_{z0} < 0$. Thus the waves are set to move upwards and to the right.

The initial intrinsic frequency of the waves situated at $z = 0$ is $\omega_0 = N_0 \cos \Theta_0$, in which $\Theta_0 = \tan^{-1}(k_{z0}/k_{x0})$. Because the flow is steady, ω_0 is constant for all time. Likewise, the horizontal phase speed measured by a stationary observer, ω_0/k_{x0}, is constant for all time. Thus the height of the critical level z_c can be determined immediately from the solution of $\bar{U}(z_c) = \omega_0/k_{x0}$. Explicitly,

$$z_c = \frac{N_0}{s_0 k_{x0}} \cos \Theta_0. \tag{6.36}$$

The actual path of the wavepacket as it approaches a critical level is found by solving (6.32) with Θ given by (3.56) and (6.33). Explicitly, the initial value problem is

$$\frac{dx}{dz} = \tan|\Theta| + \frac{k_{x0}s_0 z}{N_0 \sin|\Theta| \cos^2|\Theta|}, \quad x(0) = 0$$

$$\text{with } |\Theta| = \tan^{-1}\left(\frac{N_0^2}{(\omega_0 - k_{x0}s_0 z)^2} - 1\right)^{1/2}. \tag{6.37}$$

This is solved numerically through straightforward integration of both sides of the differential equation with respect to z.

The result is shown in Figure 6.4 for waves moving at the fastest vertical group velocity in a flow with shear strength $s_0 = 0.01N_0$. The path shown in Figure 6.4a asymptotically approaches the critical level at $z_c \simeq 81.6k_{x0}^{-1}$. At early stages during the propagation of the waves, where $\bar{U} \simeq 0$, lines of constant phase are nearly tangent to the path. This is consistent with the fact that the group velocity is oriented perpendicular to the wavenumber vector. At later times, the Doppler-shifting background wind results in a steeper angle of the phase lines compared with the slope of the path. In this calculation, after 200 buoyancy periods (in which one buoyancy period is $T_B = 2\pi/N_0$) the waves have travelled 91% of the vertical distance towards the critical level.

As the waves approach a critical level, (6.35) predicts that the amplitude will change in proportion to $[(\omega_0 - k_{x0}s_0 z)/c_{gz}]^{1/2}$. Near the critical level the waves become increasingly hydrostatic so that $c_{gz} \simeq (N_0/k_{x0}) \cot^2|\Theta|$. So, using the overturning condition (4.117), the level where internal waves break can be estimated numerically by finding the value of z where

$$\cot|\Theta(z)| = C(\omega_0 - k_{x0}s_0 z)^{1/4}. \tag{6.38}$$

Here C is a constant with respect to z that depends upon the initial wavenumber, frequency and amplitude of the waves.

This prediction does not guarantee the waves will break in reality: over sufficiently long distances the ambient flow may veer with height, in which case the

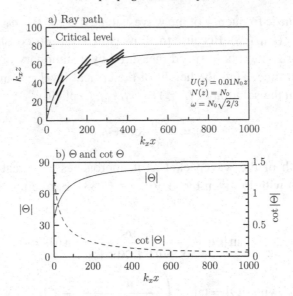

Fig. 6.4. a) The path followed by internal waves approaching a critical level in uniformly stratified fluid and in uniform shear of strength $s_0 = 0.01N_0$. The intrinsic wave frequency is set to be $\omega_0 = (2/3)^{1/2}N_0$, corresponding to waves with the fastest vertical group velocity in stationary fluid. The orientations of constant-phase lines are superimposed on the path at three positions. b) The angle Θ formed between phase lines and the vertical (solid line) and values of $\cot\Theta$ (dashed line) as the waves move along the path. The value of $\cot\Theta$ can be used to assess the stability of the waves to overturning depending upon their amplitude.

assumption of two-dimensional flow is no longer valid; over sufficiently long times the flow may no longer be steady; as the waves grow to large amplitude, weakly nonlinear effects will change their structure and may lead to instability through wave–wave interactions; if the fluid is sufficiently viscous, as may occur in laboratory experiments, the waves may broaden due to diffusion and so deposit energy to the mean flow without overturning and mixing.

The approach of internal waves to a critical level has been observed in several laboratory experiments. For example, Figure 6.5 shows internal waves launched by stratified flow over model topography (see Section 5.4) which then encounter a level in the flow above which the mean flow speed is zero, the same as that of the stationary hills. Consistent with the prediction of ray theory, the upward-propagating waves evolve to have decreasing vertical wavelength as they approach the critical level.

In Figure 6.5a the isopycnal surfaces, indicated by the dashed lines, become gradually less distorted as the waves approach the critical level. The incident waves have sufficiently small amplitude that viscosity damps the waves before they

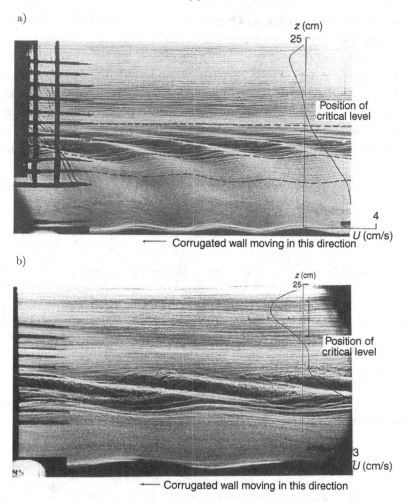

Fig. 6.5. Shadowgraph images taken from experiments in which a) small-amplitude and b) large-amplitude internal waves are launched by flow over a leftward-moving set of sinusoidal hills in a stratified shear flow. The waves approach a critical level where the velocity profile, indicated, crosses the vertical line. The dashed lines illustrate the distortion of isopycnal surfaces at different vertical levels. [Adapted, by permission of Cambridge University Press, from Figures 5 and 7 of Koop and McGee, J. Fluid Mech., **172**, 453–480 (1986).]

become overturning. In comparison, the large-amplitude incident waves shown in Figure 6.5b break turbulently below the critical level.

The second experiment demonstrates an important aspect of critical-level interactions. When waves break, they deposit momentum to the mean flow and this changes the background flow profile. Indeed, the velocity profile shown to the right of Figure 6.5b exhibits a nearly constant velocity between the critical level and the

height at which the waves break. Over time the level at which the waves break
occurs at progressively smaller heights above the source of the waves.

This interaction between waves and the mean flow has been used to explain the
essential dynamics governing the Quasi-Biennial Oscillation (or, more succinctly,
the 'QBO'). This refers to the observed zonal winds in the equatorial stratosphere
that alternately flow eastwards and westwards with a period of about two years, as
shown in Figure 6.6.

It is believed that the flow is driven by upward-propagating waves originating in
the troposphere and which deposit momentum where they break in the stratosphere,
as illustrated in Figure 6.7. Incident waves have both eastward and westward phase

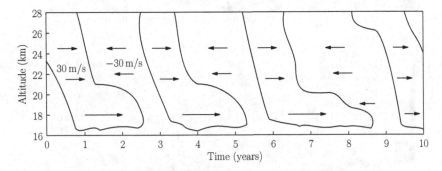

Fig. 6.6. Schematic of the alternating westward and eastward winds in the lower
equatorial stratosphere associated with the Quasi-Biennial Oscillation.

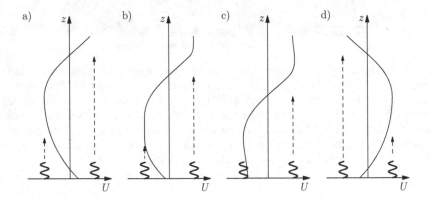

Fig. 6.7. Schematic illustrating how the absorption of internal waves at critical
levels results in the alternating westward and eastward zonal flows associated with
the Quasi-Biennial Oscillation. The solid line in each plot represents the zonal wind
profile and the left and right sinusoidal curves represent upward-propagating waves
respectively with westward and eastward zonal phase speeds. a) Westward waves
deposit momentum at a critical level which b) reduces the altitude of the level until
c) the critical level reaches the level of the source. d) The process then repeats for
eastward-propagating waves.

speeds. During the phase of the QBO when the stratospheric winds are westward, the westward-propagating waves encounter a critical level and, by depositing their momentum, they lower the altitude of the critical level. The eastward-propagating waves do not encounter a critical level and so propagate high up into the stratosphere where they dissipate or break due to other processes and so accelerate an upper-level eastward flow.

Eventually the critical level for westward waves is so low in the stratosphere that it reaches the source of the waves. Afterwards these waves can pass freely upwards through the stratosphere. Meanwhile the eastward-propagating waves encounter a critical level high in the stratosphere and, through momentum deposition, progressively lower the altitude at which this critical level is situated. The process repeats itself so that the wind in the lower stratosphere alternately flows westwards, then eastwards, then westwards again.

6.5.3 Reflection levels

Ray theory predicts that internal waves reflect from a vertical level where the extrinsic frequency matches the background buoyancy frequency. This is called a reflection level. There are two distinct idealized circumstances in which this may occur, both of which are considered here for upward-propagating incident waves. In one, the ambient flow is stationary and the stratification decreases with height. In the other, the stratification is uniform and the ambient flow decreases with height, Doppler-shifting the waves to higher extrinsic frequencies.

First we consider rightward- and upward-propagating waves in decreasing stratification with $N(z) = N_0(1 - \sigma_0 z)$, in which σ_0 is a constant. The background flow is taken to be stationary so that the intrinsic and extrinsic frequencies are equal. For waves with intrinsic frequency ω_0, the reflection level occurs where

$$z = z_r = \frac{1 - \omega_0/N_0}{\sigma_0}. \tag{6.39}$$

The path followed by these waves is found by solving (6.32) with $\bar{U} = 0$. The solution is shown in Figure 6.8a for a case with $\sigma_0 = 0.003k_{x0}$ and $\omega_0 = (2/3)^{1/2}N_0$.

As the waves approach the reflection level, the slope of the path increases with height and becomes infinite at $z = z_r$. The waves then reflect, still moving rightwards but now moving downwards along a path whose slope is negative and decreasing in magnitude.

Because there is no background flow the slope is given by (6.34): $dz/dx = -\cot\Theta$, in which Θ is the angle formed between constant-phase lines and the

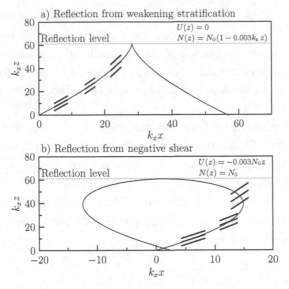

Fig. 6.8. The path followed by internal waves approaching a reflection level in a) stationary fluid whose stratification decreases according to $N = N_0(1 - 0.003\,k_xz)$ and b) uniformly stratified fluid in constant negative shear with $s_0 = -0.003\,N_0$. In both cases the initial intrinsic frequency is taken to be $\omega_0 = N_0\sqrt{2/3}$, corresponding to internal waves that move upwards at the fastest vertical group velocity in stationary fluid. In both plots, the orientations of constant-phase lines are superimposed on the path at three positions.

vertical and Θ is negative (positive) for upward- (downward-) propagating waves. Thus the slope of the phase lines matches the slope of the ray path everywhere along the path. In particular, at the reflection level phase lines are vertically oriented and the corresponding vertical wavelength is infinite.

In the second example, we consider the circumstance in which a uniformly stratified fluid has uniform but negative shear $\bar{U}(z) = -s_0 z$ with $s_0 > 0$. In this case, waves with intrinsic frequency ω_0 and horizontal wavenumber $k_{x0} > 0$ that move upwards from $z = 0$ are Doppler-shifted to increasing extrinsic frequencies $\Omega = \omega - k_x\bar{U} = \omega_0 + k_{x0}s_0z$. Eventually, the waves reach a level z_r at which the extrinsic frequency equals the buoyancy frequency. Explicitly,

$$z_r = \frac{N_0 - \omega_0}{k_{x0}\,s_0}. \tag{6.40}$$

The path followed by an internal wavepacket reflecting from a negative shear flow is shown in Figure 6.8b. As expected, the wavepacket moves upwards from the origin and reflects from the reflection level predicted by (6.40). Unlike the previous case, however, the waves follow a counter-clockwise path and approach the reflection level tangentially rather than at a cusp. During the motion the phase

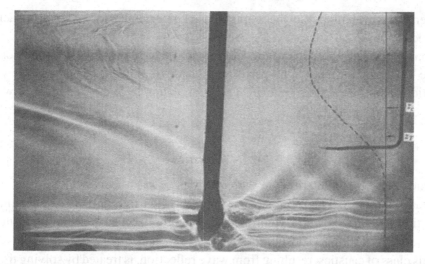

Fig. 6.9. Shadowgraph images taken from experiments in which internal waves generated by an oscillating cylinder move upwards into a leftward-moving shear flow. The leftward-propagating waves encounter a critical level; the rightward-propagating waves encounter a reflection level. [Reproduced, by permission of Cambridge University Press, from Figure 10 of Koop, J. Fluid Mech., **113**, 347–386 (1981).]

lines are not parallel with the path except at $z = 0$ where $\bar{U} = 0$. As in the previous example, the phase lines become more vertically oriented as the waves approach the reflection level, meaning that the vertical wavelength becomes infinite.

These two examples show that the energy and momentum transported by internal waves are reflected back towards the source of the waves if the waves encounter a level where their extrinsic frequency equals the background buoyancy frequency. The different behaviour of waves at a critical and reflection level is beautifully illustrated by the laboratory experiment shown in Figure 6.9. Here internal waves are generated in a uniformly stratified fluid by a vertically oscillating cylinder (see Section 5.2) and the left and right beams propagate upwards into a leftward-moving shear flow. The leftward-propagating waves encounter a critical level, their phase lines tilting towards the horizontal as they move upwards. The rightward-propagating waves encounter a reflection level resulting from their extrinsic frequency being Doppler-shifted to match the background buoyancy frequency. The shear is so strong in this case, that the looped path shown in Figure 6.8b occurs within a short vertical distance from the reflection level and so appears to be more cusp-like, as in Figure 6.8a.

In this section we have noted that as internal waves approach a reflection level their vertical wavelength becomes infinitely large. This poses a problem for ray theory, which is valid only in the limit of background variations being long compared to the vertical wavelength. Ray theory predicts its own demise for waves approaching a reflection level.

Nonetheless it is possible to model the behaviour of waves within the context of the WKB approximation by performing an asymptotic expansion of the wave equations about the reflection level. This is an example of the treatment of waves near so-called caustics.

6.5.4 Caustics

Caustics refer to singularities in the equations of ray theory which result when ray paths intersect. This occurs, for example, when internal waves encounter a reflection level as shown in Figure 6.8: the meeting at a cusp of the incident and reflected waves forms a caustic.

This class of caustics, resulting from wave reflection, is treated by solving the linearized equations of motion in a neighbourhood about the reflection level. The solutions can then be spliced together with the ray theory prediction far from the caustic to give an approximate solution for the evolution of incident and reflected waves.

To demonstrate the treatment of such caustics, we consider the evolution of incident internal waves in the lower-half plane that propagate upwards in a non-uniform shear flow with non-uniform stratification and which encounter a reflection level at $z_r = 0$. Suppose that $N(z)$ and $\bar{U}(z)$ vary continuously about $z = 0$, so that near this level we can approximate the background buoyancy frequency and horizontal velocity by linearized functions

$$N(z) = N_0(1 - \sigma_0 z) \text{ and } \bar{U}(z) = U_0 - s_0 z. \qquad (6.41)$$

We assume that $\omega_0 > 0$ and $k_{x0} > 0$ and that the constants σ_0, s_0 and U_0 are positive so that the stratification weakens with height and the background shear Doppler-shifts upward-propagating waves to higher extrinsic frequencies $\Omega(z)$. For $z = 0$ to be a reflection level, we must have

$$\Omega(z = 0) = \omega_0 - U_0 k_{x0} = N_0. \qquad (6.42)$$

The vertical structure of small-amplitude horizontally periodic disturbances in a parallel shear flow is prescribed by the Taylor–Goldstein equation (3.134). For the profiles defined by (6.41), this becomes

$$\frac{d^2\hat{\xi}}{dz^2} + k_{x0}{}^2 \left\{ \frac{[N_0(1 - \sigma_0 z)]2}{[N_0 + s_0 k_{x0} z]^2} - 1 \right\} \hat{\xi} = 0, \qquad (6.43)$$

in which we have used the reflection level condition (6.42). For conceptual convenience, the equation has been recast as a formula for the vertical displacement amplitude, $\hat{\xi}(z)$, instead of the streamfunction amplitude, $\hat{\psi}(z)$.

In (6.43) the term in curly braces is zero at $z = 0$, and the signs of s_0 and σ_0 have been chosen so that the term is positive if $z < 0$ and negative if $z > 0$. Thus solutions have an oscillatory form in the lower-half plane, consistent with propagating waves, and have a monotonically decreasing form in the upper-half plane, consistent with the structure of evanescent disturbances.

We can examine the detailed structure near the reflection level by performing a Taylor-series expansion about $z = 0$ of the term in curly braces in (6.43) and keeping only the leading-order term in z. This yields the approximate differential equation

$$\frac{d^2\hat{\xi}}{dz^2} - 2k_{x0}{}^2\sigma_0 \left(1 + \frac{s_0 k_{x0}}{N_0\sigma_0}\right) z\,\hat{\xi} \simeq 0. \tag{6.44}$$

Through a straightforward change of variables this can be converted into the canonical form of Airy's equation for the function $\xi(Z)$:

$$\xi'' + Z\xi = 0, \tag{6.45}$$

in which $Z = -\{2k_{x0}{}^2\sigma_0[1 + (s_0 k_{x0})/(N_0\sigma_0)]\}^{1/3}z$.

Generally, the solution of (6.45) is a superposition of the Airy functions Ai and Bi. However, the latter function is unbounded as $Z \to \infty$ and so is neglected. The plot of Ai(Z) is shown as the solid line in Figure 6.10a. The dashed lines represent asymptotic approximations to the Airy function:

$$\mathrm{Ai}(Z) \simeq \begin{cases} \dfrac{1}{2\sqrt{\pi}Z^{1/4}} \exp\left(-\dfrac{2}{3}Z^{3/2}\right) & Z \gg 0, \\[2ex] \dfrac{1}{\sqrt{\pi}(-Z)^{1/4}} \sin\left(\dfrac{2}{3}(-Z)^{3/2} + \dfrac{\pi}{4}\right) & Z \ll 0. \end{cases} \tag{6.46}$$

Below $z = 0$ the vertical wavelength and amplitude of the waves increases as Z increases. However, the full treatment of the Airy function shows that the vertical wavelength and amplitude do not approach infinity as $Z \to 0$. Above the reflection level the amplitude decreases exponentially as $\exp(-(2/3)Z^{3/2})/Z^{1/4}$.

The structure of the wavefield at a snapshot in time is shown in Figure 6.10b. Explicitly, the greyscale indicates values of Ai(Z) $\cos(k_{x0}x)$ with light greys indicating positive values up to 0.5 and dark greys indicating negative values as low as -0.5. As a consequence of the superposition of the incident and reflected waves,

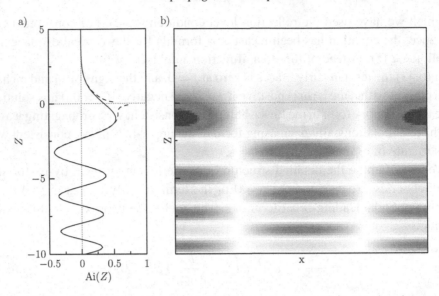

Fig. 6.10. a) Plot of Airy function (solid line) and asymptotic approximations to this function (dashed lines) for large and small Z as given by (6.46) and b) the structure of internal waves incident from below upon a reflection level where the buoyancy frequency decreases linearly with height. The greyscale indicates crests (light grey) and troughs (dark grey) of the vertical displacement field.

the disturbance field adopts a checkerboard pattern below the reflection level. The associated flux of wave action is everywhere zero, with no energy transport by the evanescent waves above $Z = 0$ and with the upward flux by incident waves cancelled by the downward flux by reflected waves below $Z = 0$.

6.6 Eckart resonance and tunnelling

Both the ocean and atmosphere are characterized by layers of strong and weak stratification. In the ocean, for example, the seasonal and main thermoclines are separated by a relatively weakly stratified region. In some circumstances internal waves can be generated in the seasonal thermocline with frequency less than the local buoyancy frequency in the region but greater than the buoyancy frequency immediately underneath. Such waves are said to be ducted, meaning that they are trapped within the stratified layer. However, depending upon their spatial scale it may be possible for the waves to transmit energy through to the main thermocline. In this sense, the seasonal thermocline acts as a leaky duct and, more generally, the waves are said to tunnel from one strongly stratified region to another through a region in which they are evanescent.

There are two approaches through which this situation can be modelled. In one, a result originally derived by Eckart in 1961, an eigenvalue problem is formulated to determine the dispersion relation of waves in a system of two strongly stratified layers separated by a weakly stratified layer. Tunnelling occurs for pairs of modes that have comparable frequency and wavenumber. In the other, a plane wave with prescribed wavenumber and frequency is assumed to be incident upon a weakly stratified layer with a strongly stratified layer beyond. This is not an eigenvalue problem but one for which transmission and reflection coefficients may be computed.

6.6.1 Eckart resonance

Here we consider the phenomenon of resonant modes periodically transferring energy between two ducts of depth H_1 and H_2 having buoyancy frequency N_1 and N_2, respectively. Explicitly, as a crude model of the seasonal and main oceanic thermoclines, we consider the wave-modes in a stationary fluid with piecewise-constant stratification given by

$$N^2(z) = \begin{cases} N_1{}^2 & -H_1 \leq z \leq 0 \\ 0 & -H_1 - D < z < -H_1 \\ N_2{}^2 & -H \leq z \leq -H_1 - D. \end{cases} \tag{6.47}$$

The depth of the lower duct is $H_2 = H - H_1 - D$ and a uniform-density region of depth D separates the two ducts. For simplicity, we assume the vertical motion is restricted at the upper and lower boundaries so that the streamfunction amplitude $\hat{\psi}(z)$ is zero at $z = 0$ and $z = -H$.

The dispersion relation and structure of internal waves with given horizontal wavenumber k_x may be found following the procedure described in Section 3.6.4. In particular, the Taylor–Goldstein equation predicts that the streamfunction amplitude in each layer has the form

$$\hat{\psi}(z) = \begin{cases} \mathcal{A}\sin(\gamma_1 z) & -H_1 \geq z \geq 0 \\ \mathcal{B}_1 e^{k_x z} + \mathcal{B}_2 e^{-k_x z} & -H_1 - D > z > -H_1 \\ \mathcal{C}\sin(\gamma_2(z+H)) & -H \geq z \geq -H_1 - D, \end{cases} \tag{6.48}$$

in which $\gamma_i = k_x[(N_i/\omega)^2 - 1]^{1/2}$ for $i = 1, 2$. If $\omega > N_i$ then γ_i is pure imaginary and the respective sine functions in (6.48) are replaced with $\imath$ times hyperbolic sine functions.

The interface conditions (2.134) and (2.135) in this stationary, Boussinesq fluid reduce to the requirement that $\hat{\psi}$ and its derivative are continuous at $z = -H_1$ and $z = -H_1 - D$. Thus we arrive at an eigenvalue problem with four equations in the four unknowns $\mathcal{A}$, $\mathcal{B}_1$, $\mathcal{B}_2$ and C. The resulting dispersion relation is

$$\tanh(k_x D)[1 + \mathcal{T}_1 \mathcal{T}_2] + \mathcal{T}_1 + \mathcal{T}_2 = 0, \tag{6.49}$$

in which

$$\mathcal{T}_i = \frac{k_x}{\gamma_i} \tanh(\gamma_i H_i), \quad i = 1, 2. \tag{6.50}$$

In the limit where the ducts are widely separated, $k_x D \gg 1$, the dispersion relation reduces to the pair of equations $k_x \tanh(\gamma_i H_i) = -\gamma_i$ for $i = 1, 2$. This is the dispersion relation for varicose waves trapped in a single duct as given by (3.155). Those waves had zero displacement at the centre of a duct of depth $2H$, which is consistent with our requirement that $\hat{\psi}$ is zero at the top and bottom of the ducts of depth H_1 and H_2, respectively.

As in the case of trapped waves in a single duct, the stratification given by (6.47) can support an infinite but countable number of modes, with the lowest modes having the broadest vertical structure. Figure 6.11 shows the dispersion relation and structure of the lowest two modes of this system in the particular case where $N_2 = N_1/2$, $H_2 = 4H_1$ and $D = 2H_1$.

Right away we see an interesting phenomenon. For long waves ($k_x H_1 \ll 1$), the streamfunction amplitude of the lowest mode (solid line) is largest within the lower duct whereas, tracking this mode to short waves ($k_x H_1 \gg 1$), it is largest within the upper duct. Conversely, the next lowest mode is largest in the upper duct for long waves and largest in the lower duct for short waves. The dispersion relations for the two modes appear to cross over between one for waves in the lower duct to one for waves in the upper duct. For the two lowest modes, this occurs in this example if their horizontal wavenumber satisfies $k_x H_1 \simeq 1$.

The phenomenon known as 'Eckart resonance' occurs as a competition between wave modes having comparable horizontal wavenumber and frequency. Consider, for example, the superposition of the two modes shown in Figure 6.11d with wavenumber $k_x = H_1^{-1}$ and frequencies $\omega_1 \simeq 0.46 N_1^{-1}$ and $\omega_2 \simeq 0.42 N_1^{-1}$. For simplicity, we choose co-ordinates so that the two modes are in phase. Hence the streamfunction resulting from their superposition is

$$\psi(x, z, t) = \hat{\psi}_1(z) \cos(k_x x - \omega_1 t) + \hat{\psi}_2(z) \cos(k_x x - \omega_2 t) \tag{6.51}$$

$$= \hat{\psi}_+(z) \cos(k_x x - \bar{\omega} t) \cos(\Delta \omega t) - \hat{\psi}_-(z) \sin(k_x x - \bar{\omega} t) \sin(\Delta \omega t),$$

in which $\bar{\omega} \equiv (\omega_1 + \omega_2)/2$, $\Delta \omega = (\omega_1 - \omega_2)/2$, $\hat{\psi}_+ = \hat{\psi}_1 + \hat{\psi}_2$ and $\hat{\psi}_- = \hat{\psi}_1 - \hat{\psi}_2$.

The superimposed streamfunction amplitude $\hat{\psi}_+$ has large (negative) value in the upper duct and relatively small value in the lower duct whereas the converse is true

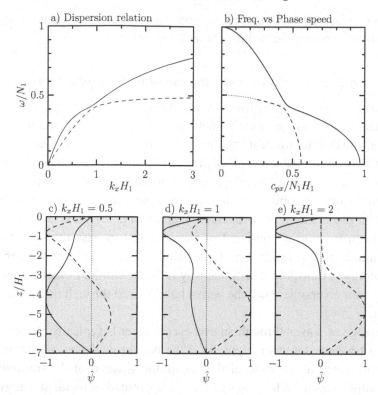

Fig. 6.11. a) The dispersion relation of the lowest (solid line) and next lowest (dashed line) modes in the stratified fluid given by (6.47) with $N_2 = N_1/2$, $H_2 = 4H_1$ and $D = 2H_1$. b) The corresponding relationship between the frequency and horizontal phase speed $c_{px} \equiv \omega/k_x$. The dotted line indicates the continuation of the second lowest mode for small c_{px} (high k_x). In reality this curve deflects to larger ω for k_x due to interactions with the third lowest mode. The streamfunction amplitude of the modes is shown for c) $k_x H_1 = 0.5$, d) $k_x H_1 = 1$ and e) $k_x H_1 = 2$. These are normalized so that $||\hat{\psi}(z)|| = 1$. The region of strong stratification in these three plots is indicated by the grey regions. In all the plots, the lowest mode is given by the solid line and the next lowest mode is given by the dashed line.

for $\hat{\psi}_-$. So at time $t = 0$ the mode is concentrated in the upper region with vertical structure given by $\hat{\psi}_+$. But after the long time $\pi/(2\Delta\omega)$ the mode is concentrated in the lower region with vertical structure given by $\hat{\psi}_-$.

This transference of energy from one strongly stratified region through a weakly stratified region to another strongly stratified region, is an example of tunnelling. The terminology brings to mind the problem of electron tunnelling across an electric potential barrier in quantum mechanics. But tunnelling in the sense of Eckart resonance is reversible. For example, in (6.52) we saw that energy shifted predominantly from one duct to another in a time $\pi/(2\Delta\omega)$. But after another half-period the energy concentrates once more in the upper duct. The energy continues to switch back and forth between the two modes with frequency $2\Delta\omega$.

6.6.2 *Transmission and reflection of incident plane waves*

Here we consider the one-way transference of energy that occurs in a manner more directly analogous with quantum tunnelling. We suppose the domain is vertically unbounded and that the ambient stratification is non-zero for large and small z. Thus internal waves can propagate vertically inwards from and outwards to infinity. The problem of internal wave tunnelling, analogous to quantum tunnelling, occurs if somewhere within the domain the stratification is so weak over a finite-depth region that incident plane waves are evanescent there. If this region is not so broad, it is possible for a fraction of the incident wave energy to pass through this barrier and transmit across the other side; the remaining fraction reflects from the barrier and the energy is transported back the way it came. Unlike Eckart resonance, the situation is not reversible: once the waves have passed through the barrier they do not return.

The passage of waves through an evanescent layer is typically characterized in terms of a transmission coefficient, T. If the background flow is stationary, this can be represented as the ratio of the magnitude-squared of the transmitted and incident amplitudes, which is the fraction of transmitted to incident energy flux by the waves. However, if the background flow at the level of the incident waves is different from that at the level of the transmitted waves, T must be defined more generally. In Section 3.4.3 we saw that the energy of waves moving in a shear flow is not conserved. Instead it is their 'wave action' or, more generally, pseudoenergy, that is conserved. The mean wave action $\langle \mathcal{A} \rangle$ of small-amplitude waves, given by (3.94), is the ratio of the energy to the extrinsic frequency, $\Omega = \omega - \bar{U}k_x$. The vertical flux of wave action is just the wave action times the vertical group velocity: $\langle \mathcal{F}_A \rangle = c_{gz} \langle \mathcal{A} \rangle$.

Thus it is physically meaningful to define the transmission coefficient for internal waves in a non-uniformly stratified shear flow to be the ratio of the transmitted to incident flux of wave action. Using the polarization relations for small amplitude, Boussinesq internal waves and the corresponding expression for c_{gz} given in terms of the wavenumber and frequency, the general formula for the transmission coefficient is

$$T = \frac{|\mathcal{A}_t|^2 \sqrt{\frac{N_t^2}{\Omega_t^2} - 1}}{|\mathcal{A}_i|^2 \sqrt{\frac{N_i^2}{\Omega_i^2} - 1}}, \tag{6.52}$$

in which $\mathcal{A}_i$, N_i, Ω_i and $\mathcal{A}_t$, N_t, Ω_t are respectively the vertical displacement amplitude, buoyancy frequency and extrinsic frequency at the level of the incident and transmitted waves. Because the reflected wave propagates in the same medium as the incident waves, the reflection coefficient is just

$$R = \frac{|\mathcal{A}_r|^2}{|\mathcal{A}_i|^2}, \qquad (6.53)$$

where $\mathcal{A}_r$ is the amplitude of the reflected waves. Unless the waves draw or deposit energy from the background shear near a critical level (a phenomenon known as 'over-reflection'), conservation of wave action requires $R + T = 1$.

In the special case in which the background flow is stationary, the extrinsic frequency above and below an evanescent layer does not change and so energy is conserved. Thus (6.52) reduces to the usual definition for the transmission coefficient:

$$T = \frac{|\mathcal{A}_t|^2}{|\mathcal{A}_i|^2}. \qquad (6.54)$$

This can be seen as the ratio of the energy of the transmitted waves relative to the energy of the incident waves. More physically, however, it represents the ratio of the outgoing to incoming energy fluxes.

For simplicity, we first consider the transmission and reflection of a small-amplitude plane internal wave in a Boussinesq fluid whose ambient stratification is prescribed by

$$N^2(z) = \begin{cases} N_0{}^2 & |z| > H/2 \\ 0 & |z| \le H/2. \end{cases} \qquad (6.55)$$

We assume the plane wave is incident from above with frequency $\omega > 0$ and horizontal wavenumber $k_x > 0$. Because the background is steady and independent of x, both ω and k_x are constant as the waves pass vertically through the non-uniformly stratified fluid. This is consistent with the ray theory prediction in Section 6.5.

From the Taylor–Goldstein equation (3.132) for stationary fluid, we know that the vertical structure of the streamfunction is given by

$$\hat{\psi}(z) = \begin{cases} \mathcal{A}_i e^{ik_z z} + \mathcal{A}_r e^{-ik_z z} & z > H/2 \\ \mathcal{B}_1 e^{k_x z} + \mathcal{B}_2 e^{-k_x z} & |z| < H/2 \\ \mathcal{A}_t e^{ik_z z} & z < -H/2. \end{cases} \qquad (6.56)$$

Here $k_z = k_x (N_0{}^2/\omega^2 - 1)^{1/2}$ is the vertical wavenumber in the strongly stratified region. We have defined it to be positive, consistent with the incident wave of amplitude $\mathcal{A}_i$ propagating downwards. Likewise, the transmitted and reflected waves of amplitudes $\mathcal{A}_t$ and $\mathcal{A}_r$ have the sign of the vertical wavenumber defined so that they propagate downwards and upwards, respectively.

Because there is no background flow and the background density is assumed to vary continuously, the interface conditions (2.134) and (2.135) require $\hat{\psi}$ and its derivative to be continuous at $z = \pm H$. Thus we have four equations relating the five unknown amplitudes. This is not an eigenvalue problem. Instead we reduce the equations so as to write four of the unknowns in terms of the fifth unknown. To this end, we assume that the incident wave amplitude is prescribed. Keeping in mind that the Taylor–Goldstein equation results from linear theory, it is understood that $|\mathcal{A}_i|$ is small, but its magnitude and phase are unrestricted otherwise.

Solving the four equations, we express $\mathcal{A}_t$ in terms of $\mathcal{A}_i$ alone. The result is then cast in terms of a transmission coefficient which, for stationary fluid, is given by (6.54). The calculation results in the following prediction:

$$
T = \left[1 + \left(\frac{\sinh k_x H}{\sin 2\Theta}\right)^2\right]^{-1}. \tag{6.57}
$$

Here $\Theta = \cos^{-1}(\omega/N_0)$ is a convenient measure of the relative incident wave frequency (3.57).

Separately, we could compute $\mathcal{A}_r$ in terms of $\mathcal{A}_i$ and so find the reflection coefficient defined by (6.53). The calculation reveals that $R = 1 - T$, consistent with energy conservation: what does not get transmitted, gets reflected.

Figure 6.12a shows values of the transmission coefficient as given by (6.57) computed for a range of incident horizontal wavenumbers and frequencies. As expected, transmission is perfect in the limit of an infinitesimally thin barrier ($k_x H \to 0$) and negligible transmission occurs if the barrier is very wide compared to the horizontal wavelength ($k_x H \to \infty$). Surprisingly, we see that even for a thin barrier, very little transmission occurs if $\omega \lesssim N_0$ or if $\omega \ll N_0$. At a fixed value of $k_x H$, maximum transmission occurs if $\omega = N_0/\sqrt{2}$, in which case $|\Theta| = 45°$.

A similar calculation can be performed to find the transmission coefficient for waves crossing a weakly stratified layer or crossing a layer of enhanced stratification. In this case we define N^2 by

$$
N^2(z) = \begin{cases} N_0{}^2 & |z| > H/2 \\ N_1{}^2 & |z| \leq H/2. \end{cases} \tag{6.58}
$$

The structure of the waves, again given by the Taylor–Goldstein equation, exhibits exponential behaviour for $|z| \leq H/2$ if $\omega < N_1$, and exhibits oscillatory behaviour

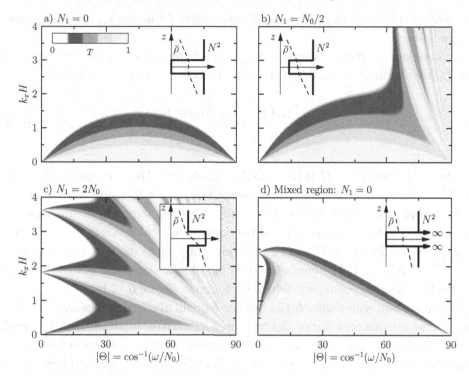

Fig. 6.12. Transmission coefficients for small-amplitude Boussinesq waves crossing a piecewise-constant stratification given by (6.58) with a) $N_1 = 0$, b) $N_1 = N_0/2$ and c) $N_1 = 2N_0$, as illustrated schematically by the solid line in the corner of each plot (the corresponding density profile is shown as the dashed line). In d) the transmission coefficient is plotted for waves crossing a localized well-mixed region whose density profile is given by (6.59). The greyscale contours for all plots are indicated in the upper left-hand corner of a).

in z otherwise. The resulting transmission coefficients are shown for $N_1 = N_0/2$ and $N_1 = 2N_0$ in Figures 6.12b and c, respectively.

In Figure 6.12b the transmission characteristics are qualitatively similar to those in Figure 6.12a if $\omega > N_1/2$ ($|\Theta| < 60°$), in which case the waves are evanescent for $|z| < H/2$. If $N_1 < \omega < N_0$, the waves are nowhere evanescent and one might expect perfect transmission even for large $k_x H$. But the figure shows that strong transmission occurs only for certain combinations of ω and k_x. These are combinations for which H is approximately a half-integer multiple of the vertical wavelengths within $|z| < H/2$. Perfect transmission requires waves to fill the cavity resonantly between the upper and lower strongly stratified regions.

Even if $N_1 > N_0$, in which case the incident waves are propagating at all depths, Figure 6.12c shows that perfect transmission does not always occur. Indeed, if $\omega \simeq N_0$ ($|\Theta| \simeq 0$), transmission occurs for a relatively small range of $k_x H$, corresponding to waves that have vertical wavelengths resonantly filling the region of enhanced stratification.

We consider one more example in detail here. Suppose that a uniformly stratified region is well mixed over a depth H so that its density profile is given by

$$
\bar{\rho}(z) = \begin{cases} \rho_0 \left(1 - \dfrac{z}{\mathcal{H}}\right) & |z| > H/2 \\ \rho_0 & |z| \le H/2, \end{cases} \tag{6.59}
$$

in which $\mathcal{H} = g/N_0^2 \gg H$ is the density scale height. The corresponding profile of N^2 is the same as that in (6.55) except that it is effectively infinite at $z = \pm H/2$ where the density jumps discontinuously. The structure of the waves is again given by (6.56). But because of the density discontinuities at $z = \pm H/2$ the interface conditions (2.134) and (2.135) now simplify to require continuity of $\hat{\psi}$ and $\hat{\psi}' - [g\bar{\rho}/(\varrho_0 c^2)]\hat{\psi}$. Nonetheless it is possible to derive an analytic solution for the transmission coefficient, which is plotted in Figure 6.12d.

In comparison with Figure 6.12a, we see a qualitatively new feature: there is a strong transmission peak that occurs for finite $k_x H$ even for small ω. This is the result of the incident internal waves resonantly exciting interfacial waves at the density jumps above and below the mixed region. When the wavenumber and frequencies of these two classes of waves match, the internal waves transmit perfectly across the mixed region.

The results of this section serve to caution those who follow the rule based upon ray theory that internal waves reflect from a region in which they are evanescent, and do not reflect if their extrinsic frequency is always smaller than N. The heuristic is valid if the vertical variations of N are long compared with the vertical wavelength of the waves, but it is not necessarily reliable otherwise.

6.7 Internal wave spectra

Throughout this book we have focused primarily upon plane waves or quasi-monochromatic wavepackets. But we have seen that oceanic and atmospheric waves are generated with a variety of spatial and temporal scales and that large-amplitude waves can become unstable or interact with each other in a way that transfers energy from one wavenumber to another, before ultimately breaking down into turbulence and dissipating.

Surprisingly, observations do not show a random distribution of internal waves. Instead, their average spectra, at least far from source regions, obey what appear to be universal distributions. In this section we will review the observations and empirical formulae that synthesize the data into conceptually convenient models.

6.7.1 Oceanic internal waves

In the open ocean, far from the continental shelves and islands, internal waves originate from a variety of sources, some of which were discussed in Chapter 5. Observations rarely point to cause-and-effect relationships linking the energetic sources to the resulting waves. Instead, measurements of temperature, velocity and shear reveal variations spanning a wide and continuous range of temporal and spatial scales.

In the open ocean, away from the surface and solid boundaries, the ensemble of waves is found to be statistically homogeneous in time and space. That is, if measurements are taken for long enough so that statistical averages and spectra are meaningful, the relationship between the amplitude of wave-induced disturbances and their frequency and wavenumber exhibits power-law behaviour.

For example, Figure 6.13 shows the displacement of isopycnals at the thermocline and their corresponding spectra. The displacement spectrum, determined from the square of the amplitudes of the Fourier transformed displacements, decreases with frequency as ω^{-2} for hydrostatic waves having frequency sufficiently smaller than the local buoyancy frequency.

Likewise the vertical structure follows an approximate power law, as shown in Figure 6.14. These data are gathered from dropped instruments that measure

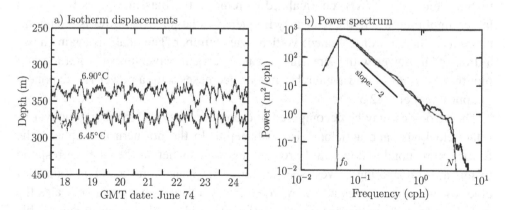

Fig. 6.13. a) Time series of isopycnal displacements about the thermocline and b) the corresponding averaged spectrum of the displacements determined from an isotherm at 6.60°C. Frequencies are plotted in cycles (not radians) per hour. The thin line in b) shows the values of a modified version of the Garrett–Munk 72 spectrum with cut-offs at the inertial frequency, f_0, and local buoyancy frequency, N. The thick line is drawn with a slope of -2. [Modified, with permission ©American Geophysical Union, from Figures 1 and 3 of Cairns & Williams, J. Geophys. Res., **81**, 1943–1950 (1976).]

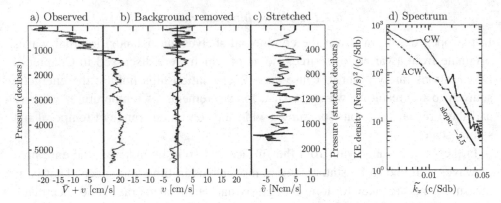

Fig. 6.14. a) Measured total northward velocity with depth, b) fluctuation north-ward velocity with mean currents removed, c) fluctuation northward velocity in WKB-renormalized co-ordinates, and d) kinetic energy spectrum against WKB-renormalized wavenumber. In stretched co-ordinates, velocity is measured in 'normalized centimetres per second' (Ncm/s) and Sdb refers to stretched decibars. Wavenumbers are given in cycles (not radians) per stretched decibar. In d) the solid (dashed) line is the spectrum of the waves that rotate clockwise (anticlockwise) with depth. The thick solid line is drawn with slope −2.5. [Adapted, with permission ©American Geophysical Union, from Figures 3 and 4 of Leaman & Sanford, J. Geophys. Res., **80**, 1975–1978 (1975).]

northward and eastward velocities. After the mean currents are removed, the fluc-tuation velocity is WKB-renormalized to remove the bias that gives the waves larger amplitude where the buoyancy frequency is larger. The power is then com-puted as a function of (rescaled) vertical wavenumber. The analysis again shows a drop-off in power with increasing rescaled vertical wavenumber. On a log–log plot the power of high wavenumber modes decreases with a power law exponent of approximately −2.5.

The velocities may be decomposed into components that, looking downwards, rotate clockwise and anticlockwise with depth. In the northern hemisphere, the former correspond to downward-propagating waves whereas the latter correspond to upward-propagating waves (see Section 3.8). That more power is found for clockwise-rotating waves at low wavenumbers indicates that on large scales the waves originate from the surface. At small wavenumbers, the power is comparable for both wave types suggesting that the waves are more isotropic in the orientation of their associated energy flux.

Observations carried out with a variety of instruments remarkably reveal similar power-law behaviour in different parts of the ocean. This universality suggests that instability as well as linear and nonlinear wave–wave interactions act to transfer energy from newly created waves across the broad spectrum between the buoyancy and inertial frequencies. Although the details of generation and interaction are poorly understood at this time, the fact that the cumulative result of these processes leads to a universal spectrum has inspired some hope that a simple model can be devised that captures the essential physics of linear and nonlinear wave–wave interactions.

Although a consistent dynamical model has yet to be constructed, an empirical model has been put together. This synthesizes the observational data from dropped, moored and towed instruments into a convenient formula known generally as the Garrett–Munk spectrum. The model was first proposed in 1972 and was later modified first in 1975, then in 1979, as a larger data set of better refined observations was made. These models are sometimes referred to succinctly as 'GM72', 'GM75' and 'GM79'. Here, we will present the formulae of the GM75 spectrum.

Several assumptions underlie its empirical formulation:

(i) The background is stationary (there are no mean currents) and the stratification decreases exponentially with depth from the top of the thermocline. Explicitly, the profile of the buoyancy frequency is given by

$$N(z) = N_\star \exp(z/H_\sigma), \tag{6.60}$$

defined for $z < 0$. The parameters in (6.60) are taken to be $N_\star = 3$ cph (5.2×10^{-3} s^{-1}), which is the value of $N(z)$ extrapolated to the surface, and $H_\sigma = 1.3$ km is the e-folding depth.

(ii) The Coriolis frequency, f, is constant. Typically, one takes $f = f_0 \simeq 7.3 \times 10^{-5}$ s^{-1}, corresponding to the inertial frequency at 30°N.

(iii) Statistically the internal waves are horizontally isotropic. Thus, the horizontal structure can be characterized by a single wavenumber $k_h = |(k_x, k_y)|$.

(iv) The waves have sufficiently small amplitude that k_h can be related to the vertical wavenumber γ_j of the jth mode and wave frequency ω by the dispersion relation deduced from linear theory. This is done by extension of (3.194) assuming, consistent with the WKB approximation, that N can be treated as constant at each vertical level. Thus

$$k_h \simeq \gamma_j \left(\frac{N^2 - \omega^2}{\omega^2 - f_0^2} \right)^{-1/2}. \tag{6.61}$$

(v) Energy is transported as effectively upwards as downwards. Thus, together with the small-amplitude assumption, the vertical structure can be represented by standing modes that extend from the ocean bottom to the virtual surface. Each mode can be characterized by a vertical mode number, $j = 1, 2, 3, \ldots$ The mode number can be related approximately to a vertical wavenumber, k_z

for sufficiently high wavenumber modes through WKB renormalization, as discussed in Section 3.5.2. Thus, the jth mode has vertical wavenumber

$$\gamma_j \simeq \frac{\pi}{H_\sigma} \left(\frac{N^2 - \omega^2}{N_\star^2 - \omega^2} \right)^{1/2} j \simeq j \frac{\pi}{H_\sigma} \frac{N}{N_\star}. \qquad (6.62)$$

In this last approximation, the waves are assumed to be hydrostatic so that $\omega \ll N_\star$. The corresponding vertical wavelength for each mode near the top of the thermocline is of the order $\lambda_{zj} \sim 2H_\sigma/j$.

(vi) The spectrum is separable, meaning that it can be represented as the product of a function of frequency with a function of (typically vertical) wavenumber.

These assumptions are made for simplicity of the model while trying to emulate at least the coarser features of the observations. One must keep in mind that this model has been developed to describe waves well below the mixed region where there are negligibly small mean currents and where N varies slowly with depth. Near the surface, observations contradict the model assumptions – the energy of low vertical modes depends upon frequency and so the spectrum is not separable there. The model also invokes linear theory even though nonlinear wave–wave interactions and instability likely play an important, if not dominant, role in explaining the universal nature of the observed spectra.

Given the above assumptions and with further constraints provided by internal wave observations, the energy spectrum per unit mass is empirically represented by

$$P(j,\omega) = E_0 H_\sigma^2 N_\star N(z) M(\tilde{\gamma}_j) W(\omega). \qquad (6.63)$$

The vertical wavenumber-dependent part of the spectrum is given by

$$M(\tilde{\gamma}_j) = (p_j - 1)\left(1 + \tilde{\gamma}_j\right)^{-p_j}, \qquad (6.64)$$

in which $\tilde{\gamma}_j \equiv \gamma_j/\gamma_{j\star} = j/j_\star$ with $\gamma_{j\star} = j_\star \frac{\pi}{H_\sigma} \frac{N}{N_\star}$ and the leading constant $(p_j - 1)$ is defined so that $\int_0^\infty M(\tilde{\gamma}_j)\, d\tilde{\gamma}_j = 1$. In the GM75 model the dependence of M upon j is defined with $p_j \simeq 2.5$ to reproduce the observed vertical wavenumber power law, $j^{-2.5}$ (or equivalently $k_z^{-2.5}$) at high wavenumbers. The characteristic mode number $j_\star$ of the energy-containing waves is introduced to ensure the spectrum does not approach infinity at low j. Here we take $j_\star = 6$ corresponding to modes having vertical wavelength of the order of 450 m near the top of the thermocline. The empirical constants p_j and $j_\star$ are chosen differently in other versions of the Garrett–Munk model. In particular, GM79 takes $p_j = 2$ and $j_\star = 3$.

The frequency-dependent part of the spectrum is given by

$$W(\omega) = \frac{2}{\pi} \frac{f_0/\omega}{\sqrt{\omega^2 - f_0^2}}, \qquad (6.65)$$

in which the leading constant, $2/\pi$, is set so that $\int_{f_0}^{N(z)} W(\omega)\,d\omega \simeq 1$. The integral exactly equals one in the limit $N(z) \to \infty$, but the approximation is good provided that $N(z) \gg f_0$. The structure of W is defined to reproduce the observed ω^{-2} power law at frequencies $\omega \gg f_0$ and also, via the $(\omega^2 - f_0^2)^{-1/2}$ term, capture the observed energy spike for waves near inertial frequencies.

The leading nondimensional constant is $E_0 = 6.3 \times 10^{-5}$. This sets the energy per unit mass of the wave field and is surprisingly consistent (to within a factor of 2) with a variety of measurements.

Finally, the dimensional terms in (6.63) give the vertical distribution of energy as it depends upon the structure of the background buoyancy frequency. Consistent with the result of WKB-renormalization, the energy changes with z in a way that is proportional to the local buoyancy frequency, $N(z)$. In particular, the energy is relatively small at great depths where the buoyancy frequency is small.

The resulting spectrum is shown in Figure 6.15. Figure 6.15a plots contours of the power as a function of frequency and mode number. The power summed over frequency and wavenumber is shown in Figures 6.15b and c, respectively. The -2.5 vertical mode power law is recovered for $j \gg j_\star$ and the -2 frequency power law is recovered for $\omega \gg f_0$ with an enhancement of energy near the inertial frequency.

The Garrett–Munk spectrum also represents the partition of energy into kinetic and potential energy. The former is given in terms of the horizontal velocity spectrum, P_u, the assumption being that the vertical velocity contributes negligibly to the kinetic energy of the wave field. This is a reasonable approximation for hydrostatic waves. Explicitly,

$$P_u(j, \omega) = E_0 \sigma^2 N_\star N(z) \left(1 + \frac{f_0^2}{\omega^2}\right) M(\tilde{\gamma}_j) W(\omega). \tag{6.66}$$

The spectrum of kinetic energy per unit mass is $P_u/2$.

The potential energy is given in terms of the vertical displacement spectrum:

$$P_\xi(j, \omega) = E_0 \sigma^2 \frac{N_\star}{N(z)} \left(1 - \frac{f_0^2}{\omega^2}\right) M(\tilde{\gamma}_j) W(\omega). \tag{6.67}$$

The available potential energy per unit mass is given by $N^2 P_\xi/2$.

From (6.66), the root-mean-square horizontal velocity is approximately $7\,\mathrm{cm/s}$ and, from (6.67), the root-mean-square vertical displacement is approximately $7\,\mathrm{m}$. Consistent with linear theory, the kinetic and potential energies are equally partitioned for waves with frequencies well above the inertial frequency.

For illustrative purposes, Figure 6.15d shows a sample vertical time series which is computed by superposing waves with amplitudes given by the square root of the power, $P_u(\gamma_j, \omega)$, and adding a random phase factor to each mode. The large

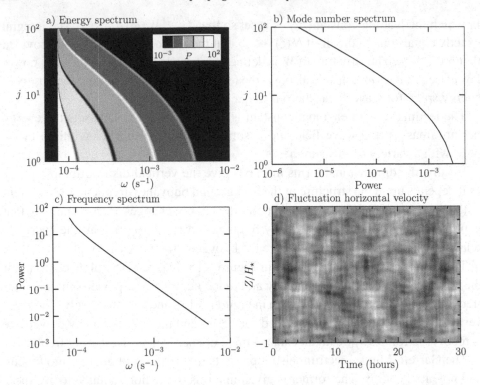

Fig. 6.15. The Garrett–Munk energy spectrum (GM75) a) as a function of frequency and mode number, b) summed over frequency and c) summed over mode number. d) shows an example of a vertical time series resulting from a wave field having the Garrett–Munk spectrum. The vertical scale is given as the WKB-renormalized co-ordinate, Z. The greyscale varies linearly from black to white as the normalized horizontal velocity varies from -1 to 1.

vertical spatial scales (of the order of mode 2) and long time scales (with periods of the order of 10 hours) stand out in this image as expected because energy is concentrated at low mode numbers and frequencies near f_0. But the presence of finer structures is also evident, indicating the non-negligible influence of smaller scale, faster frequency waves.

6.7.2 Atmospheric internal waves

Just as the spectrum of internal waves in the ocean exhibits a universal power law as a function of wavenumber and frequency, so do atmospheric internal waves.

The spectra of horizontal velocity, which is related to the kinetic energy density, and of temperature, which is related to the potential energy density, both exhibit

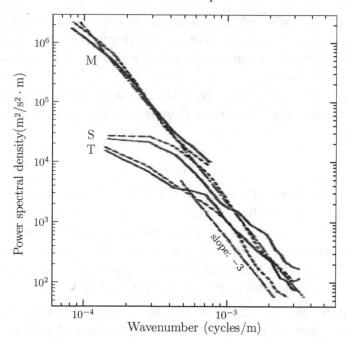

Fig. 6.16. Vertical wavenumber spectra of zonal (solid lines) and meridional (dashed lines) winds in the mesosphere (M), lower stratosphere (S) and troposphere (T). The straight lines indicate modelled slopes in the troposphere (long-dashed line) and in the mesosphere and lower stratosphere (dot-dashed line). [Adapted, with permission ©American Meteorological Society, from Figure 3 of Tsuda *et al.*, J. Atmos. Sci., **46**, 2440–2447 (1989).]

a power law dependence upon frequency whereby $\mathcal{P} \propto \omega^{-p_\omega}$, in which the measured exponent, p_ω, has values between 1 and 2 with typical values around 5/3, moderately smaller than the corresponding exponent measured in the ocean.

Vertical wavenumber spectra of velocity are peaked about a critical wavenumber $k_{z\star}$, the power falling off at higher and lower wavenumbers, as shown in Figure 6.16. The value of $k_{z\star}$ varies depending upon altitude, decreasing from approximately $2\pi/(2\,\mathrm{km})$ to $2\pi/(20\,\mathrm{km})$ going from the troposphere to the mesosphere. At all altitudes, the spectra exhibit a power-law dependence $\mathcal{P} \propto k_z^{-p_{kz}}$, in which $p_{kz} \simeq 3$, moderately larger than the corresponding exponent measured in the ocean.

The universality of a -3 power law for the vertical wavenumber spectra of velocity variance might be expected on the basis of dimensional analysis: assuming the only relevant parameters determining the spectra are N and m, then the combination N^2/m^3 is the only combination that has units of the power spectral density.

On its own, dimensional analysis does not elucidate the physics determining circumstances leading to this power law. As with the ocean, many processes may be responsible for such universality, including linear and nonlinear wave interactions, saturation and wave breaking. A variety of differing theories have been proposed, each of which provides insight but none of which on their own satisfactorily describes what is observed.

Though a cliché, it is fair enough to say that more work needs to be done to understand the nonlinear dynamics of evolving, interacting and breaking internal waves.

Exercises

6.1 Through a change of co-ordinates, $x \rightarrow x + U_0 t$, transform the differential equation (3.43) for Boussinesq internal waves in uniformly stratified, stationary fluid to describe the waves moving in a uniform flow with speed U_0 in the x-direction. Show that the resulting dispersion relation is given by $\omega = \Omega + U_0 k_x$ in which $\Omega(\vec{k})$ is the dispersion relation for internal waves in stationary fluid, given by (6.6) with ω replaced by Ω.

6.2 (a) Write the generalization for the partial differential equation that describes the motion of a wave in a background velocity field given by $(U(z), V(z), 0)$.
    (b) Fourier transform the equation in the horizontal and so show that the waves with fixed horizontal wavenumber move independently of each other.
    (c) Using the ray theory equations (6.13), show that the horizontal components of the wavenumber are invariant for a wavepacket propagating in a background whose stratification and horizontal velocity field depend only upon z.

6.3 (a) Use ray theory to derive the differential equation in terms of dx/dy for shallow water, interfacial waves in a two-layer fluid with constant upper and lower layer depths that move in a horizontal shear flow given by $U(y) = s_0 y$.
    (b) Assuming $\vec{k}_H = (k_x, k_y)$ at $(x, y) = (0, 0)$ with $k_y > 0$, describe qualitatively how k_y and the path of the waves change over time depending on whether $s_0 > 0$ or $s_0 < 0$.

6.4 Derive the differential equation (6.31) for waves in the x–z plane having z-dependent, but no x-dependent background flow and stratification.

6.5 (a) Using a Taylor-series expansion of (6.25), determine the quadratic dependence of y upon x for x near $-L$.
    (b) Repeat this calculation for long interfacial waves in a two-layer fluid. To do this, replace $gh = gs_0|x|$ in (6.25) with $g'\bar{h}$, in which $\bar{h}$ is defined in (6.28).

6.6 (a) Write a code to compute (6.25) numerically and use the code to estimate at what point on the beach, $y(0) - y_0$, incident shallow water waves will reach the shore. For this problem take $L/H_0 = 100$ (so $s_0 = 0.01$) and $k_{x0} = k_{y0} = 0.1/H_0$. Note, your answer does not depend upon the value of $\sqrt{gH_0}$.

(b) Write a code to compute the path of long interfacial waves in a two-layer fluid. As initial conditions choose the same values as in a) where $H_1 + H_2 = H_0$ and set $H_1 = 0.2H_2$. At what along-shore position do the waves intersect the bottom slope?

6.7 A shear flow given by $U(z) = U_0 - s_0 z$, with $U_0 > 0$ and $s_0 > 0$, moves over periodic hills at $z = 0$ launching waves with zero intrinsic frequency and wavenumber $k_x > 0$. Assuming the fluid is uniformly stratified with $N_0 > U_0 k_x$, sketch the path followed by the internal waves.

6.8 (a) Explicitly write the initial value problem for the path followed by anelastic internal waves starting at the origin in a uniformly stratified, uniform shear flow with $U = s_0 z$. For this problem, use the dispersion relation given by (3.182).

(b) Using a numerical solver, solve the equation for $s_0 = 0.01 N_0$, $\omega = (2/3)^{1/2} N_0$ and for $k_x H = 0.1, 1$ and 10.

6.9 Derive the condition (6.38) for overturning internal waves below a critical level in constant shear and explicitly determine the constant C in terms of the initial structure and amplitude of the waves.

6.10 Consider a non-uniformly stratified shear flow in which $N(z) = N_0(1 + \sigma_0 z)$ and $U(z) = U_0 + s_0 z$. Given the horizontal wavenumber k_{x0} and intrinsic frequency $\omega_0 < N_0$ of internal waves at $z = 0$, determine formulae for the critical level z_c and reflection level z_r. Assuming $k_{x0} > 0$ and $U_0 > 0$, for what σ_0 and s_0 do upward-propagating internal waves encounter a critical level or a reflection level?

6.11 Determine the change in variables $z \to Z$ that transforms the approximation (6.44) of the Taylor–Goldstein equation near a reflection level to the canonical form of the Airy equation (6.45).

6.12 (a) Find the dispersion relation for trapped waves in a duct with stratification given by

$$N^2(z) = \begin{cases} N_1{}^2 & -H_1 \le z \le 0 \\ 0 & z < -H_1, \end{cases}$$

assuming the domain is unbounded below and the surface at $z = 0$ is rigid.

(b) Compare your answer with the dispersion relation for varicose waves, given by (3.155), and show that a solution of (6.49) reduces to your result in the limit $k_x D \gg 1$.

(c) Determine the horizontal phase speed of the lowest wave modes in the limit of small and large k_x.

6.13   Show that the flux of wave action associated with plane internal waves near a caustic is everywhere zero. For this problem, use the definition of the wave action flux given by (3.96) and (3.90), together with the polarization relations for internal waves. Rather than working with Ai(Z) and its derivative, use the asymptotic formulae in (6.46).

6.14   Derive the formula (6.57) for the transmission coefficient, T, of internal waves passing through a uniform-density region of depth H. Also compute the reflection coefficient, R, and show that $R = 1 - T$.

6.15   Find the transmission coefficient for internal waves passing through a non-uniformly stratified, stationary fluid in which

$$N^2(z) = \begin{cases} N_0^2 & z > H/2 \\ 0 & |z| \leq H/2. \\ N_1^2 & z < -H/2 \end{cases}$$

Appendix A
Suggestions for further reading

A.1 Textbooks

Baines, P. G., 1995. *Topographic Effects in Stratified Flows*. Cambridge University Press, 482 pp. [Chapters 4 and 5 on stratified shear instability, internal wave propagation in non-uniform media and generation by hills]

Craik, A. D. D., 1985. *Wave Interactions and Fluid Flows*. Cambridge University Press, 322 pp. [Chapter 11 on wavepacket modulations]

Drazin, P. G. & Reid, W. H., 1981. *Hydrodynamic Stability*. Cambridge University Press, 525 pp. [Chapters 4 and 6 on the stability of layered and continuously stratified shear flows]

Gill, A. E., 1982. *Atmosphere–Ocean Dynamics*. Academic Press, 662 pp. [Chapter 6 on transport by small-amplitude internal waves]

Lighthill, M. J., 1978. *Waves in Fluids*. Cambridge University Press, 504 pp. [Chapter 4 on internal wave propagation in non-uniform media]

Miropol'sky, Y. Z., 2001. *Dynamics of Internal Gravity Waves in the Ocean*. Kluwer Academic Publishers, 406 pp.

Nappo, C. J., 2002. *An Introduction to Atmospheric Gravity Waves*. Vol. 85, Academic Press, 276 pp.

Turner, J. S., 1973. *Buoyancy Effects in Fluids*. Cambridge University Press, 367 pp. [Chapter 3 on finite-amplitude interfacial and internal waves]

Vallis, G. K., 2006. *Atmospheric and Oceanic Fluid Dynamics*. Cambridge University Press, 745 pp. [Chapter 2 on the derivation of equations of motion with their solution for small-amplitude internal waves]

Vlasenko, V., Stashchuk, N. & Hutter, K., 2005. *Baroclinic Tides*. Cambridge University Press, 351 pp.

Whitham, G. B., 1974. *Linear and Nonlinear Waves*. John Wiley and Sons, Inc., 636 pp. [Chapters 13, 16, 17 on modulational stability]

A.2 Review articles

Garrett, C. J. R. & Munk, W. H., 1979. Internal waves in the ocean. *Annu. Rev. Fluid Mech.*, **11**, 339–369.

Olbers, D. J., 1983. Models of the oceanic internal wave field. *Rev. Geophys. Space Phys.*, **21**, 1567–1606.

Fritts, D. C., 1984. Gravity wave saturation in the middle atmosphere, a review of theory and observations. *Rev. Geophys.*, **22**, 275–308.

Müller, P., Holloway, G., Henyey, F. & Pomphrey, N., 1986. Nonlinear interactions among internal gravity waves. *Rev. Geophys.*, **24**, 492–536.

Baines, P. G., 1987. Upstream blocking and airflow over mountains. *Annu. Rev. Fluid Mech.*, **19**, 75–97.

Wurtele, M. G., Sharman, R. D. & Datta, A., 1996. Atmospheric lee waves. *Annu. Rev. Fluid Mech.*, **28**, 429–476.

Riley, J. J. & Lelong, M.-P., 2000. Fluid motions in the presence of strong stable stratification. *Annu. Rev. Fluid Mech.*, **32**, 613–657.

Staquet, C. & Sommeria, J., 2002. Internal gravity waves: From instabilities to turbulence. *Annu. Rev. Fluid Mech.*, **34**, 559–593.

Fritts, D. C. & Alexander, M. J., 2003. Gravity wave dynamics and effects in the middle atmosphere. *Rev. Geophys.*, **41**, 1003-1–64.

Broutman, D., Rottman, J. W. & Eckermann, S. D., 2004. Ray methods for internal waves in the atmosphere and ocean. *Annu. Rev. Fluid Mech.*, **36**, 233–253.

Helfrich, K. R. & Melville, W. K., 2006. Long nonlinear internal waves. *Annu. Rev. Fluid Mech.*, **38**, 395–425.

Garrett, C. J. R. & Kunze, E., 2007. Internal tide generation in the deep ocean. *Annu. Rev. Fluid Mech.*, **39**, 57–87.

Ferrari, R. & Wunsch, C., 2009. Ocean circulation kinetic energy: Reservoirs, sources and sinks. *Annu. Rev. Fluid Mech.*, **41**, 253–282.

A.3 Journal articles

A.3.1 Anelastic and other non-Boussinesq equations

Batchelor, G. K., 1953. The conditions for dynamical similarity of motions of a frictionless perfect-gas atmosphere. *Quart. J. Roy. Meteor. Soc.*, **79**, 224–235.

Ogura, Y. & Phillips, N. A., 1962. Scale analysis of deep and shallow convection in the atmosphere. *J. Atmos. Sci.*, **19**, 173–179.

Lipps, F. B. & Hemler, R. S., 1982. A scale analysis of deep moist convection and some related numerical calculations. *J. Atmos. Sci.*, **39**, 10, 2192–2210.

Durran, D. R., 1989. Improving the anelastic approximation. *J. Atmos. Sci.*, **46**, 1453–1461.

Klein, R., 2009. Asymptotics, structure, and integration of sound-proof atmospheric flow equations. *Theor. Comput. Fluid Dyn.*, **23**, 161–195.

A.3.2 Momentum and energy transport

Eliassen, A. & Palm, E., 1961. On the transfer of energy in stationary mountain waves. *Geofys. Publ.*, **22**, 1–23.

Andrews, D. G. & McIntyre, M. E., 1978a. An exact theory of nonlinear waves on a Lagrangian-mean flow. *J. Fluid Mech.*, **89**, 609–646.

Andrews, D. G. & McIntyre, M. E., 1978b. On wave action and its relatives. *J. Fluid Mech.*, **89**, 647–664.

McIntyre, M. E., 1981. On the wave momentum myth. *J. Fluid Mech.*, **81**, 331–347.

Scinocca, J. F. & Shepherd, T. G., 1992. Nonlinear wave-activity conservation laws and Hamiltonian structure for the two-dimensional anelastic equations. *J. Atmos. Sci.*, **49**, 5–27.

Shaw, T. A. & Shepherd, T. G., 2008. Wave-activity conservation laws for the three-dimensional anelastic and Boussinesq equations with a horizontally homogeneous background flow. *J. Fluid Mech.*, **594**, 493–506.

A.3.3 Stratified shear flow stability

Miles, J. W., 1961. On the stability of heterogeneous shear flows. *J. Fluid Mech.*, **10**, 496–508.

Howard, L. N., 1961. Note on a paper by John W. Miles. *J. Fluid Mech.*, **10**, 509–512.

Holmboe, J., 1962. On the behaviour of symmetric waves in stratified shear layers. *Geofys. Publ.*, **24**, 67–112.

Drazin, P. G. & Howard, L. N., 1966. Hydrodynamic stability of parallel flow of inviscid fluid. *Adv. Applied Math.*, **9**, 1–89.

Lawrence, G. A., Browand, F. K. & Redekopp, L. G., 1991. The stability of a sheared density interface. *Phys. Fluids A*, **3**, 2360–2370.

Smyth, W. D. & Peltier, W. R., 1991. Instability and transition in finite-amplitude Kelvin-Helmholtz and Holmboe waves *J. Fluid Mech.*, **228**, 387–415.

Caulfield, C. P., 1994. Multiple linear instability of layered stratified shear flow. *J. Fluid Mech.*, **258**, 255–285.

A.3.4 Internal waves at sloping boundaries and attractors

Eriksen, C. C., 1982. Observations of internal wave reflection off sloping bottoms. *J. Geophys. Res.*, **87**, 525–538.

Maas, L. R. M. & Lam, F.-P. A., 1995. Geometric focusing of internal waves. *J. Fluid Mech.*, **300**, 1–41.

Maas, L. R. M., Benielli, D., Sommeria, J. & Lam, F.-P. A., 1997. Observation of an internal wave attractor in a confined stably stratified fluid. *Nature*, **388**, 557–561.

Grisouard, N., Staquet, C. & Pairaud, I., 2008. Numerical simulation of a two-dimensional internal wave attractor. *J. Fluid Mech.*, **614**, 1–14.

Hazewinkel, J., v. Breevoort, P., Dalziel, S. B. & Maas, L. R. M., 2008. Observations on the wavenumber spectrum and decay of an internal wave attractor. *J. Fluid Mech.*, **598**, 373–382.

A.3.5 Evolution of finite-amplitude internal gravity waves

Hunt, J. N., 1961. Interfacial waves of finite amplitude. *La Houille Blanche*, **4**, 515–531.

Whitham, G. B., 1965. A general approach to linear and nonlinear dispersive waves using a Lagrangian. *J. Fluid. Mech.*, **22**, 273–283.

Benjamin, T. B. & Feir, J. E., 1967. The disintegration of wavetrains on deep water. *J. Fluid Mech.*, **27**, 417–430.

Thorpe, S. A., 1968. On the shape of progressive internal waves. *Phil. Trans. Roy. Soc. A*, **263**, 563–614.

McIntyre, M. E., 1973. Mean motions and impulse of a guided internal gravity wave packet. *J. Fluid. Mech.*, **60**, 801–811.

Fermi, E., Pasta, J. & Ulam, S., 1974. Studies of nonlinear problems I, Los Alamos Report LA 1940, 1955. In A. C. Newell, Ed., reproduced in *Nonlinear Wave Motion* (Providence, RI, 1974). Amer. Math. Soc.

Grimshaw, R. H. J., 1975. The modulation and stability of an internal gravity wave. Technical Report 32, Department of Mathematical Sciences, University of Melbourne.

Sutherland, B. R., 2006. Weakly nonlinear internal wavepackets. *J. Fluid Mech.*, **569**, 249–258.

A.3.6 Evolution of internal solitary waves

Clarke, R. H., 1972. The Morning Glory: an atmospheric hydraulic jump. *J. Appl. Meteorol.*, **11**, 304–311.

Liu, A. K., Holbrook, J. R. & Apel, J. R., 1985. Nonlinear internal wave evolution in the Sulu Sea. *J. Phys. Oceanogr.*, **15**, 1613–1624.

Clarke, J. C., 1998. Picture of the month: an atmospheric undular bore along the Texas coast. *Mon. Weather Rev.*, **126**, 1098–1100.

Boegman, L., Ivey, G. N. & Imberger, J., 2005. The energetics of large-scale internal wave degeneration in lakes. *J. Fluid Mech.*, **531**, 159–180.

Moum, J. N. & Smyth, W. D., 2006. The pressure disturbance of a nonlinear internal wave train. *J. Fluid Mech.*, **558**, 153–177.

Van Gastel, P., Ivey, G. N., Meulerners, M. J., Antenucci, J. P. & Fringer, O., 2009. The variability of the large-amplitude internal wave field on the Australian North West Shelf. *Cont. Shelf Res.*, **29**, 1373–1383.

A.3.7 Parametric subharmonic instability of internal gravity waves

Davis, R. E. & Acrivos, A., 1967. The stability of oscillatory internal waves. *J. Fluid Mech.*, **30**, 723–736.

Mied, R. R., 1976. The occurrence of parametric instabilities in finite-amplitude internal gravity waves. *J. Fluid Mech.*, **78**, 763–784.

Drazin, P. G., 1977. On the instability of an internal gravity wave. *Proc. Roy. Soc. London*, **356**, 411–432.

Klostermeyer, J., 1982. On parametric instabilities of finite-amplitude internal gravity waves. *J. Fluid Mech.*, **119**, 367–377.

McEwan, A. D., 1983. The kinematics of stratified mixing through internal wave breaking. *J. Fluid Mech.*, **128**, 47–57.

Klostermeyer, J., 1991. Two- and three-dimensional parametric instabilities in finite amplitude internal gravity waves. *Geophys. Astrophys. Fluid Dyn.*, **61**, 1–25.

Bouruet-Aubertot, P., Sommeria, J. & Staquet, C., 1995. Instabilities and breaking of standing internal gravity waves. *J. Fluid Mech.*, **285**, 265–301.

Benielli, D. & Sommeria, J., 1996. Excitation of internal waves and stratified turbulence by parametric instability. *Dyn. Atmos. Ocean.*, **23**, 335–343.

Lombard, P. N. & Riley, J. J., 1996. On the breakdown into turbulence of propagating internal waves. *Dyn. Atmos. Oceans*, **23**, 345–355.

Benielli, D. & Sommeria, J., 1998. Excitation and breaking of internal gravity waves by parametric instability. *J. Fluid Mech.*, **374**, 117–144.

MacKinnon, J. A. & Winters, K. B., 2005. Subtropical catastrophe: Significant loss of low-mode tidal energy at 28.9°. *Geophys. Res. Lett.*, **32**, L15605–1–5.

Koudella, C. R. & Staquet, C., 2006. Instability mechanisms of a two-dimensional progressive internal wave. *J. Fluid Mech*, **548**, 165–196.

Alford, M. H., 2008. Observations of parametric subharmonic instability of the diurnal internal tide in the South China Sea. *Geophys. Res. Lett.*, **35**, L15602–1–5.

A.3.8 Internal wave interaction with its induced mean flow

Fritts, D. C. & Dunkerton, T. J., 1984. A quasi-linear study of gravity-wave saturation and self-acceleration. *J. Atmos. Sci.*, **41**, 3272–3289.

Sutherland, B. R., 2001. Finite-amplitude internal wavepacket dispersion and breaking. *J. Fluid Mech.*, **429**, 343–380.

Sutherland, B. R., 2006. Internal wave instability: Wave-wave versus wave-induced mean flow interactions. *Phys. Fluids*, **18**, 074107-1–8.

A.3.9 Interaction between internal wave beams

Teoh, S. G., Ivey, G. N. & Imberger, J., 1997. Laboratory study of the interaction between two internal wave rays. *J. Fluid Mech.*, **336**, 91–122.

Peacock, T. & Tabaei, A., 2005. Visualization of nonlinear effects in reflecting internal wave beams. *Phys. Fluids*, **17**, 061702-1–4.

Tabaei, A., Akylas, T. R. & Lamb, K. G., 2005. Nonlinear effects in reflecting and colliding internal wave beams. *J. Fluid Mech.*, **526**, 217–243.

Gerkema, T., Staquet, C. & Bouruet-Aubertot, P., 2006. Nonlinear effects in internal-tide beams, and mixing. *Ocean Modelling*, **12**, 302–318.

A.3.10 Internal wave breakdown

Fritts, D. C. & Andreassen, O., 1994. Gravity wave breaking in two and three dimensions. Part 2: Three-dimensional evolution and instability structure. *J. Geophys. Res.*, **99**, 8109–8123.

Sonmor, L. J. & Klaassen, G. P., 1997. Toward a unified theory of gravity wave stability. *J. Atmos. Sci.*, **54**, 2655–2680.

Thorpe, S. A., 1999. On internal wave groups. *J. Phys. Oceanogr.*, **29**, 1085–1095.

Yau, K.-H., Klaassen, G. P. & Sonmor, L. J., 2004. Principal instabilities of large amplitude inertio-gravity waves. *Phys. Fluids*, **16**, 936–951.

Achatz, U., 2007. The primary nonlinear dynamics of modal and nonmodal perturbations of monochromatic inertia-gravity waves. *J. Atmos. Sci.*, **64**, 74–95.

Fritts, D. C., Wang, L., Werne, J., Lund, T. & Wan, K., 2009. Gravity wave instability dynamics at high Reynolds numbers. Part I: Wave field evolution at large amplitudes and high frequencies. *J. Atmos. Sci.*, **66**, 1126–1148.

A.3.11 Internal wave generation by oscillating bodies

Mowbray, D. E. & Rarity, B. S. H., 1967a. A theoretical and experimental investigation of the phase configuration of internal waves of small amplitude in a density stratified liquid. *J. Fluid Mech.*, **28**, 1–16.

Thomas, N. H. & Stevenson, T. N., 1972. A similarity solution for viscous internal waves. *J. Fluid Mech.*, **54**, 495–506.

Stevenson, T. N., 1973. The phase configuration of internal waves around a body moving in a density stratified fluid. *J. Fluid Mech.*, **60**, 759–767.

Voisin, B., 1991. Internal wave generation in uniformly stratified fluids. Part 1. Green's function and point sources. *J. Fluid Mech.*, **231**, 439–480.

Hurley, D. G., 1997. The generation of internal waves by vibrating elliptic cylinders. Part 1: Inviscid solution. *J. Fluid Mech.*, **351**, 105–118.

Hurley, D. G. & Keady, G., 1997. The generation of internal waves by vibrating elliptic cylinders. Part 2: Approximate viscous solution. *J. Fluid Mech.*, **351**, 119–138.

Sutherland, B. R., Dalziel, S. B., Hughes, G. O. & Linden, P. F., 1999. Visualisation and measurement of internal waves by "synthetic schlieren". Part 1: Vertically oscillating cylinder. *J. Fluid Mech.*, **390**, 93–126.

Flynn, M. R., Onu, K. & Sutherland, B. R., 2003. Internal wave generation by a vertically oscillating sphere. *J. Fluid Mech.*, **494**, 65–93.

Voisin, B., 2003. Limit states of internal wave beams. *J. Fluid Mech.*, **496**, 243–293.

A.3.12 Internal wave generation by steady flow over topography

Long, R. R., 1953. Some aspects of the flow of stratified fluids. A theoretical investigation. *Tellus*, **5**, 42–58.

Drazin, P. G., 1961. On the steady flow of a fluid of variable density past an obstacle. *Tellus*, **8**, 239–251.

Miles, J. W. & Huppert, H. E., 1968. Lee waves in a stratified flow. Part 2. Semi-circular obstacle. *J. Fluid Mech.*, **33**, 803–814.

Lilly, D. K. & Kennedy, P. J., 1973. Observations of a stationary mountain wave and its associated momentum flux and energy dissipation. *J. Atmos. Sci.*, **30**, 1135–1152.

Klemp, J. B. & Lilly, D. K., 1975. Numerical simulation of hydrostatic mountain waves. *J. Atmos. Sci.*, **32**, 320–339.

Smith, R. B., 1977. The steepening of hydrostatic mountain waves. *J. Atmos. Sci.*, **34**, 1634–1654.

Lilly, D. K., 1978. A severe downslope windstorm and aircraft turbulence event induced by a mountain wave. *J. Atmos. Sci.*, **35**, 59–77.

Brighton, P. W. M., 1978. Strongly stratified flow past three-dimensional obstacles. *Quart. J. Roy. Meteorol. Soc.*, **104**, 289–307.

Peltier, W. R. & Clark, T. L., 1979. The evolution and stability of finite-amplitude mountain waves. Part II: Surface wave drag and severe downslope windstorms. *J. Atmos. Sci.*, **36**, 1498–1529.

Hunt, J. C. R. & Snyder, W. H., 1980. Experiments on stably and neutrally stratified flow over a model three-dimensional hill. *J. Fluid Mech.*, **96**, 671–704.

Smith, R. B., 1985. On severe downslope winds. *J. Atmos. Sci.*, **42**, 2597–2603.

Scinocca, J. F. & Peltier, W. R., 1989. Pulsating downslope windstorms. *J. Atmos. Sci.*, **46**, 2885–2914.

Scinocca, J. F. & Peltier, W. R., 1994. The instability of Long's stationary solution and the evolution toward severe downslope windstorm flow. Part II: The application of finite-amplitude local wave-activity flow diagnostics. *J. Atmos. Sci.*, **51**, 623–653.

Aguilar, D. A., Sutherland, B. R. & Muraki, D. J., 2006. Generation of internal waves over sinusoidal topography. *Deep-Sea Res. II*, **53**, 96–115.

A.3.13 Internal solitary wave generation by flow over topography

Farmer, D. M. & Armi, L., 1999a. The generation and trapping of solitary waves over topography. *Science*, **283**, 5399, 188–190.

Farmer, D. M. & Armi, L., 1999b. Stratified flow over topography: the role of small scale entrainment and mixing in flow establishment. *Proc. Roy. Soc., Series A*, **455**, 3221–3258.

Armi, L. & Farmer, D. M., 2002. Stratified flow over topography: Bifurcation fronts and transition to the uncontrolled state. *Proc. Roy. Soc., Series A*, **458**, 513–538.

Cummins, P. F., Vagle, S., Armi, L. & Farmer, D. M., 2003. Stratified flow over topography: Upstream influence and generation of nonlinear internal waves. *Proc. R. Soc. Lond. A*, **459**, 1467–1487.

Klymak, J. M. & Gregg, M. C., 2004. Tidally generated turbulence over the Knight Inlet sill. *J. Phys. Ocean.*, **34**, 5, 1135–1151.

Stastna, M. & Peltier, W. R., 2004. Upstream-propagating solitary waves and forced internal-wave breaking in stratified flow over a sill. *Proc. Roy. Soc., Series A*, **460**, 3159–3190.

Klymak, J. M., Pinkel, R., Liu, C. T., Liu, A. K. & David, L., 2006. Prototypical solitons in the South China Sea. *Geophys. Res. Lett.*, **33**, 11, L11607.

A.3.14 Inertia gravity wave generation by tidal flow

Haury, L. R., Briscoe, M. G. & Orr, M. H., 1979. Tidally generated internal wave packets in Massachusetts Bay. *Nature*, **278**, 312–317.

Balmforth, N. J., Ierley, G. R. & Young, W. R., 2002. Tidal conversion by subcritical topography. *J. Phys. Oceanogr.*, **32**, 2900–2914.

Smith, S. G. L. & Young, W. R., 2002. Conversion of the barotropic tide. *J. Phys. Oceanogr.*, **32**, 1554–1566.

Legg, S. & Adcroft, A., 2003. Internal wave breaking at concave and convex continental slopes. *J. Phys. Oceanogr.*, **33**, 2224–2246.

Simmons, H. L., Hallberg, R. W. & Arbic, B. K., 2004. Internal wave generation in a global baroclinic tide model. *Deep-Sea Res. II*, **51**, 3043–3068.

Munroe, J. R. & Lamb, K. G., 2005. Topographic amplitude dependence of internal wave generation by tidal forcing over idealized three-dimensional topography. *J. Geophys. Res.*, **110**, C02001.

Nash, J. D., Kunze, E., Lee, C. M. & Sanford, T. B., 2006. Structure of the baroclinic tide generated at Kaena Ridge, Hawaii. *J. Phys. Oceanogr.*, **36**, 1123–1135.

Peacock, T., Echeverri, P. & Balmforth, N. J., 2008. An experimental investigation of internal tide generation by two-dimensional topography. *J. Phys. Oceanogr.*, **38**, 235–242.

A.3.15 Internal waves at critical and reflection levels

Bretherton, F. P., 1966. The propagation of groups of internal gravity waves in a shear flow. *Quart. J. Roy. Meteorol. Soc.*, **92**, 466–480.

Booker, J. R. & Bretherton, F. P., 1967. The critical layer for internal gravity waves in shear flow. *J. Fluid Mech.*, **27**, 513–539.

Bretherton, F. P. & Garrett, C. J. R., 1969. Wavetrains in inhomogeneous moving media. *Proc. Roy. Soc. A.*, **302**, 529–554.

Plumb, R. A., 1977. The interaction of two internal waves with the mean flow: Implications for the theory of the quasi-biennial oscillation. *J. Atmos. Sci.*, **34**, 1847–1858.

Plumb, R. A. & McEwan, A. D., 1978. The instability of a forced standing wave in a viscous, stratified fluid: A laboratory analogue of the quasi-biennial oscillation. *J. Atmos. Sci.*, **35**, 1827–1839.

Brown, S. N. & Stewartson, K., 1980. On the nonlinear reflexion of a gravity wave at a critical level. Part 1. *J. Fluid Mech.*, **100**, 577–595.

Koop, C. G., 1981. A preliminary investigation of the interaction of internal gravity waves with a steady shearing motion. *J. Fluid Mech.*, **113**, 347–386.

Fritts, D. C., 1982. The transient critical-level interaction in a Boussinesq fluid. *J. Geophys. Res.*, **87**, 7997–8016.

Koop, C. G. & McGee, B., 1986. Measurements of internal gravity waves in a continously stratified shear flow. *J. Fluid Mech.*, **172**, 453–480.

Winters, K. B. & D'Asaro, E. A., 1989. Two-dimensional instability of finite amplitude internal gravity wave packets near a critical level. *J. Geophys. Res.*, **94**, 12709–12719.

Lamb, K. & Pierrehumbert, R. T., 1992. Steady-state nonlinear internal gravity-wave critical layers satisfying an upper radiation condition. *J. Fluid Mech.*, **238**, 371–404.

Winters, K. B. & D'Asaro, E. A., 1994. Three-dimensional wave instability near a critical level. *J. Fluid Mech.*, **272**, 255–284.

Dörnbrack, A., Gerz, T. & Schumann, U., 1995. Turbulent breaking of overturning gravity waves below a critical level. *Appl. Sci. Res.*, **54**, 163–176.

Sutherland, B. R., 2000. Internal wave reflection in uniform shear. *Q.J.R.M.S.*, **126**, 3255–3287.

Yu, Y., Hickey, M. P. & Liu, Y., 2009. A numerical model characterizing internal gravity wave propagation into the upper atmosphere. *Adv. Space Res.*, **44**, 836–846.

A.3.16 Internal wave ducting and tunnelling

Eckart, C., 1961. Internal waves in the ocean. *Phys. Fluids*, **4**, 791–799.

Fritts, D. C. & Yuan, L., 1989. An analysis of gravity wave ducting in the atmosphere: Eckart's resonances in thermal and Doppler ducts. *J. Geophys. Res.*, **94**, D15, 18455–18466.

Sutherland, B. R. & Yewchuk, K., 2004. Internal wave tunnelling. *J. Fluid Mech.*, **511**, 125–134.

Smith, R. B., Jiang, Q. & Doyle, J. D., 2006. A theory of gravity wave absorption by a boundary layer. *J. Atmos. Sci.*, **63**, 774–781.

Yu, Y. & Hickey, M. P., 2007. Numerical modeling of a gravity wave packet ducted by the thermal structure of the atmosphere. *J. Geophys. Res.*, **112**, A06308–1–12.

Nault, J. T. & Sutherland, B. R., 2007. Internal wave tunnelling across a mixed region. *Phys. Fluids*, **19**, 016601–1–8.

Brown, G. L., Bush, A. B. G. & Sutherland, B. R., 2008. Beyond ray tracing for internal waves. Part II: Finite-amplitude effects. *Phys. Fluids*, **20**, 106602–1–13.

Broutman, D., Eckermann, S. D. & Rottman, J. W., 2008. Practical application of two turning-point theory to mountain-wave transmission through a wind jet. *J. Atmos. Sci.*, **66**, 481–494.

A.3.17 Internal waves in transient and veering flows

Broutman, D. & Young, W. R., 1986. On the interaction of small-scale oceanic internal waves with near-inertial waves. *J. Fluid Mech.*, **166**, 341–358.

Shutts, G. J., 1995. Gravity-wave drag parameterization over complex terrain: the effect of critical level absorption in directional wind shear. *Quart. J. Roy. Meteorol. Soc.*, **121**, 1005–1021.

Sun, H. & Kunze E., 1999. Internal wave–wave interactions. Part I: The role of internal wave vertical divergence. *J. Phys. Oceanogr.*, **29**, 2886–2904.

Broad, A. S., 1999. Do orographic gravity waves break in flows with uniform wind direction turning with height? *Quart. J. Roy. Meteorol. Soc.*, **125**, 1695–1714.

Hasha, A., Bühler, O. & Scinocca, J., 2008. Gravity wave refraction by three-dimensionally varying winds and the global transport of angular momentum. *J. Atmos. Sci.*, **65**, 2892–2906.

Vanderhoff, J. C., Nomura, K. K., Rottman, J. W. & Macaskill, C., 2008. Doppler spreading of internal gravity waves by an inertia-wave packet. *J. Geophys. Res.*, **113**, C05018.

A.3.18 Oceanic internal wave spectra

Garrett, C. J. R. & Munk, W. H., 1972. Space-time scales of internal waves. *Geophys. Fluid Dyn.*, **3**, 225–264.

Cairns, J. L., 1975. Internal wave measurements from a midwater float. *J. Geophys. Res.*, **80**, 299–306.

Sanford, T. B., 1975. Observations of the vertical structure of internal waves. *J. Geophys. Res.*, **80**, 3861–3871.

Leaman, K. D. & Sanford, T. B., 1975. Vertical energy propagation of inertial waves: A vector spectral analysis of velocity profiles. *J. Geophys. Res.*, **80**, 1975–1978.

Garrett, C. J. R. & Munk, W. H., 1975. Space-time scales of internal waves: A progress report. *J. Geophys. Res.*, **80**, 291–298.

Cairns, J. L. & Williams, G. O., 1976. Internal wave observations from a midwater float, 2. *J. Geophys. Res.*, **81**, 1943–1950.

Munk, W., 1981. Internal wave and small-scale processes. In B. A. Warren & C. Wunsch, Eds., *Evolution of Physical Oceanography* (Cambridge, MA, 1981), pp. 264–291. MIT Press.

Henyey, F. S. & Pomphrey, N., 1983. Eikonal description of internal-wave interactions: A non-diffusive picture of "induced diffusion", *Dyn. Atmos. Oceans*, **7**, 189–208.

Pinkel, R., 1984. Doppler sonar observations of internal waves: The wavenumber-frequency spectrum. *J. Phys. Oceanogr.*, **14**, 1249–1270.

Henyey, F. S., Wright, J. & Flatté, S. M., 1986. Energy and action flow through the internal wave field: an eikonal approach. *J. Geophys. Res.*, **91**, 8487–8495.

Allen, K. A. & Joseph, R. I., 1989. A canonical statistical theory of oceanic internal waves. *J. Fluid Mech.*, **204**, 185–228.

Garrett, C., 2001. What is the "near-inertial" band and why is it different from the rest of the internal wave spectrum. *J. Phys. Oceanogr.*, **31**, 962–971.

Levine, M. D., 2002. A modification of the Garrett-Munk internal wave spectrum. *J. Phys. Oceanogr.*, **32**, 3166–3181.

Polzin, K., 2004. A heuristic description of internal wave dynamics. *J. Phys. Oceanogr.*, **34**, 214–230.

Alford, M. H. & Zhao, Z., 2007. Global patterns of low-mode internal-wave propagation. Part I: Energy and energy flux. *J. Phys. Oceanogr.*, **37**, 1829–1848.

Pinkel, R., 2008. The wavenumber-frequency spectrum of vortical and internal-wave shear in the Western Arctic Ocean. *J. Phys. Oceanogr.*, **38**, 277–290.

A.3.19 Atmospheric internal wave spectra

VanZandt, T. E., 1982. A universal spectrum of buoyancy waves in the atmosphere. *Geophys. Res. Lett.*, **9**, 575–578.

Dewan, E. & Good, R. E., 1986. Saturation and the "universal" spectrum for vertical profiles of horizontal scalar winds in the atmosphere. *J. Geophys. Res.*, **91**, 2742–2748.

Smith, S. A., Fritts, D. C. & VanZandt, T. E., 1987. Evidence for a saturated spectrum of atmospheric gravity waves. *J. Atmos. Sci.*, **44**, 1404–1410.

Tsuda, T., Inoue, T., Fritts, D. C., van Zandt, T. E., Kato, S. and Fukoo, S. 1989. MST Radar observations of a saturated gravity wave spectrum. *J. Atmos. Sci.*, **46**, 2440–2447.

Weinstock, J., 1990. Saturated and unsaturated spectra of gravity waves and scale-dependent diffusion. *J. Atmos. Sci.*, **47**, 2211–2225.

Hines, C. O., 1991. The saturation of gravity waves in the middle atmosphere. Part II: Development of Doppler-spread theory. *J. Atmos. Sci.*, **48**, 1360–1379.

Eckermann, S. D., 1999. Isentropic advection by gravity waves: quasi-universal M^{-3} vertical wavenumber spectra near the onset of instability. *Geophys. Res. Lett.*, **26**, 201–204.

Chunchuzov, I., 2002. On the high-wavenumber form of the Eulerian internal wave spectrum in the atmosphere. *J. Atmos. Sci.*, **59**, 1753–1774.

Hines, C. O., 2002. Theory of the Eulerian tail in the spectra of atmospheric and oceanic internal gravity waves. *J. Fluid Mech.*, **448**, 289–313.

Wang, L., Geller, M. A. & Alexander, M. J., 2005. Spatial and temporal variations of gravity wave parameters. Part I: Intrinsic frequency, wavelength, and vertical propagation direction. *J. Atmos. Sci.*, **62**, 125–142.

Pulido, M., 2005. On the Doppler effect in a transient gravity-wave spectrum. *Quart. J. Roy. Meteorol. Soc.*, **131**, 1215–1232.

Klaassen, G. P. & Sonmor, L. J., 2006. Does kinematic advection by superimposed waves provide an explanation for quasi-universal gravity-wave spectra? *Geophys. Res. Lett.*, **33**, L23802-1–5.

A.3.20 Drag parameterization in general circulation models

Lindzen, R. S., 1981. Turbulence and stress owing to gravity wave and tidal breakdown. *J. Geophys. Res.*, **86**, 9707–9714.

Palmer, T. N., Shutts, G. J. & Swinbank, R., 1986. Alleviation of a systematic westerly bias in general circulation and numerical weather prediction models through an orographic gravity drag parameterization. *Quart. J. Roy. Meteor. Soc.*, **112**, 1001–1039.

McFarlane, N. A., 1987. The effect of orographically excited gravity wave drag on the general circulation of the lower stratosphere and troposphere. *J. Atmos. Sci.*, **44**, 1775–1800.

Hines, C. O., 1997. Doppler-spread parameterization of gravity-wave momentum deposition in the middle atmosphere. Part 2: Broad and quasi-monochromatic spectra, and implemenation. *J. Atmos. Sol.-Terr. Phys.*, **59**, 387–400.

McLandress, C., 1998. On the importance of gravity waves in the middle atmosphere and their parameterization in general circulation models. *J. Atmos. Sol.-Terres. Phys.*, **60**, 1357–1383.

Alexander, M. J. & Dunkerton, T. J., 1999. A spectral parameterization of mean-flow forcing due to breaking gravity waves. *J. Atmos. Sci.*, **56**, 4167–4182.

Warner, C. D. & McIntyre, M. E., 2001. An ultrasimple spectral parameterization for nonorographic gravity waves. *J. Atmos. Sci.*, **58**, 1837–1857.

McLandress, C. & Scinocca, J., 2005. The GCM response to current parameterizations of nonorographic gravity wave drag. *J. Atmos. Sci.*, **62**, 2394–2413.

Index

Bold face denotes a primary entry or an extended discussion. *Italics* denote a definition.

Printed in the United States
By Bookmasters